Risk, Systems and Decisions

Series Editors

Igor Linkov
U.S. Army ERDC, Vicksburg, MS, USA

Jeffrey Keisler
College of Management, University of Massachusetts
Boston, MA, USA

James H. Lambert
University of Virginia, Charlottesville, VA, USA

Jose Figueira
University of Lisbon, Lisbon, Portugal

More information about this series at http://www.springer.com/series/13439

Ivan Damnjanovic • Kenneth Reinschmidt

Data Analytics for Engineering and Construction Project Risk Management

 Springer

Ivan Damnjanovic
Texas A&M University
College Station, TX, USA

Kenneth Reinschmidt
College Station, TX, USA

ISSN 2626-6717   ISSN 2626-6725   (electronic)
Risk, Systems and Decisions
ISBN 978-3-030-14253-7   ISBN 978-3-030-14251-3   (eBook)
https://doi.org/10.1007/978-3-030-14251-3

Library of Congress Control Number: 2019934720

This Springer imprint is published by the registered company Springer Nature Switzerland AG
The registered company address is: Gewerbestrasse 11, 6330 Cham, Switzerland

Preface

I returned, and saw under the sun, that the race is not to the swift, nor the battle to the strong, neither yet bread to the wise, nor yet riches to men of understanding, nor yet favor to men of skill; but time and chance happeneth to them all. (Ecclesiastes, 9:11)

This book is about time and chance as they affect projects. More specifically, the main objective of this text is to provide foundations for the assessment of uncertainty and risks on engineering projects of all types. It deals with the spectrum of uncertainty, from the variability in construction operations to the risks in unique, complex, first-of-a-kind projects. In looking at field operations or other project activities, we use probabilities to try to describe the natural variability of work, and we are concerned with answering the question: Is the reported performance on this activity merely reflecting these inherent variations, or is it sending a message that the activity is about to go seriously out of control? In looking at major risks on complex projects, we use data, information, and knowledge about the underlying behavior to express the confidence in our risk estimates. In general, the text places emphasis on building data-driven models, and these models are of necessity mathematically inspired. As the British physicist William Thomson (Lord Kelvin) said (1883), "when you cannot measure it, when you cannot express it in numbers, your knowledge is of a meager and unsatisfactory kind." And "meager and unsatisfactory" is a good description of many project risk assessments.

It may be argued that project managers should place their confidence in experience, judgment, and gut feel, not mathematics. However, learning about risk by the trial-and-error method can take a long time, and the lessons can prove to be very expensive. The view here is that it is much easier for a project manager to have confidence in a decision after he/she has examined a risk model from all possible views and played out a number of scenarios, alternates, and options, than to rely on his/her judgment alone. One of the reasons why project risk assessment and management has become such an active topic for research as well as education in the past few years is that experience and judgment alone have been inadequate. Therefore, the principle used here is that model building should be informed by, and consistent with, judgment and experience, but any model beats no model nine times out of ten.

The text covers risk identification and assessment methods for projects that are already defined in terms of objectives, specifications, resource plans, sequencing, and work breakdown structure. In other words, the methods we cover deal with assessment of risks, not making decisions in response to the risks. This is deliberate as the text is looking to be general, not be specific to owners, subcontractors, and contractors, their risk attitude, or the type of contracting strategy and project delivery method being used.

The content is considered suitable for graduate students in engineering, construction, or project management, as well as practitioners aiming to develop, improve, and/or simplify corporate project management processes. This book is based on the class lecture notes developed from Fall 2001 to Spring 2018 and taught as a part of the graduate course in Project Risk Management at Texas A&M University. This text is perhaps more mathematical than many other texts, and this is deliberate. The mathematics contains nothing beyond what an engineering graduate is expected to know: some algebra, a little calculus, a little statistics, and, especially, undergraduate-level understating of the probability theory.

The field of project risk assessment and management is actively evolving, and we may anticipate that better methods will continue to be developed. This text is an attempt to provide a bridge from the qualitative and anecdotal to the quantitative and analytical way of thinking. The authors encourage students and practitioners to make their own contributions to the advancement of project risk management.

Alea iacta est. (Gaius Julius Caesar, 49 BC)

College Station, TX, USA   Ivan Damnjanovic
October 2018   Kenneth Reinschmidt

Acknowledgments

We would like to thank our families on their patience and support as well as express gratitude to many colleagues and students on helpful suggestions.

In Memoriam

Kenneth Reinschmidt (March 26, 1938–December 31, 2018)

It was Spring 2014, when Ken and I embarked on this writing endeavor. By then, Ken had been teaching Project Risk Management for more than 10 years, while I had just began. Over the next 5 years, across countless afternoons, we discussed the content, combined the notes and problem sets, and drafted and revised the chapters. When Ken fell sick, the settings of our meetings changed, but Ken remained as focused and devoted as ever. Unfortunately, as this book was headed to publication, on December 31, 2018, Ken passed away. At this juncture therefore, I would like to take few moments to reflect on Ken's remarkable career and share the influence he has imparted both personally on myself and on the field in general.

Ken was a military veteran, dedicated public servant, industry visionary, and brilliant researcher and engineer. Over his long and esteemed career, Ken achieved excellence in all areas he pursued. He entered military active duty in January 1966, while on a leave of absence from MIT, and led the team development of an integrated computer-based system for planning and management of military operations. He was honorably discharged, in December 1967, at the rank of captain. Further dedicating his energy to the public, he chaired multiple National Research Council committees and initiatives, provided testimony to Congress on Electric Power System Reliability, and served on numerous other committees. During his time working in industry, Ken raised through the ranks to be become the Elected President and Chief Executive Officer of Stone & Webster Advanced Systems Development Services, Inc. and through that work he impacted the nuclear industry as a whole. Finally, in academia, as a professor at both MIT and TAMU, Ken was fundamental in developing of what is now known as Building Information Modeling (BIM), use of Artificial Intelligence and expert systems in engineering and construction, and implementation of advanced computing methods to project management. In essence, Dr. Kenneth Reinschmidt was instrumental in introducing the application of computer to Civil Engineering. His contributions have been acknowledged by his peers, culminating in 1991, in which Ken was elected to the National Academy of Engineering and became a Fellow of the American Association for the Advancement of Science.

To conclude, although the time that we collaborated was relatively short in light of Ken's long career, I am grateful for the opportunity and honor to have Dr. Kenneth Reinschmidt as a friend, colleague, and co-author.

Contents

Part I
Introduction

Chapter 1
Introduction to Uncertainty and Risk

Abstract In this chapter we discuss the concept of uncertainty and risks from the two different viewpoints – the theoretical and the project management viewpoint. We provide an overview of uncertainty classification that extends beyond the two typical approaches and discuss the types of uncertainty project managers are often challenged with – volitional, agonistic, and dialectic uncertainty. Further, we provide evidence of a divergence in approaches adopted by project managers and decision theorists as well as of the link that allows us to develop a holistic approach to project risk management.

Keywords Uncertainty · Project manager · Managerial attitude · Probabilities and decisions

1.1 Viewpoints on Uncertainty

The differences in what people mean when they say "uncertainty" depends on a philosophical position they take. One school of thought says that the universe is based on chance, and uncertainty is essentially characterized by relative frequencies of observed phenomena. Others, starting from the time of Plato, say that the universe must be deterministic, and all uncertainty is caused by our feeble capabilities to measure it or to understand it ("God does not play dice with the universe" – Albert Einstein). However, this strictly deterministic view was laid to rest by quantum mechanics: at the most fundamental level, the universe appears to be random. In other words, even when one controls all causal factors, some outcomes of the experiment will vary randomly. For example, see Heisenberg's Uncertainty Principle, which shook the foundations of classical physics at the beginning of the twentieth century.

Corresponding to these two philosophical viewpoints we classify uncertainty into: (a) aleatory uncertainty, from the Latin word alea, die; plural aleae, dice, and therefore referring to gambling (an aleator is a gambler); this type of uncertainty is characterized by variability in repeated experiments, such as flipping a coin or

© Springer Nature Switzerland AG 2020
I. Damnjanovic, K. Reinschmidt, *Data Analytics for Engineering and Construction Project Risk Management*, Risk, Systems and Decisions,
https://doi.org/10.1007/978-3-030-14251-3_1

rolling dice; and (b) epistemic uncertainty, from the Greek word επιστημη (skill, understanding, experience, or knowledge) is due to lack of information or only partial knowledge of the phenomena on the part of the observer. Epistemic uncertainty is subjective, not objective; it does not exist independent of the observer, and can differ between observers.

However, from the project viewpoint these two categories can be limiting in expressing the uncertainty as we *experience* it on projects. What is being missed here are the conditioning and feedback, confidence in assumptions, and the capacity to define and evaluate events beyond already materialized and studied phenomena. For example, in project setting the likelihood of a safety incident could be reduced just by a project team being aware of it; technical risks could materialize due to erroneous assumptions that were never explicitly stated or considered; similarly, project cost overruns could occur due to events that contradict broadly accepted knowledge. In a strict theoretical interpretation this ignorance can be considered epistemic uncertainty, but from the practical perspective this uncertainty is typically not part of a deliberate risk assessment process. The general state of knowledge about the systems' phenomena and project behavior rarely, if ever, show up in project documentation. Therefore, for practical reasons it is useful to highlight its importance explicitly. We will refer to this class of uncertainty as agnostic uncertainty, from the Greek word άγνωστος (ignorant, not knowing). It is not uncertainty about the knowledge, rather it is uncertainty about our ignorance; often, far more dangerous of the two.

The classification of uncertainty in aleatory, epistemic and agnostic components could be considered analogous to the popular classification of project uncertainties into knowns and unknowns. The uncertainty behind known knowns could be considered of aleatory nature if it is based on validated theoretical foundation (e.g. Newton's laws). Unknown knowns, on the other hand, relate to a spectrum of aleatory or epistemic uncertainties that are derived from either large data sets or rooted in deterministic assumptions with unknown parameters; finally, unknown unknowns, unforeseen, unimaginable, surprise, black swans and white ravens' events result from our overconfidence in assumptions and understanding of the phenomena and the system behavior in general.

1.1.1   Aleatory Uncertainty

Aleatory uncertainty is measured or characterized by relative frequencies: the number of times a particular event occurs out of N repeated experiments. In gambling, for example, with dice, there is complete knowledge about the potential states to be encountered (in a modern die, the integers 1–6, and with two dice, the integers 2–12). This type of uncertainty is objective, it is a characteristic of the real world, and it can be measured, at least approximately. This type of uncertainty is the business of statisticians. It is external to and independent of the observer; e.g., a radioactive cesium atom is presumed to decay with a certain probability whether or not anyone observes it. Presumably any number of experimenters or observers of identical experiments would observe the same relative frequencies in the long run. The relative frequencies, either derived from observations or from deductive principles, are then called probabilities. A probability is a relative frequency taken to the limit. For example, let the relative frequency of some event j be given by the ratio $f_j = n_j/N$, in which n_j is the number of observations of event j and N is the total number of observations. Then the probability of event j is $p_j = \lim_{N \to \infty} f_j = \lim_{N \to \infty} \left[n_j / N \right]$. But we can never do an infinite number of observations, so these relative frequencies or probabilities are actually based on the outcomes of experiments that are never run. Or, in observing a die, we may say the probability of each side coming up is (or, rather, should ideally be) 1/6 without making any observations of the actual relative frequencies; this set of probabilities (the probability distribution) is assumed to be characteristic of some ideal die, even if may be a poor assumption for any die you happen to be betting on.

We may subdivide this classification of aleatory uncertainty in the following subtypes.

Type 1, in which we know the form of the probability distribution and the parameters of the probability distribution. For example, we say the probability distribution for the flip of a fair coin is the binomial distribution, and the probability of obtaining a head on one flip is 1/2. Similarly for the six-sided die: the probability for each side is said to be equal, and lacking any better information, equality implies the probability of each is 1/6. This is the type of probability commonly covered in probability courses and with which most people are familiar.

Type 2, in which the form of the probability distribution is known (or believed to be known based on some theory), but the parameters are unknown. This might be the case with a coin that may have been tampered with, or a pair of dice that we may have suspicions about. There is no question about the form of the probability distribution, but determining the actual values of the parameters (in the case of the dice, the actual relative frequencies of getting each of the 11 outcomes) is a point at issue.

Type 3, in which the parameters (relative frequencies) are known (through observations) but the form of the probability distribution is unknown. This often occurs when one has a lot of data, including relative frequencies of various events, but has no theory to tie them together. Here the reasoning is inductive: given these observations, what probability function best describes the data? We may either fit some probability distribution to the data, or else use the observed relative frequencies themselves as the empirical probability distribution.

Type 4, in which neither the form of the probability distribution nor any values for the parameters are known. Hence, if the form is not known, even the identity of the parameters is not known. Our job, then, may be to determine both the form of the distribution and its parameters simultaneously.

1.1.2   Epistemic Uncertainty

Epistemic uncertainty is the uncertainty that does not derive from variability in repeated experiments. It arises because of our limitations; reality is thought to be deterministic but we cannot see or know the true reality, only shadows of it. Pushed to the limit, this position would say that the future could be predicted without error if we only had enough information. For example, from this viewpoint, the classical gambling problem that motivated the development of probability theory could finally be solved: the outcome of shooting dice would not be random, because, given knowledge of the initial positions and orientations of the dice, and all the forces acting on them during the roll, one could predict with certainty how the dice would come up. Thus, we use the roll of dice as a standard for randomness simply because we cannot (the casino won't let us) or do not choose to make all these measurements and computations. If we could, of course, no one would ever again shoot dice.

Epistemic uncertainty is not based on relative frequencies, as the "experiments" of interest are not repeated, and an event may happen only once or not at all. As such, it is the business of managers and engineers. Although some people object to the use of subjective probabilities, the issue in practice typically is whether to use subjective probabilities or no probabilities at all; that is, to stay with deterministic analysis (which is equivalent to probabilistic analysis with all probabilities either zero or one). One may adopt the viewpoint that we would use relative frequencies if we had any data, but as typically we do not, and as managers we must nevertheless make decisions, we may have use subjective probabilities as better than nothing.

1.1.3   Agnostic Uncertainty

Agnostic uncertainty is about our confidence and completeness of knowledge and assumptions upon which we develop probabilities and judgments. In a strict sense it cover both frequentist and epistemic viewpoints as they both are associated assumptions. We may further subdivide agnostic uncertainty in the following subtypes.

Volitional uncertainty is a type of uncertainty in the area of projects, in which the subjective probability distributions are not objective or "out there," but are capable of being manipulated by the very people who develop them and use them. (Volitional derives from the Latin *uolo*, I wish for something, want something, or choose something with free will – cf. *uoluntarie*, of one's own accord.). This uncertainty does not derive from that fact that we may not know the probabilities on the roulette wheel the casino is using, but from the possibility that the casino may be consciously capable of affecting these probabilities in its favor, based on the distribution of bets. This may sound like a nonscientific position, but people and institutions often modify their behavior based on what they perceive to be (or are told to be) probabilities of various outcomes, and these changes in behavior in turn change the probabilities. Therefore, the problem of predicting behavior under uncertainty is not one of linear extrapolation; it involves nonlinear feedback loops, psychology, and game theory.

Suppose you were asked to give the probability that you will spend this evening doing homework, as opposed to going to the movies, watching TV, or playing poker. You may give the odds for each, but these are not odds that someone else would care to bet on, as the actual outcome depends entirely on your volition. Someone placing a bet on what you will do this evening would have to consider your psychological state, peer pressure (to play poker), and whether you stand to profit from the bet.

As another example, suppose that a contractor is informed by an objective, unbiased risk analyst that the probability that the contractor will overrun his schedule and thereby forfeit a substantial incentive payment is 50%. We may expect that this contractor will take whatever steps he can to change this probability. In fact, the amount that this probability changes due to contractor actions may depend on the amount of the incentive payment relative to the costs to the contractor of getting earlier completion. We would expect the contractor to be able to change this probability by any of a number of actions, until the probability of overrunning is much less than 50%, but if he/she had not been informed of the 50% probability of overrunning, he might have taken no action, and the 50% would have continued to be true. Conversely, if the contractor had been informed that the probability of overrunning on this job was 1%, the contractor would probably place his attention and resources on other jobs, and the probability of overrunning might increase as a result.

Under such conditions, what does the term "probability" mean? It certainly has nothing to do with relative frequency. This job will ultimately finish; the contractor will get incentive payment or won't; there will be no repetitions of this experiment from which to gather statistics.

A true probabilist might argue that the apparent issue arises because the problem is miss-specified. What was called a probability is really a conditional probability; the quoted 50% probability is really a value that is conditional on everything remaining the same, that is, the contractor taking no action, or being unaware of the probability statement. There are then multiple conditional probability distributions, for the probability of overrunning conditional on the contractor taking no action, some action, moderate action, a lot of action, etc. That is, volitional risk is not a property of an event or a project, but rather is something that can be mitigated or managed through actions of people. Thus, if X is some risk event, we may characterize the uncertainty about whether the event X will occur in one of the following ways:

- There is an aleatory probability $P[X]$ that is characteristic of X and that can be estimated objectively from the relative frequency of past occurrences of X.
- There is an epistemic or subjective probability $P[X]$ that reflects the observer's degree of belief that X will occur, whether or not X has ever occurred in the past.
- There are volitional risks $P[X|\text{no mitigation}]$, $P[X|\text{mitigation of type 1}]$, $P[X|\text{mitigation of type } n]$ etc., depending on what somebody chooses to do after obtaining some information about $P[X]$.

Theory of Evidence and Dempster-Shafer Theory, which deal with degrees of belief and how to change them based on new evidence (Dempster 1976; Pearl 2014). Thus, when the hypothetical unbiased observer says that the probability of over-running is 50%, what he means is that his degree of belief that the project will overrun, based on observed conditions, past history, and other factors, is about midway between no chance at all and dead certainty. This degree of belief may well change if the contractor takes some action that changes the observer's belief in the outcome. We might then inquire, what actions could the contractor take that most cost-effectively change the observer's degree of belief that the project will overrun? If there is more than one observer, with different degrees of belief, how can they be combined into one? Or should they be?

Fuzzy Sets, which assign quantitative values (membership functions) to linguistic terms, such as "risky." Fuzzy sets are super sets of classical set theory or Boolean logic, in which a proposition is either true or false (1 or 0). In fuzzy set theory and fuzzy logic, propositions may take on any values in the interval from 0 to 1. Therefore, fuzzy set theory is said to be more general than classical set theory, although Boolean logic is general enough to power all digital (really, binary) computers (Klir and Yuan 1995). Subjective or Bayesian probabilities do not derive from classical set theory, but rather are a part of the mathematical field of measure theory. Therefore, fuzzy sets and subjective probabilities are two completely different, incommensurate ways of dealing with uncertainty. A simplified example of the difference is the following. When categorizing some object, in fuzzy set theory the parameters of the object are known with perfect certainty, but classification is difficult because the boundaries of the classes (sets) are vague or ambiguous (i.e., fuzzy), due to the limitations (ambiguity, lack of precision) of language. In the probabilistic approach, the boundaries of the classes are known with perfect precision, but classification is difficult because the values of the parameters (measurements) are uncertain.

In addition to the above-mentioned, there are many other theories and methods to represent uncertainty in decision process including possibility theory, interval probabilities, entropy, ambiguity, and info-gap theory. In attempt to obtain more realistic representation of uncertainty in the system behavior, researchers and scientist will continue to look for new methods.

1.2.1   Process Variability

Project outcomes and data can also be analyzed probabilistically from a time-series perspective. In fact, this text also examines the dichotomy of uncertainty suggested by the methods of Statistical Quality Control (SQC), Statistical Process Control (SPC), and, most recently, Six-Sigma. Here, adopting the viewpoint of Statistical Process Control, we make a distinction between processes (that is, project time series) that are in control and those that are out of control.

Stating that a process is in statistical control does not mean that it doesn't vary, or that is completely deterministic. Rather, this term recognizes that there are uncertainties or variability in the process, due to the action of common causes, but the process stays within the limits of this variability. This variability or uncertainty is inherent to the nature of the process (Thompson and Koronaki 2002). As long as the variability in the process stays within bounds, the process is said to be under control. In the language of statistical quality control or Six-Sigma, σ is the natural or inherent standard deviation of the variability of the process, and as long as the output of the process is randomly variable but remains within the bounds defined by the mean plus or minus 3σ, it is considered to be in control.

However, the fact that a project or process is in statistical control does not mean that it is acceptable. The natural variability may be larger than we can tolerate. The outputs may be in control, but they may not meet our requirements. In this case, the process as defined is not capable of producing acceptable work. For example, suppose that we are concerned with the total cost (or duration) of a project. Each work package in the project has some expected cost, and also some uncertainty or variability due to the nature of the work, local conditions, etc. However, it may happen that, when one examines the variability or uncertainty in the total project cost, which is in some sense the sum of the cost uncertainties in all the work packages, the likelihood of exceeding the project budget is unacceptably large. The solution to this problem is to change the process. Some or many of the work packages may have to have their work processes changed in order to reduce the expected cost or the variability (uncertainty) in the costs to an acceptable level.

Conversely, a project or process may go out of control due to some extraordinary external or internal cause. The solution to this problem, if it occurs, is to track down the cause and eliminate it. A better solution is to identify such causes before they happen and take steps to assure that they don't happen, or that their impact is much reduced. This is the objective of risk analysis and mitigation.

In this text, the first kind of uncertainty is that associated with the natural variability of processes, even those under control. Incremental risks include risks that are not major in themselves but can accumulate to constitute a major risk. For example, a cost overrun in one subcontract may not in itself constitute a risk to the project budget, but if a number of subcontracts overrun simultaneously, due to coincidence or to some common cause, then there may be a serious risk to the project budget. Individually, such risks may not be major or difficult to identify; the risk really lies in the combination of a number of them, and the lack of recognition that these could occur simultaneously. These incremental risks are typically analyzed using probability distributions, in the form of either probability density functions or probability mass functions. Often we use probability distributions because they can represent our lack of information or state of ignorance by relatively few parameters (usually two or three), and who wants to have to specify a lot of parameters to express one's ignorance? In this case, the variability in each activity or process is considered to be incremental or differential; it is the combination of all the activities that is of concern.

In the second case, attention is focused on specific, discrete events that are uncertain in that they may or may not occur. These catastrophic risks include risks that could be major threats to the project performance, cost, or schedule. Such risks might include dependence on critical technologies that might or might not prove to work, scale-up of bench-level technologies to full-scale operations, discovery of waste products or contamination not expected or not adequately characterized, dependence on single suppliers or sources of critical equipment, etc. These risks are typically discrete events, and must be individually managed. This is the area usually considered to be risk mitigation or risk management, but in fact both these cases require management. In the first case, the project manager must identify the control limits and design and manage the project execution process such that it stays within the required limits relative to time, cost, and quality. In the second case, he/she identifies and manages discrete risks.

Further, there are situations when variations and changes in trend are due to transitions in internal behavior, rather than external risk; projects, in fact, often experience changes in trend over time and exhibit non-stationarities and tipping points. This dynamic nature of uncertain systems' behavior can be represented using stochastic/probabilistic processes. However, as this is rather a very broad area of probability theory, for practical purposes, we limit the scope of this text to time-series, process control, and few fundamental processes such as Poisson process.

1.2.2   Probabilities and Decisions

Uncertainty assessment and risk analysis are generally done for the purpose of making risk-informed decisions. Hence, it is upon the decision-maker to interpret the result of the analysis. Figure 1.1 illustrates this process. The two typical paths are illustrated with a blue and orange colored lines; the former shows when system/project data is available (blue line – frequentist approach), and the latter shows when the data is not readily available and when the experts' judgements are required (orange line – epistemic approach). In both cases the probability estimates are based on the analyst's assumptions. However, there is a path, illustrated with a maroon colored line, that doesn't involve the same level of assumptions. It feeds processed data about the system directly to the decision-maker. This type of data is representative of the system behavior, but not expressed in terms of outcome probabilities.

There are two reasons why one may want to consult other-than probability measures when making decisions. First, probabilities are sometimes hard to interpret and/or can be misleading; for example, one may be tempted to replace flight data with the cockpit indicator that reports the results of a risk analysis for a stall probability. However, from the pilot's perspective (i.e. decision-maker), this information would be counterproductive as flight data i.e. the position of an aircraft and speed are much informative and easier to map into decisions than the estimates of the event probabilities. How much one should be one worried if this probability is 1%? Would 50% probability warrant an immediate mitigation decision? Pilots regularly make maneuvers that significantly increase this probability, yet they are

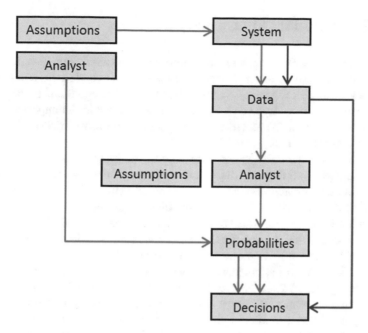

Fig. 1.1 Uncertainty and risk assessment decision process

in full control. This doesn't imply that providing such probabilities is wrong; in fact all commercial aircrafts have a similar feature i.e. an alarm, but it stresses the fact that proper decisions cannot be made in isolation of system-level data. Second, by relying only on probability measures one allows for convolution of assumptions. For example, we regularly make decisions about distribution type, sample independence, and stability of the underlying process that has generated the data. Similarly, subject-matter experts assign probability based on their experience about the systems behavior but based on many conscious and subconscious assumptions and biases (Damnjanovic and Aven 2018). The bottom line is – How valid are these assumptions?

To overcome this issue project managers supplement the probability measures with indicators developed using system-level data that have no, or only few embedded assumptions. Here the system feature is used as an implicit indicator of outcome probabilities; this is similar, or perhaps the same as leading indicators. For example, task completion times are highly sensitive to resource availability; if the tasks share common resources, then such set up is more risky from the perspective of completion time, as any event affect the common resource will be propagated to all tasks. Hence, having a measure of resource-task dependences is informative about the risk, but not defined in terms of probabilities. Similarly, as mentioned before, project managers often use scope index to assess potential risks with scope creep, and feedback loops in design to reflect design rework risk. Some may refer to these as explanatory variables in a propensity function. But then again, this would imply adding assumptions such as type of the function, which is precisely what we try to avoid.

1.3   Is Uncertainty Always Bad?

In general, most people prefer less uncertainty to more uncertainty. In fact, they may take many actions to buffer or protect themselves from uncertainties and their consequences. Also, risk is generally a pejorative term. Although there are people who like to take risks (e.g., skydivers), the intent is still to avoid the unpleasant consequences of the risk; if the risk actually happens (the parachute doesn't open), it is generally considered bad.

However, in many cases uncertainty may mean opportunity. In fact, one may say there are no opportunities without uncertainties (if there were, someone would have found them already). Even Chinese language symbol for risk reflects its dual nature 危機; it encompasses a symbol reflecting danger and a symbol representing opportunity. Hence, if one wants to make a lot of money on Wall Street, one has to look at the stocks that are volatile (have large variability, called). However, this is also the way to lose a lot of money on Wall Street.

The judicious use of alternates or options may add value to projects even when (or especially when) these options have uncertainties (or risks). As an introductory example to what will follow, consider a manager on some project with the following conditions: if, after engineering is complete, the construction cost estimate less than $8,000K, then he is a hero; if it is more than $10,000K then he is the goat. Suppose that he will know the cost with sufficient certainty to make a go or no-go decision after the end of the detailed design phase. Suppose also that there are two alternate technologies. Technology 1 is conventional and straightforward, with an a priori (before the design phase) expected value of $9000K and a standard deviation of $1000K. Assuming these cost uncertainties are approximated by the well-known Normal probability distribution, then the project manager calculates that the probability of exceeding the maximum allotted budget if he chooses this design option is about 16%, which is not bad, but the probability that he will get positive recognition for coming in less than the lower target is also not better than 16%. See Figs. 1.2 and 1.3 for the probability density functions and the cumulative distribution functions.

Even though the probability distribution is symmetric, the project manager's view of the outcomes is not necessarily symmetric. He/she may feel that being a hero is good, but the utility of being a hero is a lot less than the disutility of being a goat. That is, he may be risk averse.

The project manager also has available an alternate process, technology 2, with a higher *a priori* expected value of $10,000K and a much higher standard deviation of $4,000K. Technology 2 is not only expected to cost more than technology 1, it is far riskier. Using the Normal assumption, the project manager calculates the chances of overrunning the maximum budget with method 2 to be 50%. On the other hand, the chances of being a hero with this technology are 31%. Assume that the project manager has to make a decision on the technology before the design starts; he/she does not have enough time in the schedule to perform one design and then do the other if the first one does not come out satisfactorily. Probably under these conditions the project manager chooses technology 1; the chances of winning

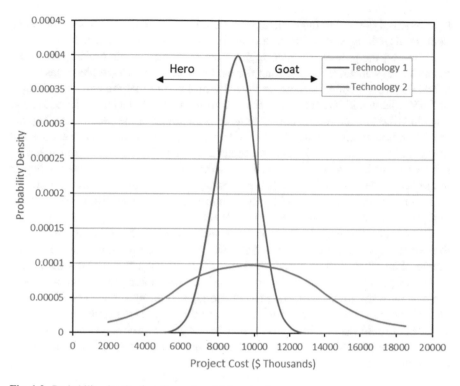

Fig. 1.2 Probability density functions of project cost outcome

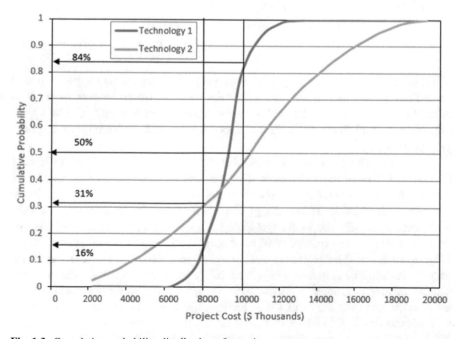

Fig. 1.3 Cumulative probability distributions for project cost outcome

big are small but so are the chances of losing big. Technology 2 has an attractive likelihood of doing very well, but too much chance of doing really badly.

However, the project manager might consider designing the project using both technologies concurrently, choosing the best one after the design phase has established the construction cost. In this case, the probability that the maximum budget of $10,000K would be exceeded is the likelihood that both technologies cost more than that (the project manager always chooses the lower of the two). Assuming that the technologies are independent (a higher than expected cost for one does not imply a higher than expected cost for the other), then the probability of exceeding the maximum budget is just the probability that both technologies result in higher than acceptable cost estimates. More personally, the probability that the project manager is a goat is the probability that technology 1 exceeds $10,000K and technology 2 exceeds $10,000K = (0.16)(0.50) = 0.08.

Conversely, the probability of beating the minimum target is the probability that either technology is less than $8000K, or 1 – the probability that both are greater than $8000K. This comes to $[1.0 - (1.0 - 0.16)(1.0 - 0.31)] = 0.42$. That is, by designing using both technologies and then exercising the option to choose the one with the lower cost, the project manager reduces the chance of looking like a goat by one-half, from 16% to 8%, while at the same time increasing his chances of looking like a big hero from 16% to 42%. In probability notation, let x_1 represent the construction cost using technology 1, and let x_2 be the construction cost using technology 2. Then,

$$P[PM = Goat] = P\left[(x_1 > 10000) \cap (x_2 > 10000)\right] = (0.16)(0.50) = 0.08$$
$$P[PM = Hero] = P\left[(x_1 \le 8000) \cup (x_2 \le 8000)\right]$$
$$= 1 - P\left[(x_1 > 8000) \cap (x_2 > 8000)\right]$$
$$= 1 - (1.00 - 0.16)(1.00 - 0.31) = 1 - (0.84)(0.69) = 0.42$$

Of course, there is no free lunch; designing using both technologies will cost more than designing just one, and additional analysis would be needed to determine if it is worth it; here the point is that variability or uncertainty can be a good thing if it is used to create additional opportunities and decision options for the project manager.

So, is variability good or bad? It could be either. Generally speaking, variability is more valuable or more desirable if it is optional, or constitutes an alternative that can be used or taken advantage of by the project manager, and less desirable when it is inescapable. If technology 2 were the only option available, then the high variance of this alternative would be undesirable, but really makes little difference, because the probability of overrunning the maximum allowable cost is 1/2, regardless of the uncertainty in the cost of technology 2. However, the presence of technology 2 makes a high variability in technology 1 desirable. If the project manager decides to pursue both design options simultaneously, he can never do worse than the results with technology 1, and he might do a lot better with technology 2. Of course, real projects are not as simplistic as this example, but in any case it is up to the project manager to identify options that can make variability an advantage and add value to this project. As in the old saw, if fate hands you a lemon, make lemonade.

1.4   Managerial Attitudes to Risk and Uncertainty

Project managers are concerned with ways to manage projects in the face of uncertainty, to analyze risks and to mitigate them. Projects continually face new risks, which must be identified, analyzed, and understood in order to develop a framework for selecting projects and successfully executing them. However, the term *risk* has different meanings to different people. For example, to economists and financial analysts risk and uncertainty are synonyms; the smaller the variations, the lower the risks. In engineering and decision-theory, on the other hand, risk is often defined as expectation over a set of unfavorable outcomes. Many books have been written in an attempt to define the terms *risk* and *uncertainty* precisely. That will not be attempted here. The difficulty with precise academic definitions of these terms is that no one in project management feels obliged to use these definitions. Perhaps it would be less ambiguous if they did. Perhaps, on the other hand, academic decision theorists should learn to deal with ambiguity. But under the circumstances, in order to promote improved communication between all participants in a project, it is necessary to use the terms as they are used; that is, vaguely. It is clear from observation that academic decision theorists (those who know how decisions ought to be made) and project managers (those who make decisions) use the terms risk and uncertainty with very different meanings (March and Zur 1987). In fact, to most managers "risk is not primarily a probability concept." Therefore, risk is not evaluated on the basis of uncertainty or probability distributions, as used in decision theory.

Risk is Exposure.   Managers often look at risk as their exposure to loss. That is, the term risk is usually applied to negative events, although a large variance simply means large variability in either direction around the mean. As a result, one might hear statements such as "You have a high risk of a heart attack" but utterances such as "You are at a high risk of winning the state lottery" are rare. Some quantification of the corporate exposure may be made by such means as scenario analysis, but identification of exposure is more commonly the response to "what if" type questions. Risk assessment may largely consist of meetings in which participants try to think of "what if" questions that would lead to organizational financial or other exposure. Lawyers are particularly good at this exercise. It is a useful exercise, but it is often arbitrary and inconsistent. Risks are weighted by outcomes and not likelihoods, so that the process can be highly conservative, given that relatively low probability events can be considered high risks if the consequences are great. Equally, it can be very unconservative, if major risk factors are overlooked or forgotten.

It is clear that, if risk means exposure, then risks are not additive. One may say, "Our investment in this project is $10,000,000, and if Event X happens, our exposure is the loss of our entire investment of $10,000,000." And similarly, "If Event

Y happens, then our exposure is the loss of our entire investment of $10,000,000." But one cannot then say, "If both events *X* and *Y* happen, then our exposure in this project is twice our investment." When deciding to undertake a project or not, management may perform a maximization calculation, in which the largest of all exposures due to all credible events is taken as the measure of the risk of the project.

In this approach, risk is typically denominated in dollars, the financial exposure of the firm if something goes wrong with the project. The risk may be equivalent to the cost of getting out of the project once one is in. Thus, a manager may say, "Our risk in this project is $10,000,000," meaning that the firm has invested that amount or more (in cash or in some other way) in the project and will not be able to recover that amount if the project fails. If the project has some salvage value, then the exposure is the difference between the initial investment and the price for which it can be sold if it fails.

This focus of managers on exposure, or the consequences of failure, does not mean that they are oblivious of probabilities. They may not like to emphasize probability because that is related to gambling, and that implies that the managers are not in control, but rather subject to chance. But, even if managers may not calculate probabilities objectively and mathematically, they may have general subjective, qualitative ideas of what constitutes greater or lesser risk. This qualitative assessment or feel for probability or likelihood is then merged with the numerical assessment of exposure, as in, "For this kind of project, we need to keep our risk (that is, exposure) under $1,000,000." *This kind of project* is, based on the manager's experience, a member of a class of projects that have similar subjective probabilities of going bad. In some way, this may relate to relative frequencies: "Of the last six of *this kind of project* done, two have been outright failures." But in many cases, the probabilities are obtained from a sample of one: "We did something like this before, and we won't do one again."

It is common for decision theorists to denigrate this *gut feel* for project risks as not being quantitative or objective, but this does not prove that managerial experience is not valuable. Many knowledge-based (expert) systems have tried to capture this expertise about how to assess projects. It is not at all dissimilar to a physician's experience in diagnosing diseases: some do it better than others; in general, more experience means better results. Some progress has been made in automating diagnosis, but no one knows how physicians do it, and good diagnosticians continue to be in high demand.

One may indeed explain *gut feel* in Bayesian terms: initially, inexperienced managers have little knowledge about projects, and therefore have prior probability distributions on project outcomes that have very high variance. Each project constitutes a new set of information, which modifies the manager's prior distribution into a posterior distribution. Over a number of years, this distribution evolves to one with less variance. When a prior distribution has high variance, any data point has a great influence on the posterior distribution, but when the prior has a low variance, a new data point has very little effect on the posterior distribution. Hence, managers may seem set in their opinions to outsiders, but to themselves they have simply learned from many experiences.

Risk is Multidimensional Concept.   Another difference between the decision theoretic view and the managerial view is that theorists like to summarize project risks
into a single risk probability, whereas managers have very limited desire to reduce
the risk to aggregate probabilities. Instead, managers look on risks as multidimensional, with a maximum exposure considered for each risk category. The combination of all risks in a single a priori number is useful in a decision whether or not to
proceed with the project. Obviously, this decision is important, but it is only one
part of total risk management. The academic focus on reduction of risks to a single
number has actually had a deleterious effect on the development of methods for risk
assessment and management. People refer to risk assessment as if it were nothing
more than the simplification of many risks and circumstances into a single number,
often by Monte Carlo simulation, and much of the available software reflects this
simplified viewpoint. In this area, project managers are much more sophisticated
than the decision theorists, in that they are able to consider and balance a number of
risk factors and circumstances independently. In fact, this is the essence of holistic
risk management as opposed to simplified risk assessment: holistic risk management deals with the total risk environment of the project, not just at the initial decision, but also throughout the life of the project.

In fact, the holistic view on risk management requires not just the synthesis or
integration of risks into a single number, but the analysis, or differentiation, of risks
into manageable parts. Only when the risks are identified and differentiated can
management come up with policies on how to deal with them.

Moreover, and possibly the most serious objection, the focus on combining risks
into some single risk index or number diverts attention from the most serious issue
at the core of risk assessment: the *inference problem.* What are the probability distributions, where do they come from, and how do we estimate them, in the almost
total absence of relative frequency data?

Taking Risk Depends on Situations.   Higher-level organisms as well as organizations survive by taking only reasoned risks and avoiding excessive risks. Large
organizations protect themselves against unwarranted risks by internal and external
reviews and signoffs. For example, a major commercial aircraft manufacturer is said
by one of its engineers to require a huge number of signoffs on all engineering
changes – a policy that is frustrating to those engineers who feel that they have great
new ideas, but no doubt is reassuring to all the passengers flying in its products,
who, if asked, would probably feel that corporate restraints on risk-taking in aircraft
design are a very good idea. On the other hand, time plays a big role in distributed
decision-making, and risks often need to be taken at the level where the problem
arises. For example, if a new technical challenge appears on the site, valuable time
would be lost if the decisions are not made promptly. Hence, there is a balance
between too much risk-taking and too-little risk taking at different levels of the
organizational structure.

In general, managers typically warrant risk taking when faced with likely failure.
By this principle, contractors will take more risks (for example, by submitting very
low bids to "buy" jobs) when business is bad and their survival is under threat. Also,

by this principle, project managers would be more likely to take risks on a bad project than on a project that is going well. As an example, consider two similar projects: for one, Earned Value Analysis indicates that the Cost at Completion will be $1,000,000 over budget; for the other, $1,000,000 under budget. So in the first case, the contractor would show a loss of $1,000,000, and in the second case a profit of $1,000,000. Suppose that there were some new process that had some chance of reducing costs by $1,000,000. Which project would take the risk of using this process: the first one, which might reduce the loss from $1,000,000 to break-even, or the second, which might increase the profit from $1,000,000 to $2,000,000? In line with the principle stated here, most people would probably say the project showing the loss would take the risk – even though the monetary gain would be the same for either. In fact, this behavior may be entirely rational, and even optimal.

Risks are Perceived Controllable. Managers much like other people find patterns and underlying rules in sequences of events that are in fact completely random. The difficulty here is, what if the set of events is really indeterminate, and the fact that the project manager imposes his/her logic on them does not make them determinate?

This principle is very important in assessing the actual behavior of managers when assessing and managing risks. Senior managers in particular have arrived at senior positions by making good decisions or having good luck (which may be hard to distinguish). Naturally, they prefer to think it is good decision-making. Therefore, they are led by the very circumstances of their positions in the organization to believe that they control events, and not that events control them. Successful corporate executives may be like generals with a string of victories, who come to believe that they are – think of Napoleon invading Russia. Unfortunately for him, subsequent events did not follow Bonaparte's logic, and the result was a disaster for the French and their allies.

While the empirical managerial approaches to risk show a sharp contrast with the decision-theoretic viewpoints, the comparison is somewhat misleading. Just as we may admire the nerve of a tightrope walker, without seeing the safety net stretched below him, or of a steelworker on a tall building, while failing to notice the safety belt tying him off, so we may think managers are taking risks when they have the skills derived from experience to mitigate them, avoid them, or hedge them, in ways not immediately apparent. This text is concerned with how to acquire some of these skills without spending so many years to do it.

In summary, the empirical managerial approach to risk is:

- Break down the total risk into its components
- Analyze the risk for each component, largely in terms of its maximum exposure for loss, in the total context of the project, the environment, and historical performance
- If any risk is unacceptable, take steps to reduce it, manage it, and control it
- Revise the project definition until all risk dimensions are acceptable before commitment

The more experienced and successful a manager is, the more he/she believes that he/she can control risks because the manager has gotten where he/she is by succeed-

Table 1.1 Decision-theoretic vs. managerial viewpoint on risk

Decision-theoretic view	Managerial view
Sees risks as probabilities	Sees risks as exposure
Synthesizes individual risks into one risk factor	Breaks out risks into individual components for mitigation
Quantifies risks numerically	Characterizes risks verbally and qualitatively
Looks at probability distributions over (ideally) all possible outcomes	Looks at relative few possible outcomes.
Sees risks as random events	Sees risks as avoidable or controllable
Finds optimal solutions under uncertainty	Incrementally moves to intuitively satisfactory solutions ("satisfices")

ing in previous circumstances; he/she does not attribute this success to luck. Therefore, project managers are more willing to accept risks if they have more experience with successful previous projects. Conversely, project managers may be unwilling to accept risks if they have not had experience successfully managing projects under relevant conditions of public scrutiny, regulations, outside stakeholder influence, tight budgets, fixed price contracts, adversarial relations with contractors, etc. Successful project managers may not always be correct in their assumptions that they can control risks, and making a mistake in this regard can have serious consequences.

We can summarize the differences in the two approaches to risk in Table 1.1.

At this point, the reader may wonder, with all these managerial approaches to risk assessment, why should one consider probability theory and decision analysis? The reason, of course, is that the common or intuitive management approach does not necessarily give good results. There are some managers who are very good at risk assessment and management. Unfortunately, there are too few of them. The method of education of managers in this field is essentially one of apprenticeship: junior managers observe more senior ones and apply what they have learned to projects of their own, until they either rise in the organization and replace their mentors, or they fail. Unfortunately, too many fail. In Table 1.1 the skills on the left can be taught; the skills on the right can be learned, but cannot be taught. Therefore, industry is interested in better methods for risk management that are more consistent, more objective, and reproducible; that can be formally taught; and also that give better results. This text is intended to try to help meet that need by bridging (or at least straddling) the gap between the decision theoretic approach and the managerial approach.

1.5 Holistic Approaches to Risk

Although the viewpoints on risk summarized above are different, they are not mutually exclusive. As noted above, managers take risks when they have sufficient experience to understand the nature of the risks involved, when to take risks and when not, and how to control and manage risks. Conversely, project managers who may not have enough experience with fixed-price contracts, project management, and budget and schedule control to have developed the confidence to take (i.e., to accept responsibility for controlling) risks may be perceived, and may perceive themselves, as risk averse.

One solution lies in integrating the analytical and experiential approaches to risk described above. By identifying, objectifying, quantifying, and estimating risks, by inferring appropriate probabilities, and by assessing these individual risks through simulation, scenario analysis, decision analysis, and other techniques, project managers should be able to overcome lack of experience by means of analytics. By synthesizing the managerial approach to risk with decision theoretic and analysis methods, project managers should be able to take appropriate risks because the analysis should quantify the risks and simulation should indicate how they can be controlled.

This text is concerned with the use of the probabilistic approach to examine managers' assumptions and methods, to try to determine to what extent managers may actually conform to decision theoretic methods and principles. That is, we will use some of the theory to try to analyze what managers are really doing, to determine to what extent decision theory is descriptive and not merely prescriptive. Considering the long history of projects, it would be remarkable of project managers, contractors, and others involved had not developed techniques for buffering themselves from the effects of uncertainty. These risk mitigation or safety factors may be so ingrained that they are not actually highly visible, but they may be there. In this process, we will try to build some models of managerial approaches to risk assessment and management and to compare these with our beliefs about the operations of the real world of projects.

References

Damnjanovic I, Aven T (2018) Critical slowing-down framework for monitoring early warning signs of surprise and unforeseen events. In: Knowledge in risk assessment and management. Wiley, Hoboken, pp 81–101

Dempster G (1976) A mathematical theory of evidence. Princeton University Press, Princeton

Gelman A, Carlin J, Stern H, Dunson D, Vehtari A, Rubin D (2013) Bayesian data analysis. Chapman and Hall/CRC, New York

Klir G, Yuan B (1995) Fuzzy sets and fuzzy logic, vol 4. Prentice hall, Princeton

March JG, Zur S (1987) Managerial perspectives on risk and risk taking. Manag Sci 31(11):1404

Pearl J (2014) Probabilistic reasoning in intelligent systems: networks of plausible inference. Elsevier, Amsterdam

Thompson JR, Koronaki J (2002) Statistical process control: the Deming paradigm and beyond, 2nd edn. Chapman & Hall, Boca Raton

Chapter 2
Project Risk Management Fundamentals

Abstract In this chapter we present the fundamentals of project risk management. We provide an overview of the overall process including risk identification, qualitative and quantitative risk assessment, and risk mitigation/treatment. We discuss different approaches to modeling project risks and provide a context for the materials we cover in the following chapters.

Keywords Project uncertainty and risks · Risk management framework · Representation methods

2.1   Uncertainty and Risks in Projects

It is often not obvious why projects may have large amounts of uncertainty. In many straightforward projects, both duration and cost are largely determined by the quantity of work units to be done and the unit rates. In placing concrete, for example, the number of yards of concrete is determined from the design drawings. The number of yards that can be placed in a day by a crew determines the duration, and the cost per yard for concrete and formwork, materials and labor, determines the cost. In this type of situation, uncertainty may be introduced primarily by lack of foreknowledge of external factors such as weather, which may affect productivity.

Excavation can be very similar to concrete placement, in that the gross volume is known. But in addition to being weather dependent, often uncertainty is introduced by lack of knowledge of the quantities of various materials that may not be known until the work is actually performed, due to inadequate subsurface sampling. The differences in time and cost to excavate rock, compared to loose material, can introduce risks.

The risks described above, due to weather, subsurface conditions, etc., would seem to be well understood and quantifiable. There are, however, many types of projects, the duration and cost of which are not primarily determined by a fixed quantity of work to be done multiplied by a unit rate. These include, for example, engineering projects that must execute multiple design cycles, iterating until some

© Springer Nature Switzerland AG 2020
I. Damnjanovic, K. Reinschmidt, *Data Analytics for Engineering and Construction Project Risk Management*, Risk, Systems and Decisions,
https://doi.org/10.1007/978-3-030-14251-3_2

design objectives are met. The number of cycles may depend on the difficulty of meeting these objectives, which is not really known until it is done. In general, this type of uncertainly applies to all kinds of projects that consist of a number of steps with acceptance tests or quality control tests at the end of each step. In these projects, if the acceptance test is successfully passed, then the project enters a new step or phase, but if the acceptance test is not passed, the previous step (or even a number of previous steps) must be repeated until the acceptance criteria are met. Construction projects may fall into this category if the quality assurance requirements are stringent compared to the level of quality achieved, as in the case of pipe welding, for which welds that do not pass radiographic examination must be cut out and redone until they do. In such a case, the time and cost of producing an acceptable weld are not determined by unit rates so much as by the reject rates; higher quality means shorter durations and lower costs, whereas lower quality means the opposite.

Similar examples abound in a number of fields. Software development is an area in which attempts to predict time and cost based on quantity of work, that is, estimated number of lines of code, have proven to be unreliable, and various organizational structures have been tried, such as prototyping and spiral development model, to reduce the probability of rework and recycling. In the construction examples cited earlier, even concrete placement can be driven by rework considerations if the acceptance criteria are stringent, either for high strength or dimensional accuracy. Uncertainty is increased by high reject rates, and these are often characterized by acceptance criteria that are at the margin or the boundary of the processes being used. In such cases it may be useless or even counterproductive to focus on unit rates and unit costs; to reduce uncertainty one must focus on methods to achieve higher quality or adoption of alternate improved processes, in order to increase the acceptance rates, or relax the acceptance criteria.

Many projects may fall into this category including most notably: (a) Scientific and R&D projects. In cases in which new science is being brought along from theory to bench-scale laboratory tests to pilot plants to full-scale operations, there may be a number of places in which acceptance tests need to be applied. These points may be addressed in the form of readiness-to-proceed reviews or Critical Decision points. If performance, for example, is not adequate at the end of one step or phase, then that step should be prolonged or repeated until it is acceptable. Pushing forward into the next step without a readiness review or before the acceptance conditions in the prior step have been achieved, in order to meet predetermined project schedules or budgets, almost invariably generates poor results; and (b) Environmental restoration projects. Some projects may adopt an incremental approach to cleanup, in which restoration proceeds in steps based on permit requirements, characterization of the pollutants, cleanup technologies, etc. For example, permits may require that cleanup technology developed in one phase be demonstrated to be at a certain level before proceeding with the next phase. Therefore, the time and cost for complete cleanup depends on the probability that these acceptance tests will be successful and the readiness reviews will be passed.

Some of the risks that might apply to projects are listed below. As no such list can be comprehensive you may want to start with this list and add your own favorites.

In addition to thinking about risks that affect project cost and duration, do not forget to include risks associated with the mission, use, business case, or economic viability of the project.

- Accidents
- Availability of third-party, nonrecourse financing
- Changes in owner's need for facility
- Client/user scope changes
- Competitive factors (reduced priced, reduced sales) compared to economic projections
- Construction cost increases
- Construction delays
- Contractor default
- Contractor inexperience or incompetence
- Costs of borrowing money
- Costs of termination
- Decommissioning and cleanup costs
- Economic and business conditions (for example, recession)
- Enforceability of contracts
- Engineering changes and design development
- Environmental factors
- External influences: delays, changes
- Failure of equipment to perform to specifications
- Failure of technology to perform as required
- Force majeure
- High bids
- Inability to accept product
- Inability to deliver raw materials
- Inaccuracy of operating cost estimates
- Inaccuracy of construction cost estimates
- Inadequate number of bidders, inadequate competition
- Inflation
- Interest rate increases
- Late delivery of equipment
- Late start due to upstream conditions
- Low availability, reliability, or throughput
- Low bidder unreasonably low
- Maintenance costs higher than expected
- Managerial experience
- Mismatch of technology and project conditions
- Operating costs higher than expected
- Operational accidents, equipment failures
- Operational performance: productivity, efficiency, availability, reliability
- Permitting and licensing delays or rejections
- Poor quality of construction; failure to meet specifications

- Reduced patronage of the facility (compared to economic projections)
- Regulatory or legal challenges
- Service or economic life less than projected
- Scale-up of technology from laboratory or pilot plant
- Site conditions
- Spills or leaks of toxic materials
- Startup problems
- Strikes and work stoppages
- Unavailability of skilled labor
- Use of new or unproved technology
- Waste characterization
- Weather

It is rather unsatisfying to have one large incomplete list of risks, from both practical and theoretical viewpoint. So, one may be tempted to structure this list around some common principles. For example, first we may think of classifying risks that relate to different stakeholders' objectives; therefore, we typically have:

Performance, scope, quality, or technological risks. These include the risks that the project when completed fails to perform as intended or fails to meet the mission or business requirements that generated the need for it. Performance risks can lead to schedule and cost risks if scope creep is permitted to increase the time and cost of the project.

Schedule risk. This is the risk that the project takes longer than anticipated or scheduled. Schedule risk may lead to cost risks, as longer projects always cost more, and to performance risk, if the project is completed too late to perform its intended mission fully.

Cost risk. This is the risk that the project costs more than budgeted. Cost risk may lead to performance risk if cost overruns lead to scope reductions to try to stay within the baseline budget. Cost risk may also lead to schedule risk if the schedule is extended due to lack of funds to accomplish the project with increased costs.

All of these risks may come in two varieties:

Incremental risks. These include risks that are not major in themselves but can accumulate to constitute a major risk. For example, a cost overrun in one subcontract may not in itself constitute a risk to the project budget, but if a number of subcontracts overrun simultaneously, due to coincidence or to some common cause, then there may be a serious risk to the project budget. Individually, such risks may not be serious or difficult to identify; the risk really lies in the combination of a number of them, and the lack of recognition that these could occur simultaneously.

Catastrophic risks. These include risks that could be major threats to the project performance, cost, or schedule. Such risks have included dependence on critical technologies that might or might not prove to work, scale-up of bench-level technologies to full-scale operations, discovery of waste products or contamination not expected or not adequately characterized, dependence on single suppliers or sources of critical equipment, etc.

We can continue further with the classification process and distinguish risks based on systems notation (i.e. internal or external to the system); when the risk is introduced and when it realizes in the project life-cycle (planning, design, execution); what type of operation it is linked to (welding, compaction, assembly, logistics, drilling), is it technical, product related, or non-technical, process related, and so forth. This and similar classification methods are useful from the perspective of trying to provide structured input for future analysis; however, there is a danger in over-classification and ignoring the fact that many of them are interdependent. For example, technological risks could easily affect the schedule, and vice versa; or the same risk could affect multiple activities e.g. weather, change of regulation, or material shortages could affect the outcomes of a number of activities.

To summarize, uncertainty and risk in projects is driven not only by external factors such as weather and site condition affecting well-defined operations (i.e. concrete placing or excavation), but also by a combination of internally-defined conditions and requirements, organizational structure, and distributed decision-making which makes the outcomes highly uncertain and the list of risks that apply to the project rather long. Faced with this challenge it is critical to develop a method to "break-down" this list into categories, yet making sure that the interdependencies among them are fully accounted for.

2.2 Risk Management Framework

Managing risks is one of the most important functions of the owner in making any major project successful. In general, the owner initially owns all of the risks, as it is the owner's decision to execute the project. Of course, it is also true that not executing the project may entail risks, to the ability of the owner to perform its mission.

Risk management *is not* a function the owner can delegate to contractors. Contractors and consultants can play major roles in identification and assessment of risks, but there remains an essential role for the owner that cannot be delegated: the identification, mitigation, acceptance, and management of the owner's risks.

The definition of the major steps to be taken in the process of analyzing and managing risks is somewhat inconsistent among different organizations. International Organization for Standardization (ISO) defines four main processes that constitute risk management process: assessment, treatment, monitoring, and communication. Society for Risk Analysis (SRA), on the other hand, has a different classification approach. The SRA defines the overarching term as risk analysis which then covers assessment, characterization, communication, management, and policy. Finally, Project Management Institute (PMI) has its own definition of steps required to make risk-informed decision-making: identification of risks, qualitative risk analysis, followed by a quantitative risk analysis, then planning for risk responses, and finally controlling risks. This is in addition to establishing the context for doing risk management activities to start with (PMI 2008).

As previously discussed, there is really no point in debating definition of the term "risk" and which of the project risk management definition is more elaborate and encompassing as project managers are not compelled to adopt any. Hence, our approach does not emphasize either definition; rather, it is focused on typical activities associated with understating the impact of risk and uncertainty in project decision-making; although, it is closest to the PMI classification. To this aim, one needs to do the following: (a) identify risks, (b) assess their impact using qualitative and quantitative methods, and (c) develop transfer and mitigation strategies; in other words – treat the risks. These three steps are performed at various levels during project life-cycle phases including planning, design and execution. Next, we discuss these three steps in more details.

2.2.1 Risk Identification

The owner may not be in a position to identify all the risks of a project unassisted, due to lack of familiarity with similar projects, but it is the responsibility of the owner's representative to make sure that all the significant risks are identified. The actual identification of risk may be carried out by the owner's representatives, by contractors, and by internal and external consultants or advisors. The risk identification function should not be left to chance, but should be explicitly covered in the Statement of Work (SOW) for the project, the project Work Breakdown Structure (WBS), the project budget, the project schedule, and the Project Execution Plan.

Risk identification is one of the most important functions of the integrated project team (IPT), and is one major reason why IPTs should be formed very early in the project and should meet face-to-face as soon as possible. Members of the integrated project teams should be selected on the basis of their ability to bring breadth of experience and viewpoints to the risk identification process. Ample examples exist of ill-advised projects that have gone forward because only the viewpoints of those with vested interests in the project were ever heard. Participation of all the members of the IPT is necessary to make sure that all significant project risks are identified. The owner's representative should be present at all such meetings.

There are a number of methods in use for risk identification. Typically, they involve brainstorming sessions by the IPT or a significant subset of it. In general, personal contact and group dynamics are involved in successful risk identification. Assigning the risk identification process to a contractor or individual member of the project staff is rarely successful, and may be considered to be a way to achieve the appearance of risk identification without actually doing it. However, objective, impartial external consultants and advisors may provide useful inputs on risk identification. In fact, risk identification should be specified as one of the major functions and contributions of the IPT. Projects should be required to include all risks identified in the project risk assessments. In the risk identification process, it is essential first to elicit all possible risks, without necessarily analyzing them. As in any brainstorming process, no idea should be rejected and every participant should be encouraged to expand on the ideas of others.

Although risk identification is a process that should be performed early in the project life cycle (starting even before the project is committed), and which should be formalized by project management, risk identification should not stop after this phase. Risk identification is not perfect, and therefore should be an ongoing process throughout the project life cycle, especially as new people or contractors are added to the project and may bring different experiences and viewpoints to the risk identi-fication. For this reason, the project risk management plan should provide at least for periodic updates.

2.2.2   Risk Assessment (Qualitative)

Following the initial risk identification phase, the project should have a working list of risks that have been identified as potentially affecting the project. From this list, the project should screen out those that require follow up and those that seem minor and do not require further attention. This process requires some qualitative assess-ment of the magnitude and seriousness of each identified risk. There are various methods to facilitate this. One common method is based on the well-known Failure Modes and Effects Analysis (FMEA), which was developed to assess failures in equipment and systems, but which has also been applied in one form or another to project risks. This type of analysis goes one step beyond risk identification to include a qualitative assessment, typically based on a subjective assessment of the magnitude of the impact of the risk event on the project (often on a scale from one to ten) multiplied by the likelihood that the risk event will occur (often on a scale from one to ten). We can also including a third parameter – the degree of warning that the project will have regarding the actual occurrence of the risk event (also on a scale from one to ten). This third parameter may give some support for the project establishing early warning indicators for specific serious risks, which might not otherwise have been done.

This form of risk assessment is qualitative and relative, not quantitative and absolute. It is primarily for screening out the identified risks that require follow-up, because of high impact or high likelihood, or both, from the risks that do not appear to require follow-up, because of both low impact and low likelihood (see Fig. 2.1). However, due to changes in project conditions or perceptions, risks that appear to have low impact or low likelihood at one time may appear differently at another. Therefore, the project has to re-evaluate risks periodically to assure that some risk previously considered negligible has not increased in either impact or likelihood to the level requiring management attention.

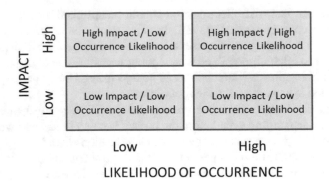

Fig. 2.1 Qualitative screening of risks

2.2.3   Risk Assessment (Quantitative): Low Impact, High Probability

Low impact, high probability risks are those largely due to uncertainties – about the actual costs of materials and labor, the actual durations of activities, deliveries of equipment, productivity of the work force, changes due to design development or owner's preferences, etc. These uncertainties are normally considered to lie within the natural variability of the project planning, design, construction, and startup process. Each of these uncertainties, taken alone, would have little impact on the project. However, taken together, there is the possibility that many of the estimates of these factors would prove to be too optimistic, leading to cumulative effects such as loss of performance, schedule overruns, and cost overruns.

There are basically two methods for addressing this type of uncertainty: (a) Apply project contingencies to cover the uncertainties; and (b) Change the process to one in which there is less uncertainty (variability).

The second approach is no doubt preferable, but is not always used. The use of lump-sum, fixed-price contracts instead of cost-plus contracts is one obvious way to reduce the cost uncertainty for the owner, although it also certainly increases the cost uncertainty for the contractor. But lump-sum, fixed-price contracts may increase the tendency for the development of adversarial relations between the owner and the contractor, whose interests are not completely aligned. Teaming, partnering, and incentive-based contracts in general may be regarded as efforts to align the interests and objectives of both the owner and the contractors and thereby to reduce the uncertainties and risks that may be associated with misalignment of interests. Often, however, the preference among contractors is to cover increased uncertainty by application of higher contingencies, which are limited by what competitors do in the bidding process. That is, contingencies are in effect set by competition and market forces rather than risk analysis.

Due to the incremental nature of this type of uncertainties, they may often be covered by contingency. Contingency is an amount included in the schedule or budget that is not identified with specific factors, but is included to cover a reasonable amount of process uncertainty. To be effective, contingency must be held at a high level in the project. The frequent practice of assigning contingencies to work packages of contracts only assures that the contingencies will be expended; there will be little opportunity to transfer contingency allowances from one work package or contract to another unless these funds are controlled at the project level. For this reason, contractors may be highly disposed to the assignment of contingencies to contracts.

Contingencies should be controlled and managed at the same level at which changes are controlled. Any change approved should be offset against the remaining contingency. The quantitative determination of the proper amounts of schedule and cost contingencies can be made through the application of simple probability theory, in which probabilities are used to represent uncertainties.

2.2.4   Risk Assessment (Quantitative): High Impact, Low Probability

By definition, high-impact, low-probability events are rare occurrences, and therefore it is very difficult to assign probabilities to them. Data do not exist, and subjective estimates of probabilities may be unreliable due to the lack of experience of personnel with such rare events. However, the objective is not the assignment of accurate probabilities to rare events, but the determination of what management actions should be taken to mitigate and manage them. If a certain specific risk is identified as plausible, and if management determines that this risk should be explicitly managed if it had a likelihood of more than 1 in 100 of occurring, then the only issue is whether it is more than 1 in 100 or less than 1 in 100; a determination that the probability is 1 in 50 is irrelevant.

High-impact, low-probability events in general cannot be covered by contingencies. The computation of the expected loss for an event as the product of the loss given the event occurs times the probability of the event is largely meaningless. For example, suppose a certain project is expected to cost $1,000,000 if a certain event does not occur, and $50,000,000 if it does. One will certainly not assign a contingency of $50,000,000 to a $1,000,000 project. If the probability of this event is estimated as 0.02, the expected loss due to the event is $1,000,000. One will not assign this number as a contingency either. If one did, the estimated cost with contingency would rise 100% to $2,000,000. If the event occurs, the contingency of $1,000,000 is completely inadequate to cover it. If the event never occurs, experience shows that the extra $1,000,000 is likely to be spent anyway.

The only way to deal with high-impact, low-probability events is to mitigate them, by reducing the impact or reducing the likelihood, or both. However. risk mitigation and management certainly is not cost-free. In the simple illustration above, it might be worth it to the owner to expend as much as $1,000,000 more to mitigate the $50,000,000 risk (and perhaps more than $1,000,000, if he owner is very risk-averse). (If the risk is the owner's; no contractor is going to expend $1,000,000 for risk mitigation on a $1,000,000 project.) To mitigate the high-impact, low-likelihood risks, it is necessary to identify specific risk mitigation activities. These activities should be estimated and scheduled, and should be included in the project budget and the project network schedule. This means that risk mitigation activities will be tracked and managed just as other critical project activities are.

2.2.5   Risk Mitigation and Treatment Approaches

Insurance Many of the risks associated with projects, especially with construction projects, are insurable. The insurer operates on the same statistical principles as discussed in this course. Insurance spreads out the risk. The insurer sells a large number of insurance policies to various customers. If all these risks are independent, by the law of large numbers the relative combined risk to the insurer declines, compared to the individual risks. If the risks are not actually independent, the insurer has a problem.

Insurance may then be used for those aspects of a project that are common to many projects but independent of other projects, and hence for which the insurer has a large number of customers. Some organizations that do projects have organized their own insurance companies, partly on the belief that they know the risks better than general insurance companies, and can therefore carry the risk at lower premiums.

Risk Transfer Risk transfer is like insurance in that the person or organization holding the risk transfers it to someone else, for a fee of some kind, although not in this case to an insurance company. The risk transfer is like any other transfer in a marketplace, but rather than transferring *goods* in exchange for other goods or money, the risk market transfers *bads*, accompanied by money or other considerations.

For example, the owner may seek to transfer some risk to a contractor, and a general contractor may seek to transfer some risk to subcontractors. In general, the owner starts with the risk, as it is the owner's project before any contracts have been let. Of course, the owner has the prerogative to eliminate project risk by not undertaking the project at all, although this leaves the owner with the risks attendant on *not* doing the project.

Risk transfer can be entirely appropriate when both sides fully understand the risk and the rewards. The side that assumes the risk may do it on the basis that it has knowledge, skills, or other attributes that will reduce the risk, compared to the risk if the owner assumes the risk. If this is true, then it is equitable and economically efficient to transfer the risks, as each party believes to be better off after the exchange than before. This means that net project value has increased by the risk transfer.

Symbolically, we can say that, before any risk transfer, the owner has risk R and amount of money M, and his value placed on this is $V_o(R, M)$. The contractor originally has no risk and no rewards associated with this project, so his value is

$V_c(0,0) = 0$. The owner transfers to the contractor some of his risk, say R_T, along with money M_T, and retains residual risk R_o. The contractor receives money M_T and assumes risk R_c. The total value of the project, after the transfer, increases: $V_o(R_o, M - M_T) + V_c(R_c, M_T) > V_o(R, M) + V_c(0,0)$. Note that it is not claimed that $R_o + R_c = R$, as the risks may not be additive.

As rewards are quantitative (that is, dollars), proper understanding of the risk-reward tradeoff on both sides depends on a quantitative assessment of the risk. Also, in a perfect market with free flow of information, each side would know not only his quantitative assessment of the risk, but also the other side's assessment.

Unfortunately, all too often the risks are not quantitatively assessed and one or both sides may seek to gain an advantage over the other side by concealing his own risk assessment from the other. Such attempts lead to competition and secrecy (or even misrepresentation). As a result of one side trying to gain advantage over the other, the value of the project is not maximized. That is, the total value of the project does not attain the value that could have been reached by full disclosure.

Sometimes owners try to coerce contractors, through market power, to accept risks they would not otherwise assume. If there is a buyer's market in construction, for example, owners can shop around for some contractor willing to accept the owner's risk at less reward than anyone else. Sometimes these attempts are not legally enforceable – such as owners requiring contractors to accept the risk of the owner's own negligence. Even if legal, the outcome may be bad for all concerned, even the owner. For example, the contractor may believe that the combination of risk and reward provide by the owner is unacceptable – the risk is too high or the reward is too low. By taking this contract, the contractor may see that he would face a risk of going bankrupt. However, the contractor may see another risk – the risk of going bankrupt if he does not take the contract. That is, perhaps $V_c(R_c, M_T) < 0$ but $V_c(R_c, M_T) > V_c(0,0)$. The contractor accepts the risk on the basis that going bankrupt in the long term is better than going bankrupt in the short term, and, besides, the risk may never happen. This method may work as long as the risk in question never materializes, but if the critical risk does occur, the contractor goes bankrupt and the project is jeopardized. One would hardly say that this is a win for the owner.

Risk Buffering   Risk buffering or risk hedging is the establishment of some reserve or buffer that can absorb the effects of many risks without jeopardizing the project. A contingency is a buffer. A large contingency reduces the risk of running out of money before the project is complete. If two people go to Las Vegas, one with $10,000 in his pocket and the other with $1000, it is clear that, although the odds may be identical for both, the one with the larger stake has much better chance of survival.

Buffering applies to time as well as budget. It can also apply to the provision of reserves of manpower, machines, or other resources used by the project. Contingency is useful when the risks are incremental and additive, and no single risk is dominant. Note that, as discussed elsewhere, contingency is not necessarily a good way to manage risks of very high impact coupled with low probability.

Although contingency can be (a part of) a successful risk strategy, in the end, adding contingency simply means increasing the project budget (or schedule). Therefore, increasing contingency should not be the first resort; unless all the risks are small, efforts to avoid, transfer, or otherwise mitigate risks should be investigated before recourse is had to increased contingency.

Risk Avoidance   Risk avoidance is the elimination or avoidance of some risk, or class of risks, by changing the parameters of the project. It is related to the question, "How can this project be redesigned so that the risk in question disappears (or is reduced to an acceptable value?" The solution may be engineering, technical, financial, political, or whatever else addresses the cause of the risk. Often, redesign to avoid risks results in a much-improved project. One must take care, however, that avoiding one known risk or set of risks does not lead to taking on unknown risks of even greater consequence.

Risk avoidance is an area in which quantitative, even if approximate, risk assessments are needed. For example, the project designers may have chosen some solution, say A, over the alternative, say B, because the cost of A is estimated or quoted to be less than the cost of B – on a deterministic, single point basis. However, quantitative risk analysis might show that the least-cost approach A is much riskier than the alternative, B. The function of quantitative risk assessment is to determine if the predicted reduction in risk by changing from alternative A to alternative B is worth the cost differential.

Risk Control   Risk control refers to assuming a risk but taking steps to reduce, mitigate, or otherwise manage its impact or likelihood. Risk control can take the form of installing data gathering or early warning systems that provide information to assess more accurately the impact, likelihood, or timing of a risk. If warning of a risk can be obtained early enough to take action against it, then information gathering may be preferable to more tangible (and expensive) actions.

Risk control, like risk avoidance, may not be free, or even inexpensive. If the project is about developing a new product, and competition presents a risk, then one might buy out the competitors, but this could be expensive. A less-expensive alternative might be to team up with one's major competitor to develop and market a joint product. Another solution might be to accelerate the development project, even at some considerable cost, to reduce market risk by beating the competition to market.

Options and Alternatives   Options and alternatives refer to changes in the project to create optional courses of action. For example, if technical risks related to some new technology are of concern, one could set up parallel development teams to pursue different technological options concurrently. This might be expensive, but necessary to increase the likelihood that one would succeed. In the Manhattan Engineering District project, nuclear physicists were unsure whether an enriched uranium device or a plutonium device would work, so they developed both. It turned out that both worked, but the additional cost was considered to be justifiable. Of course, the ultimate risk, that this project was intended to forestall, that the enemy would develop the weapon first, turned out not to exist at all.

There are many places in which options can be inserted to deal with virtually any project technical, market, financing, or other risks. The use of these options may, however, require some imagination and changes from the usual methods and practices. Many options involve the creation or purchase of information, because risk is uncertainty and information reduces uncertainty. It must be stressed, however, that creating options to generate new information is not the same as simply postponing decisions to wait for some new data to materialize.

Organizational Structures Sometimes organizational restructuring can reduce risks. Whether the best organization is tight, with central control and high accountability, or loose and decentralized, with decisions made primarily at the local level, depends on the project and the nature of the risks. That is, the proper organizational structure is *contingent* on the situation; there is no universally best form of project organizational structure.

Risk Assumption Risk assumption is the last resort. It means that, in the end, if risks remain that cannot be avoided, transferred, insured, mitigated, eliminated, or controlled, then they must simply be accepted in order that the project may proceed. Presumably, this implies that the risks associated with going ahead are nevertheless less than (or more acceptable than) the risks of not going forward.

In summary, in this text we use the term project risk management as the overall framework consisting of three distinct activities: (a) identification, (b) assessment, and (c) transfer and mitigation. While this somewhat departs from ISO and SRA approaches, it is done for the purpose of fitting risk analysis into project management process, rather than vice versa. In fact, many project managers look at identification process as a unique task not part of the larger assessment process. Nevertheless, these differences are superficial as they only to terminology rather than substance of the process.

2.3   Representation of Project Risks

As previously mentioned we classify the quantitative risk assessment into two different categories: Low Impact – High Probability and High Impact – Low Probability risk assessments. The representation of Low Impact – High Probability risks follows the general process for representing uncertainty associated with random outcomes such as probability density distributions, stochastic processes, and others. Here the outcome can be defined as any element of design or construction process where multiple incremental and often hard-to-distinguish risks result in output variability. For example, this output is typically defined as duration and cost of elements in project's WBS such as activities, tasks, and work packages, but is could also be defined as the arrival rate of rework items and design changes. Any elementary textbook in statistics and probability theory would provide a good reference point to survey commonly used models. This text is has no ambition of being such reference.

In practice the representation of High Impact – Low Probability risks is often limited to the outputs from qualitative risk identification and assessment methods such as FMEA. As previously mentioned, this is typically a form of risk score that is based on categorical data and ordinal numbers for likelihood and consequence; for example Very High likelihood = 5 and Moderately-severe consequences = 4. The risk score is then calculated by a product or similar arithmetic operation between the numbers assigned to the likelihood and the consequence (e.g. Risk Score of 20 = 5×4). While this score measure is simple for communication purposes and useful for determining if the risk warrants a mitigation strategy, it persistent use in projects as a quantitative measure is unfortunate. This is because it is mathematically incorrect, and dangerously misleading to decision-makers.

In order to provide a High Impact – Low Probability representation of that can be used a basis for quantitative assessment and analysis we feel the need to provide the link between general principles of engineering design risk analysis and project risk management.

A typical approach in engineering design is to decompose risk into a triple defined by the hazard, vulnerability, and consequence (HVC) components (McLaughlin 2001; Mander et al. 2012). For example, given a known location of future building and the exposure of such location to seismic risks (i.e. hazard) a structural engineer design a building with features that can provide structural responses (i.e. vulnerability) such that it minimizes the overall expected damage over the facility's life-cycle (i.e. consequences). Here, there is first a hazard component defined by a model that maps frequency or probability to some intensity measure. Then, there is a structure response model that maps how the structure would respond to different level of hazard intensity, and it is followed by a function that maps structural responses to damage.

These HVC components can be viewed as more general and decomposed representation of the individual risk. This representation is flexible and can be further reduced to more aggregate forms based on the available data and applicability. For example, in projects (in contrast to engineering design) it is not always possible to obtain data on hazards and vulnerability separately; or furthermore, it is often not feasible as hazard and vulnerability components are not independent.

Hence in this kind of situation one can combine hazard and vulnerability components into a single measure i.e. probability and create a risk representation that is now commonly used in project management: Risk = Probability × Consequence.

Figure 2.2 illustrates the four-step process that defines the risk by modeling hazard, vulnerability, and consequence functions separately. These functions are typically exponential; hence we can use log-log transformation to show their relationship in a close-to-linear form. This is only for the purpose of illustration and the process can be applied to any function type.

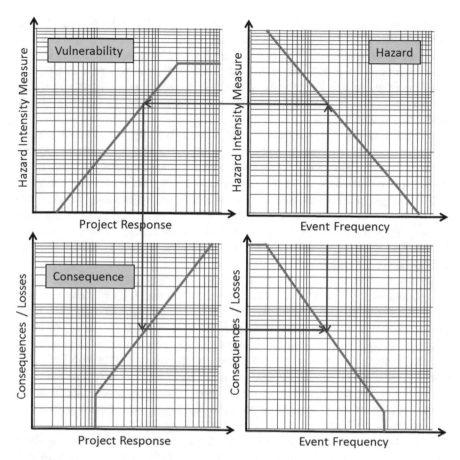

Fig. 2.2 HVC representation of risk

In the upper right corner one defines the hazard in terms of the frequencies and intensity. In many instances in projects this is, arguably, the key source of uncertainty; for example, the occurrence of a significant weather event that will shut down the construction site. However, it is natural to think that weather events differ in their intensity; hence the graph in upper right quadrant represents frequency vs. intensity of the hazard. Structures, teams, projects, and organizations respond differently to the same hazard intensity. In other words they show different vulnerability to the condition. The graph in upper left quadrant shows a functional relationship between project response and the hazard intensity. The vulnerability can also be referred to as exposure to hazard and it is often a decision variable; in order words, we make project or design decisions that will expose our project or structure/product to the hazard at different levels. One would be tempted to say that we should address this by avoiding vulnerability in the system, but this is often not feasible, or if it is, then it is often too costly.

The next step in this HVC risk modeling approach is to account for the consequences given how the project or the process responds to it. This is shown in a bottom left quadrant. Finally, based on the available information in the first three quadrants, one can derive the probability distribution of consequences given the frequency/probability. This is shown in a bottom right quadrant. Hence, the integral over the function defined in a bottom right quadrant is also an expected value of the risk.

As shown in Fig. 2.2, the four-step HVC modeling approach to risks requires separate functional representation of hazards, responses, and consequences. As such it can be extended to account for different types of responses and outcomes or different types of variables i.e. continuous or discrete. By doing this one can assess correlation and dependency between the activities and objectives. However, note that the only source of uncertainty in Fig. 2.2 is the hazard intensity – frequency function; response and consequence functions are deterministic. This is also not a limitation as the representation can include model uncertainty in vulnerability and consequence components as well.

Risk defined in FMEA or in a similar risk identification methods can also be represented using Condition-If-Then constructs (Garvey 2008). The condition event represents the early warning sign or the root cause; the risk events are probabilistic events that may occur because the condition is present, while the consequence(s) events represent the impact of the risk event on the objectives. Figure 2.3 illustrates this construct. Suppose the condition is that high traffic volume is present in and around the project site. A risk event might be that the access to the site is inadequate, and the consequences of the risk event include delays, which could cause an increase in the required resources, namely construction management and labor.

The Condition-If-Then representation provides a logical framework for defining and monitoring risk and is consistent with HVC representation of risk; Condition Event represents Hazard, Risk Event represents Vulnerability, and Consequence remains Consequence. From Fig. 2.3 we can also see that Condition-If-Then representation provides a causality-based relationship that can easily be

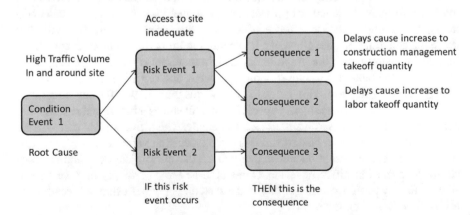

Fig. 2.3 The condition-if-then risk representation

integrated with Fault Tree Analysis (FTA), Event Trees (ET), and probability models including Bayesian networks.

To illustrate how Low-Impact High-Probability and High-Impact Low-Probability risk representation fits into project networks consider Fig. 2.4. At the lowest level of detail when incremental risk can be capture as the overall variability, see the top network (Fig. 2.4a), the risks is captured as probability distributions of activity durations and associated costs. To account of High-Impact Low-Probability risk consider figure at the next level (Fig. 2.4b); if the available data allows for identifying independent risk factors, the risk events can be modeled using probabilities

Fig. 2.4 Network representation of risk. (**a**) Agregrate representation (low impact, high probability). (**b**) Discrete risk representation (high impact, low probability). (**c**) Discrete risk representation with resource loaded schedule (high impact, low probability)

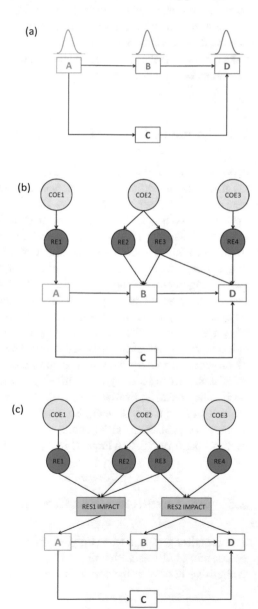

(RE node) and the consequence on the particular activity (activities). This RE representation can be further expanded to use the typical Condition-If-Then network for each risk. Finally, if a resource-loaded schedule is available one can link Condition-If-Then network to resources as shown in Fig. 2.4c. The key difference between this and the previous network representation is the intermediate layer between the risk event and project activities. This layer defines project response given different kind of project structures, typically defined in terms of resource bundle required to complete the activity (Govan and Damnjanovic 2016).

One can observe that the two bottom network formulations have risks with common conditions. In other words, the risks connected with the common cause will be inherently correlated. Further, one also observe common dependency not only on based on the condition, but also on a common resources bundle. This implies that risks can be correlated without having a common cause event, but having common resources i.e. vulnerability.

2.4 Scope of This Text

From what we said until this point even novices can see that project risk management covers not only technical concepts coming from statistics, probability and decision theory, but also principles that define how individuals, companies and organizations in general perceive, process, and respond to uncertainty. Attempting to cover all aspects of project risk management in details in a single textbook hence require providing relevant background that we feel may distract the reader from the original objective – introduce holistic and data analytics based approach to project risk management. Perhaps the reader looking to find the ultimate guidebook on the theory and practice of project risk management may find this limitation in scope unsatisfactory, but in our experience such reference would be difficult to assemble anyways. There is just too much "variance" in how projects are planned and executed across different industry segments. Therefore we here focus on the elements of project risk management theory and practice that are in common and relatively consistent across industry segments – quantitative assessment of low impact high probability risks, and variance in project activities in general. In our viewpoint, the majority of texts on project risk management already provide content on high impact low probability risks, while rigorous treatment of incremental low impact high-probability risk to a large degree is missing.

2.5 Organizations of the Content

The remaining of this text is organized as follows. In the next part (Part II: Risk Assessment in Project Planning – Chaps. 3, 4, 5, 6, 7, 8, and 9) we provide a comprehensive review of the formulations for cost and duration functions, evaluation of functions of random variables using second moment approach including estimating mean and variances from data and/or expert judgements, modeling the effect of independence and correlations, and estimating management reserves and contingencies. In Part III: Risk Monitoring and Reassessment in Project Execution (Chaps. 10, 11, 12, 13, and 14) we focus on applying Bayesian revision to managing contingencies, forecasting project completion using S-curves, implementing statistical process control methods for earned-value analysis, and using learning curves for forecasting.

References

Garvey PR (2008) Analytical methods for risk management: a systems engineering perspective. Chapman and Hall/CRC, Boca Raton

Govan P, Damnjanovic I (2016) The resource-based view on project risk management. J Constr Eng Manag 142(9):04016034

Mander JB, Sircar J, Damnjanovic I (2012) Direct loss model for seismically damaged structures. Earthq Eng Struct Dyn 41(3):571–586

McLaughlin SB (2001) Hazard vulnerability analysis. American Society for Healthcare Engineering of the American Hospital Association, Chicago

Project Management Institute (2008) A guide to the project management body of knowledge, 4th edn. Newtown Square, Pennsylvania

Part II
Risk Assessment in Project Planning

Chapter 3
Second Moment Approach for the Probability Distribution of Project Performance Functions

Abstract In this chapter we present the second moment methods for evaluating project performance functions where random variables can be continuous and/or discrete. We provide a comprehensive review of the method in context of its accuracy when compared to the results from the Monte Carlo simulation. Furthermore, we analyze the effect of correlations among the random variables and the linearization of the project performance functions.

Keywords Method of moments · Project performance functions · Correlations

3.1 Introduction

Quantitative assessments of project risks require use of linear or nonlinear project performance models. For example, the total project cost as one of the most important project indicators is simply the linear sum of work package costs; similarly, project duration is the sum of activity durations on the critical path. On other hand, there are several instances when quantitative risk assessment requires nonlinear models; for example, parametric cost models or productivity-based models are often nonlinear. But, as will be seen below, some approximate methods may be used to linearize even these functions, without the need for numerical methods such as Monte Carlo simulation. There are many instances in which Monte Carlo simulation is the best, or even only, method, especially when the model is discontinuous, involves decisions, or cannot be readily expressed in closed mathematical form. In this chapter, however, it will be seen that simple second moment methods give essentially the same answers as Monte Carlo simulation.

The second moment approach does not deal with full probability distributions but uses only the means and variances (the first two moments) to characterize uncertainty. Of course, the first two moments are actually measures of central tendency and say little about the probabilities in the tails of the distributions. The second moment approach given here is based on some simplifying approximations about the forms of the probability distributions, and these assumptions define the tails of the distributions. However, in most cases of project risk assessment the

© Springer Nature Switzerland AG 2020
I. Damnjanovic, K. Reinschmidt, *Data Analytics for Engineering and Construction Project Risk Management*, Risk, Systems and Decisions,
https://doi.org/10.1007/978-3-030-14251-3_3

probability distributions used to quantify uncertainty are largely subjective and based on judgment and experience rather than hard data. If there are historical data, they are typically sparse, and probably give little or no information about the tails of the distributions anyway. Therefore, it may be unnecessary to perform long simulations or detailed calculations when the inputs are at best known only to perhaps two significant digits.

The second moment approach may be used when one is interested in total project variability or uncertainty. Using this approach, one may wish to find the risk function or probability distribution of, for example, the total cost of a project. This probability distribution may then be used to determine if the process is capable, in the statistical process control sense, of meeting the requirements. For example, suppose that the user's specification or requirement is that a certain project be executed for a total budget of $10,000,000, including contingencies, and that the likelihood of overrunning this budget should be less than 10%. Suppose that the second moment analysis gives a total cost probability distribution such that the likelihood of exceeding $10,000,000 is 5%. Then the project as structured is capable of meeting the specification. On the other hand, if the analysis shows that the probability of exceeding $10,000,000 is 20%, then the project is not capable of meeting the sponsor's requirement. Either the process that generates this project must be reengineered or managed such that the probability of exceeding the budget is reduced to the acceptable number, or the sponsor must agree to add more budget or contingency, or the project will be cancelled.

There is another reason for performing this analysis, in addition to determining whether or not the project is too risky as it stands. It was stressed earlier that one major function of the project manager is to manage the project risks, and the risks are typically associated with individual project elements (work packages, activities, contracts, etc.). However, given that a complex project may have many work packages, and that the project manager has limited time and resources, how does he/she know which work packages should get his/her attention? As will be shown, the variability or uncertainty in the total cost of a project can be estimated, and if this variability is too high, then those work packages that make the greatest contribution to this uncertainty can be identified, and these should receive the most attention from the project manager. That is, the project manager can use these results for time management.

One variable of interest in all projects is cost. But there may be other variables related to project performance. For example, in aircraft design, performance considerations may place an upper limit on the weight of the aircraft. The weight of the aircraft is the sum of the weights of all the systems and components comprising the aircraft. Before the aircraft is designed, engineers make estimates of the weights of each system and component, but these estimates may have significant uncertainties, depending on how much this aircraft differs from previous models. If the analysis of the *a priori* probability distribution on the weight of the total aircraft indicates that there is too much chance of exceeding the weight limit, then some actions need to be taken to reduce the weights of some components, or to reduce the uncertainty associated the weights. Therefore, the project manager for the aircraft design process should be interested in which components are making the greatest contributions to the total weight uncertainty.

Software development projects may have performance criteria such as execution time. Electronic products, consumer products, medical equipment, and many other types of development projects have performance criteria as well. Depending on the state of the technology used, the final product performance, and the engineering cost to meet this performance, may be highly uncertain before the project begins.

This chapter presents a simple method for answering these questions, without the need for Monte Carlo simulation. It is applicable to project risks that meet the definitions given here, of differential risks or uncertainties. That is, it applies to processes that are in statistical control, in which the variability is due to common causes, but that may or may not be capable of meeting the specifications. This method is not necessarily applicable to processes that are out of control, that is, for which there is some unique external cause with very low probability of occurrence but very high impact on the project if it does occur.

3.2 Formulation of the Second Moment Method

Suppose there are p work packages or other cost elements and suppose that experts are available to estimate some parameters of the subjective probability distributions for each work package or activity. The usual practice in engineering analysis as well as project cost estimation is the *bottoms-up* method: the parameters of the lowest level elements (activities, work packages, or line items) are estimated, and then these are combined based on known principles to find the parameters of the total system. The logic behind this exercise is that different specialists or cost estimators may be able to estimate some specific processes, work packages or subcontracts, but none of them is qualified by knowledge or experience to estimate the statistical properties of the total project all at once. It may be the case that whoever makes the cost estimate (or bid) for a work package is most qualified to estimate the probability distribution on that cost. For example, suppliers or subcontractors may make initial cost estimates for equipment or subcontracts before sufficient design information is available to make firm offers. As more engineering information becomes available, the uncertainty in these estimates can get smaller. Therefore, it seems reasonable to perform the estimate for each work package separately and then combine the results in some way using probability concepts.

Methods for making subjective judgments about probability distributions are discussed elsewhere. Here, it is assumed that for each work package j there is a cost X_j, assumed to be uncertain but characterized by some probability distribution, and that some experts have estimated some parameters, from which the mean value $\mu_j = E[X_j]$ and the standard deviation $\sigma_j = \sqrt{E\left[\left(X_j - \mu_j\right)^2\right]}$ can be derived for all work packages. The goal is to determine the probability distribution for the sum of these terms over all line items.

The common assumption that all the work package costs are statistically independent may be a poor assumption and may result in very unconservative estimates of the total risk. Therefore, some experts familiar with the interaction between work packages j and k can estimate the correlation coefficients $\rho_{j,\,k}$, where $\rho_{j,\,k}$ is a number between -1 and $+1$. The meaning of $\rho_{j,\,k}$ can be interpreted as follows: suppose that there are two work packages, j and k, with variances σ_j^2 and σ_k^2, and correlation coefficient $\rho_{j,\,k}$. The quantity $\rho_{j,k}^2$ is the fraction of the variance of X_j that is explained or removed by knowledge of X_k. That is, suppose the project manager initially does not know the true values of X_j or X_k, but attributes to them the variances σ_j^2 and σ_k^2, respectively. Suppose then that work package k finishes, and the project manager now knows the true value of X_k. The project manager's uncertainty in work package j is now $\sigma_j^2 - \rho_{j,k}^2 \sigma_j^2 = \sigma_j^2 \left(1 - \rho_{j,k}^2 \right)$. If $\rho_{j,k}^2 = 0$, then knowledge of the true value of X_k provides no information about the value of X_j, because the estimate of the variance does not change. If $\rho_{j,k}^2 = 1$, then the project manager's estimate of the variance of work package j is zero; knowledge of the true value of X_k provides complete information about the value of X_j. Clearly, most cases are somewhere between these limits. The critical question is this: If knowledge of the true value of the cost of work package k would cause you to revise your estimate about work package j, then you believe that activities j and k are correlated. Conversely, if knowledge of the true value of the cost of work package k would not cause you to revise your estimate about work package j, then you believe that activities j and k are independent (for more about estimating correlation coefficients using expert judgments see Chap. 6).

Using the correlation coefficients, the covariances are computed as $\rho_{j,\,k}\sigma_j\sigma_k$. For the general case:

$$\text{Mean total cost} = \mu_T = E\left[\sum_{j=1}^{p} X_j \right] = \sum_{j=1}^{p} E\left[X_j \right] = \sum_{j=1}^{p} \mu_j$$

Here, the symbols $E\left[X_j \right]$, $\mu_j, \bar{X}_j$ all have the same meaning, the average value, the mean value, or the expected value of the unknown variable X_j

$$\text{Variance of the total cost} = \sigma_T^2 = \sum_{j=1}^{p}\sum_{k=1}^{p} \rho_{j,k}\sigma_j\sigma_k$$

That is, the variance of the sum is the sum of all the *variances and covariances*. This may be proved, but the proof is given elsewhere and is omitted here. Noting that symmetry requires that $\rho_{j,\,k} = \rho_{k,\,j}$, then the previous equation may be rewritten as:

$$\text{Variance of the total project cost} = \sigma_T^2 = \sum_{j=1}^{p}\sigma_j^2 + 2\sum_{j=1}^{p-1}\sum_{k=j+1}^{p} \rho_{j,k}\sigma_j\sigma_k$$

In this last equation note that the variance of the sum of the costs is the sum of the variances of all the individual work package costs *plus* the sum of all the off-diagonal covariances. From this it is easily seen that, if the correlations are generally positive, that is, $\rho_{j,k} > 0$, then assuming independence when the variables are not independent is equivalent to neglecting all of the off-diagonal covariance terms, which would result in a low value for σ_T^2 and would underestimate, perhaps grossly underestimate, the true value of the total variance.

In this derivation it is not assumed that all the probability distributions for all the work packages are Normally distributed, or follow any other particular distribution. In fact, the work package costs may well not be Normally distributed, as they certainly cannot be negative. There are a number of reasons for believing that the work package costs are nonnegative, asymmetric, and skewed to the right. To put it another way, the probability distributions for the individual work package costs may have third, fourth, fifth, and even higher moments, but the method given here uses only the first and second moments (the mean and the variance respectively). We are not assuming that the moments higher than the second are zero; we are just not using them. This is, of course, an approximate method, and the justification for these approximations will be discussed below.

It is readily apparent that all of the calculations given above can be easily implemented. To express these equations in compact matrix form, define the covariance matrix as the p-by-p matrix $\mathbf{V}$, in which the elements are:

$$v_{j,k} = \sigma_{j,k} = \rho_{j,k}\sigma_j\sigma_k \quad \text{for all } j \text{ and } k, \ 1 \le j \le p, 1 \le k \le p$$

with the definitions

$$\rho_{j,j} = 1, \ \sigma_{j,j} = \sigma_j\sigma_j = \sigma_j^2$$

Then the covariance matrix is, taking advantage of symmetry:

$$\mathbf{V} \equiv \begin{pmatrix} v_{11} & v_{12} \cdots & v_{1p} \\ v_{21} & v_{22} \cdots & v_{2p} \\ v_{p1} & v_{p2} \cdots & v_{pp} \end{pmatrix} \equiv \begin{pmatrix} \sigma_{11} & \sigma_{12} \cdots & \sigma_{1p} \\ \sigma_{21} & \sigma_{22} \cdots & \sigma_{2p} \\ \sigma_{p1} & \sigma_{p2} \cdots & \sigma_{pp} \end{pmatrix} \equiv \begin{pmatrix} \sigma_1^2 & \rho_{12}\sigma_1\sigma_2 \cdots & \rho_{1p}\sigma_1\sigma_p \\ \rho_{12}\sigma_1\sigma_2 & \sigma_2^2 \cdots & \rho_{2p}\sigma_2\sigma_p \\ \rho_{1p}\sigma_1\sigma_p & \rho_{2p}\sigma_2\sigma_p \cdots & \sigma_p^2 \end{pmatrix}$$

In general, the covariance matrix may be fully populated. Also, it should be assumed that all that the work package cost correlation coefficients are nonzero, unless it is explicitly shown otherwise.

Consequently, Monte Carlo simulation is unnecessary to find the mean and variance of the total cost or duration of a project; one simply has to sum all the work package means to find the mean of the total, and to sum all the variances and covariances to find the variance of the total. This method has the following two mnemonics: (a) the mean of the sum is the sum of the means; and (b) the variance of the sum is the sum of the *covariances*.

This method gives only the first two moments (the mean and the variance) of the total cost and does not give a specific probability distribution on the total, nor does it account for skewness, kurtosis, and other higher moments. Therefore, it may be argued, Monte Carlo simulation is necessary to determine the form of the probability distribution and the values in the tails.

First, Monte Carlo Simulation gives useful results in the tails only when a large number of repetitions is performed. Second, the Central Limit Theorem indicates that the probability distribution of a sum of independent random variates, drawn from any probability distribution, approaches the Normal. Of course, the Central Limit Theorem is not strictly applicable to the general case considered here, as independence is not assumed. However, "any linear combination of normally distributed variables, not necessarily independent, is normally distributed" (Denrell 2004) by the replication property of the Normal distribution. The issue concerns linear combinations of variables in general, not necessarily independent and not necessarily Normal. An extensive set of Monte Carlo experiments have been performed with linear combinations (that is, sums) of individual work package costs drawn from symmetric or highly skewed asymmetric distributions, with negative, zero, or positive correlation coefficients. From these experiments, it can be concluded empirically that the probability distribution of the sum is nearly Normal, with mean μ_T and variance σ_T^2 as computed from the expressions given above, regardless of the skewness or correlation of the individual terms in the summation. In other words, empirically, the Central Limit Theorem seems to give good approximations even when the variates are not independent. And good approximations are all that we need here, because the probability distributions of the individual work package costs are subjective to start with.

Therefore, the conclusion is that a Normal distribution, with mean and variance computed as above, is a reasonable approximation, which allows us to determine the entire probability distribution, such as the quantiles in the upper tail, using the tables of the standardized unit Normal distribution. Some confirmations of this principle are given next.

These conclusions apply to variates that are the sum (or the weighted sum) of dependent or independent random variates. Therefore, this method can be used to approximate the uncertainty in the duration of a path through a project network, as the path length is the sum of the durations of the individual activities along the path. Of course, for project durations using the critical path (the longest path from start to finish), this method requires that the critical path does not change for random variations in all of the network activities. If, for some values of the random variables, different paths can become the critical path, the total duration will not be the sum of a fixed set of activity durations. Application of the maximization operator to the set of all possible paths is not a linear operation, and other methods, for example Monte Carlo simulation, become necessary.

Of course, populating the complete covariance matrix may be necessary, but not easy. Estimating the mean values requires one estimate per work package, for a total of p means. Estimating the variances, or the main diagonal terms, requires one estimate per work package, or p variance estimates. However, estimating the dependencies, or off-diagonal terms, requires estimating $\dfrac{p(p-1)}{2}$ correlations. As the number of variances increases as p but the number of covariances to estimate increases as p^2, the estimation process is not trivial if p is large. Nevertheless, experience has shown that it is quite feasible for engineers and constructors in industry to estimate these correlation coefficients on the basis of experience.

Example 3.1

An actual power plant under construction some years ago is used as an example application. The project was large, but the example is small; all the calculations here can be done by spreadsheet or even by calculator. The project costs were summarized into 18 cost accounts. For each of these accounts, the cost engineers on the project estimated three points on the probability distribution for the cost. These three points corresponded to the 10th, 50th, and 90th percentiles, x_{10}, x_{50}, x_{90}. However, other percentiles could be used by suitably modifying the expressions below. The cost accounts and the estimated values are given in the table below. Note the following:

- Account 1 represents the costs expended up to the date of the estimate.
- Account 44, the largest single account, is interest on funds used during construction and therefore depends on the other costs.
- Accounts 51, 53, 55, 57, and 59 represent additional costs if the schedule slips.
- The accounts vary considerably in size.

The second moment method described earlier requires moments, and these moments are computed from the engineers' estimates for three points for each account, using equations developed by Keefer and Bodily (1983).

The values for the computed means (μ) and standard deviations (σ) are given in the last two columns of Table 3.1.

Summing the mean values in this table gives an expected value for the total project cost of $3043 million. Assuming the cost accounts are all independent, the square root of the sum of the variances of all accounts gives a standard deviation for the total project cost of $97 million.

However, the cost engineers on the project also estimated the correlation coefficients, which are given in the matrix below.

Account	1	10	12	14	16	18	22	24	26	28	32	44	45	51	53	55	57	59
1	1	□	□	□	□	□	□	□	□	□	□	□	□	□	□	□	□	□
10	□	1	0.6	□	□	0.6	0.6	0.6	0.6	□	□	0.8	□	□	□	□	□	□
12	□	0.6	1	□	□	□	□	□	□	□	□	□	□	□	□	□	□	□
14	□	□	□	1	□	□	□	□	□	□	□	□	□	□	□	□	□	□
16	□	□	□	□	1	0.8	0.8	0.6	□	□	□	0.8	□	□	0.8	□	□	□
18	□	0.6	□	□	0.8	1	□	0.6	□	□	□	0.8	□	□	0.8	□	□	□
22	□	0.6	□	□	0.8	□	1	0.6	□	□	□	0.8	□	□	□	□	□	□
24	□	0.6	□	□	0.6	0.6	0.6	1	0.6	□	□	0.8	□	□	□	□	□	□
26	□	0.6	□	□	□	□	□	0.6	1	□	□	0.8	□	□	□	□	□	□
28	□	□	□	□	□	□	□	□	□	1	□	0.8	□	□	□	□	□	□
32	□	□	□	□	□	□	□	□	□	□	1	0.8	□	□	□	□	□	0.3
44	□	0.8	□	□	0.8	0.8	0.8	0.8	0.8	0.8	0.8	1	□	□	□	□	□	□
45	□	□	□	□	□	□	□	□	□	□	□	□	1	□	□	□	□	□
51	□	□	□	□	□	□	□	□	□	□	□	□	□	1	0.4	0.4	0.4	0.4
53	□	□	□	□	0.8	0.8	□	□	□	□	□	□	□	0.4	1	0.4	0.4	0.4
55	□	□	□	□	□	□	□	□	□	□	□	□	□	0.4	0.4	1	0.4	0.4
57	□	□	□	□	□	□	□	□	□	□	□	□	□	0.4	0.4	0.4	1	0.4
59	□	□	□	□	□	□	□	□	□	□	0.3	□	□	0.4	0.4	0.4	0.4	1

Table 3.1 Three-point cost estimates (in $1,000,000); means and standard deviations

Account no	10th percentile x_{10}	50th percentile x_{50}	90th percentile x_{90}	Work package name	Mean (μ)	Std Dev (σ)
1	1035	1036	1037	Expenses prior to 1 July	1036	0.78
10	135	185	220	Manual labor	180	33.3
12	50	76	100	Construction services	75	19.5
14	1	2	10	Equipment	3.4	4.36
16	10	14	30	Bulk materials	17.7	8.29
18	20	36	60	Distributable materials	38.1	15.8
22	20	91	100	Construction services (labor)	71.7	34.4
24	100	134	175	Engineering	136	29.3
26	100	123	150	Project direction	124	19.5
28	50	67	100	Quality control	71.9	19.9
32	125	144	150	Owner's project management	140	10.2
44	1050	1083	1135	Interest during construction	1089	33.4
45	−15	−5	10	Escalation	−3.4	9.82
51	−20	20	120	AFUDC with slippage	38.7	56.4
53	−1	1	10	Distributable materials slippage	3.2	4.6
55	−3	3	20	Engineering with slippage	6.4	9.34
57	−3	3	20	Project direction with slippage	6.4	9.34
59	−4	4	25	Owner's costs with slippage	8	11.7

Using these correlations and the standard deviations for the account costs, the covariance matrix can be easily computed using the equations given earlier. Summing all the terms in the covariance matrix gives the variance of the total cost; the square root of this is the standard deviation of the total cost, $170 million. Note that this is considerably higher than the $97 million computed under the assumption of independence. The difference, $73 million, is not trivial. (Moreover, there is some reason to believe that the correlations should be generally higher that those given.)

Looking at the comparison another way, the mean and standard deviation can be used to compute the 90th percentile for the total project cost; this is the value that has a 10% likelihood of being exceeded. For the two cases, independent and correlated, the 90th percentile costs are estimated to be:

- Independent: $3168 million
- Correlated: $3261 million

That is, the project manager's estimate for the owner of the cost such that there would be a 90% likelihood that the estimate would not be exceeded, would be $93 million low if the figure based on the independence assumption were used.

Figures 3.1 and 3.2 show the probability density function for the costs, obtained from this computation.

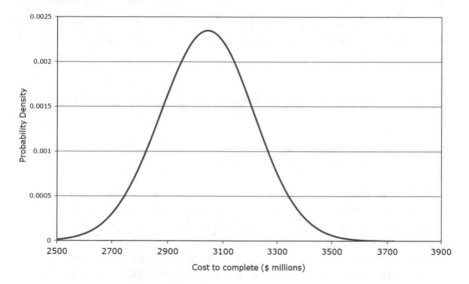

Fig. 3.1 Probability density function for cost to complete

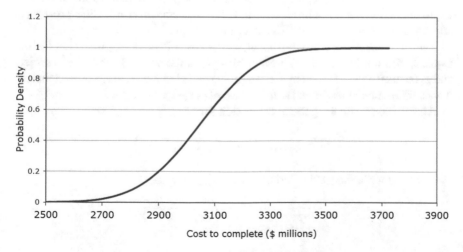

Fig. 3.2 Cumulative density function for cost to complete

3.3 Second Moment Approximations for Nonlinear Functions

Second moment approximations can be obtained for many nonlinear functions or combinations of variables, not necessarily independent. Suppose that $G(x_1, x_2, \ldots, x_m)$ is some known function of $x_1, x_2, \ldots, x_m$ and we wish to find the mean and variance of this function given that the means and variances of the random variates, $x_1, x_2, \ldots, x_m$ have been estimated. Let $\bar{x}_1 = E[x_1], \bar{x}_2 = E[x_2], \ldots, \bar{x}_m = E[x_m], \bar{G} = E[G(x_1, x_2, \ldots, x_m)]$. Howard (1971) gives a second moment approximation for a nonlinear function of m variables, $G(x_1, x_2, \ldots, x_m)$:

$$\bar{G} = E\left[G(x_1, x_2, \ldots, x_m)\right] \cong G(\bar{x}_1, \bar{x}_2, \ldots, \bar{x}_m) + \frac{1}{2}\sum_{i=1}^{m}\sum_{j=1}^{m}\frac{\partial^2 G}{\partial x_i \partial x_j}\text{cov}\left[x_i, x_j\right]$$

$$\text{var}[G] = \sigma_G^2 \cong \sum_{i=1}^{m}\sum_{j=1}^{m}\frac{\partial G}{\partial x_i}\frac{\partial G}{\partial x_j}\text{cov}\left[x_i, x_j\right]$$

All the partial derivatives $\partial G/\partial x_1, \partial G/\partial x_2, \ldots, \partial G/\partial x_m$ are evaluated at the mean values $x_1 = \bar{x}_1, x_2 = \bar{x}_2, \ldots, x_m = \bar{x}_m$. If analytic derivatives of the function $G(x_1, x_2, \ldots, x_m)$ are not available, but the output $G(x_1, x_2, \ldots, x_m)$ can be computed, the partial derivatives $\partial G/\partial x_1, \partial G/\partial x_2, \ldots, \partial G/\partial x_m$ may be approximated by taking finite differences over multiple values of the output. Note that Howard's formulation has a term in the approximation such that the expected value of the function is dependent on the values of the second derivatives of the function.

Suppose, as an example, we are considering a project where its future cost can be approximated using a nonlinear cost function that considers a similar project and adjusts the estimate based on differences in the scope.

Let a represent a factor adjustment, x_1 the cost of previous project, x_2 the unit capacity of a similar project, x_3 the unit capacity of the currently considered project, and x_4 the fixed cost of the item that was only needed in the previous project then we have a difference in project cost defined as: $a[x_1(x_2 - x_3) - x_4]$

Also we can write the general expression for $G(x_1, x_2, \ldots, x_m)$ as:

$$G(x_1, x_2, x_3, x_4) = a\left[x_1(x_2 - x_3) - x_4\right]$$

We can now write the partial derivatives of $G(x_1, x_2, x_3, x_4)$:

$\dfrac{\partial G}{\partial x_1} = a(x_2 - x_3)$	$\dfrac{\partial G}{\partial x_3} = -ax_1$
$\dfrac{\partial G}{\partial x_2} = ax_1$	$\dfrac{\partial G}{\partial x_4} = -a$

These derivatives are to be evaluated at the mean values of the variables. Similarly, the second derivatives are:

$\dfrac{\partial^2 G}{\partial x_1^{\;2}} = 0$	$\dfrac{\partial^2 G}{\partial x_2^{\;2}} = 0$	$\dfrac{\partial^2 G}{\partial x_4^{\;2}} = 0$
$\dfrac{\partial^2 G}{\partial x_1 \partial x_2} = a$	$\dfrac{\partial^2 G}{\partial x_1 \partial x_3} = -a$	$\dfrac{\partial^2 G}{\partial x_1 \partial x_4} = 0$
$\dfrac{\partial^2 G}{\partial x_2 \partial x_3} = 0$	$\dfrac{\partial^2 G}{\partial x_2 \partial x_4} = 0$	$\dfrac{\partial^2 G}{\partial x_3 \partial x_4} = 0$

We may now write the approximation for the expected value of the annual revenue as:

$$\bar{G} = E\left[G\left(x_1,x_2,x_3,x_4\right)\right] \cong a\left[\bar{x}_1\left(\bar{x}_2 - \bar{x}_3\right) - \bar{x}_4\right] + \frac{a}{2}\sum_{i=1}^{4}\sum_{j=1}^{4}\frac{\partial^2 G}{\partial x_i \partial x_j}\text{cov}\left[x_i,x_j\right]$$

Where the covariance may be written as $\text{cov}[x_i, x_j] = \rho_{i,j}\,\sigma_i\,\sigma_j$.

And the double summation can be written as the sum of all the terms in a 4 by 4 matrix:

$$\frac{1}{2}\sum_{i=1}^{4}\sum_{j=1}^{4}\frac{\partial^2 G}{\partial x_i \partial x_j}\text{cov}\left[x_i,x_j\right] = \frac{1}{2}Sum$$

Here, *Sum* is the summation of the terms in the following matrix

$$\begin{bmatrix} & 1 & 2 & 3 & 4 \\ 1 & 0 & \rho_{1,2}\sigma_1\sigma_2 & -\rho_{1,3}\sigma_1\sigma_3 & 0 \\ 2 & \rho_{1,2}\sigma_1\sigma_2 & 0 & 0 & 0 \\ 3 & -\rho_{1,3}\sigma_1\sigma_3 & 0 & 0 & 0 \\ 4 & 0 & 0 & 0 & 0 \end{bmatrix}$$

We can now substitute the values for the partial derivatives into the general expression for the variance:

$$\text{var}[G] = \sigma_G^{\;2} \cong \sum_{i=1}^{4}\sum_{j=1}^{4}\frac{\partial G}{\partial x_i}\frac{\partial G}{\partial x_j}\text{cov}\left[x_i,x_j\right] = a^2 Sum$$

Here *Sum* is the summation of all the terms in the following matrix:

$$
\begin{bmatrix}
 & 1 & 2 & 3 & 4 \\
1 & \left(\bar{x}_2-\bar{x}_3\right)^2\sigma_1^2 & \bar{x}_1\left(\bar{x}_2-\bar{x}_3\right)\rho_{1,2}\sigma_1\sigma_2 & -\bar{x}_1\left(\bar{x}_2-\bar{x}_3\right)\rho_{1,3}\sigma_1\sigma_3 & -\left(\bar{x}_2-\bar{x}_3\right)\rho_{1,4}\sigma_1\sigma_4 \\
2 & \bar{x}_1\left(\bar{x}_2-\bar{x}_3\right)\rho_{1,2}\sigma_1\sigma_2 & \bar{x}_1^2\sigma_2^2 & -\bar{x}_1^2\rho_{2,3}\sigma_2\sigma_3 & -\bar{x}_1\rho_{2,4}\sigma_2\sigma_4 \\
3 & -\bar{x}_1\left(\bar{x}_2-\bar{x}_3\right)\rho_{1,3}\sigma_1\sigma_3 & -\bar{x}_1^2\rho_{2,3}\sigma_2\sigma_3 & \bar{x}_1^2\sigma_3^2 & \bar{x}_1\rho_{3,4}\sigma_3\sigma_4 \\
4 & -\left(\bar{x}_2-\bar{x}_3\right)\rho_{1,4}\sigma_1\sigma_4 & -\bar{x}_1\rho_{2,4}\sigma_2\sigma_4 & \bar{x}_1\rho_{3,4}\sigma_3\sigma_4 & \sigma_4^2
\end{bmatrix}
$$

Note that: The equations given earlier are approximations, and are not exact under all circumstances. The validity of the approximations can be gauged by comparing the approximation results with results from Monte Carlo simulation.

Example 3.2
Assume that the means and standard deviation of the four variables are as given in Table 3.2.

To simplify the presentation, take $a = 1$.

Case 3.1 As a first comparison, we assume that all the variables are independent. That is, we take the correlation matrix to be:

$$
\begin{bmatrix}
 & 1 & 2 & 3 & 4 \\
1 & 1 & 0 & 0 & 0 \\
2 & 0 & 1 & 0 & 0 \\
3 & 0 & 0 & 1 & 0 \\
4 & 0 & 0 & 0 & 1
\end{bmatrix}
$$

We now compute the answers in two ways:

- Compute the approximate mean $\bar{G}$ and variance $var[G]$ from the equations above.
- Simulate the process by Monte Carlo, compute the histograms of the results, and compute the mean and variance from the computed values for $[x_1(x_2 - x_3) - x_4]$.

The Table 3.3 shows the results for this case. The Monte Carlo simulation used 32,000 random trials.

Table 3.2 Example 3.2 data

Variable	Mean (μ)	Standard deviation (σ)	Coefficient of variation (COV)
x_1	100	20	0.20
x_2	50	10	0.20
x_3	40	10	0.25
x_4	500	10	0.02

Table 3.3 Case 3.1 results

Method	Mean	Standard deviation	Skewness
Approximation	500.00	1428.32	NA
Monte Carlo simulation	502.58	1463.49	0.16

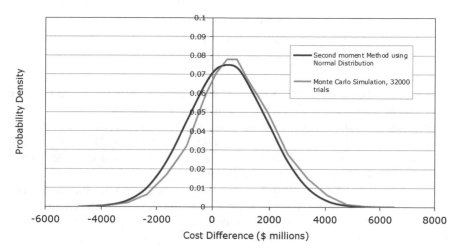

Fig. 3.3 Comparison of Monte Carlo with second moment method

Note the large value for the standard deviation, nearly three times the mean, computed by both methods. The largest coefficient of variation of any of the input variables was 25%, but the coefficient of variation of the resulting function is 293%. This results in a large probability that the cost difference is negative.

Figure 3.3 plots the probability density functions computed from the results. The Monte Carlo curve is the histogram computed by the simulation, the curve for the approximate method is a Normal distribution using the computed mean and variance. The approximate method determines two moments of the resulting distribution, but does not determine the entire distribution. Therefore, a Normal distribution is used.

From these comparative results, we can observe that:

- The difference in the means is negligible.
- The difference in the standard deviations is negligible.
- The Monte Carlo simulation histogram is very close to Normal. The computed skewness coefficient shows minor skew to the right, but this is negligible.
- There is a large probability of negative cost difference.

Therefore, we can conclude that for this comparison, with no correlations among the input variables, the approximate method gives results indistinguishable from Monte Carlo simulation with 32,000 trials.

Case 3.2 Here, we consider a situation in which the correlation matrix is taken to be:

$$
\begin{bmatrix}
 & 1 & 2 & 3 & 4 \\
1 & 1 & 0 & 0 & 0 \\
2 & 0 & 1 & 0.8 & 0 \\
3 & 0 & 0.8 & 1 & 0 \\
4 & 0 & 0 & 0 & 1
\end{bmatrix}
$$

We now compute the answers in the same two ways to obtain the results (the Monte Carlo simulation used 32,000 random trials) (Table 3.4):

Note that the standard deviation computed by both methods has decreased considerably compared to the case with all independent variables. The coefficient of variation of the resulting cost difference has fallen to 133%, much less than in Case 3.1, but still very large compared to the inputs.

Figure 3.4 plots the probability density functions computed from the results. The Monte Carlo curve is the histogram computed by the simulation, the curve for the approximate method is a Normal distribution using the computed mean and variance. Note that the scale on the abscissa is not the same as in the figure accompanying Case 3.1.

From these comparative results, we can observe that:

- The difference in the means is negligible.
- The difference in the standard deviations is negligible.
- The Monte Carlo simulation histogram is very close to Normal. The computed skewness coefficient shows a slightly larger skew to the right, but this is negligible, and not readily visible on the plot.
- There is a large probability of negative cost difference, but smaller than in Case 3.1.

Therefore, we can conclude that for this comparison, with positive correlation between x_2 and x_3 the approximate method gives results indistinguishable from Monte Carlo simulation with 32,000 trials.

Case 3.3 Here, we consider a situation in which the correlation matrix is taken to be (blank cells signify 0.0):

$$
\begin{bmatrix}
 & 1 & 2 & 3 & 4 \\
1 & 1 & -0.8 & 0 & 0 \\
2 & -0.8 & 1 & 0 & 0 \\
3 & 0 & 0 & 1 & 0 \\
4 & 0 & 0 & 0 & 1
\end{bmatrix}
$$

Table 3.4 Case 3.2 results

Method	Mean	Standard deviation	Skewness
Approximation	500.00	663.40	NA
Monte Carlo simulation	505.94	679.59	0.33

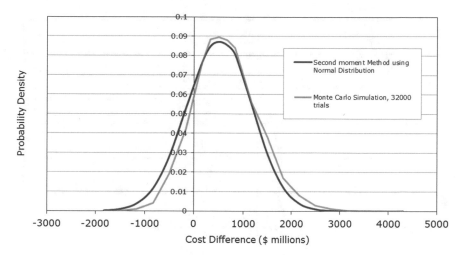

Fig. 3.4 Comparison of Monte Carlo with second moment method

We now compute the answers in the same two ways to obtain the results (the Monte Carlo simulation used 32,000 random trials) (see Table 3.5):

Note that in this case the means computed by each method have decreased from 500 to about 340, due to the correlation specified. We can see from this that ignoring correlation and assuming all variables to be independent result in errors not only in the variances but also in the mean values. The standard deviation computed by both methods, however, is large, almost as large as in the independent case (Case 3.1). The coefficient of variation of the resulting revenue is now 386%, more even than in Case 3.1, due to the high standard deviation and lower expected value. The skewness computed from the Monte Carlo simulation is now negative, indicating a skew to the left, toward lower cost difference, which can be seen in the figure below.

Figures 3.5 and 3.6 plot the probability density functions and the cumulative probability distribution functions computed from the results. The Monte Carlo curve is the histogram computed by the simulation, the curve for the approximate method is a Normal distribution using the computed mean and variance. Note that the scale on the abscissa is not the same as in the figures given before.

From these comparative results, we can observe that:

- The difference in the means is negligible.
- The difference in the standard deviations is negligible.
- The Monte Carlo simulation histogram deviates from the Normal. The computed skewness coefficient shows a visible skew to the left. Note that the second

Table 3.5 Case 3.3 results

Method	Mean	Standard deviation	Skewness
Approximation	340.00	1311.53	NA
Monte Carlo simulation	338.00	1345.87	−0.59

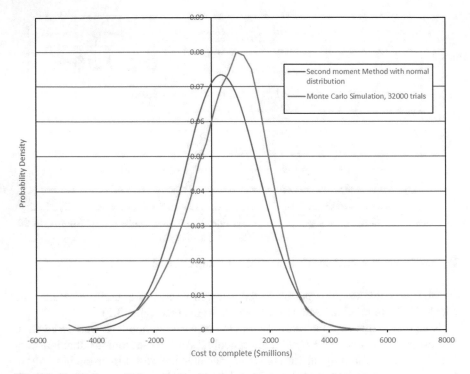

Fig. 3.5 Comparison of Monte Carlo with second moment method – histogram

moment approximation cannot match results that are skewed, as skewness is a third moment property, under the assumption that the result is Normal. Some other assumption would have to be made regarding the third moment.

- However, the differences between the two solutions on the cumulative probability distribution plot are small. The probability of negative revenues, for example, is larger for the second moment method using the Normal curve than for the Monte Carlo simulation, 0.398 versus 0.328, a difference of about 0.070 at the maximum.

Therefore, we can conclude that, for this comparison, with negative correlation the approximate method gives results that are conservative compared to Monte Carlo simulation with 32,000 trials. However, this conclusion is not necessarily valid for other conditions involving nonlinear combinations of variables. More research needs to be done on suitable second moment approximations for typical cases used in practice.

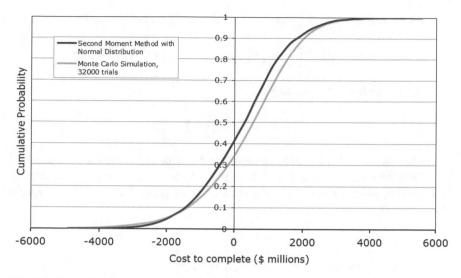

Fig. 3.6 Comparison of Monte Carlo with second moment method – cumulative

3.4   Discrete Random Variables in Linear and Nonlinear Models

In some project performance functions discrete random variables must be combined with continuous. Therefore, we discuss here how to get the first two moments of these types of variables.

Consider, for example, some cost element of a project that may be incurred – or may not. We might be seeking a building permit, for example, and there may be some factor that may or may not be required by the permitting agency. The regulator might require a flood control system that we do not feel is necessary, but the regulator could nevertheless impose it. Call this factor y. Let p represent the probability that this factor will be required (that the regulator insists on it in order to issue a building permit). Then we have the situation that either:

(i) The value of y is zero, with probability $1 - p$.
(ii) The value of y is Y with probability p. Here we assume that Y is known with no error.

Then the mean value of y is:

$$\mu_y = E[y] = pY + (1-p)0 = pY$$

The expected value of y^2 is:

$$E[y^2] = pY^2 + (1-p)0 = pY^2$$

Then the variance of y is given by:

$$\sigma_y^2 = Var[y] = E[y^2] - (E[y])^2 = pY^2 - (pY)^2 = p(1-p)Y^2$$

And the standard deviation of y is:

$$\sigma_y = Y\sqrt{p(1-p)}$$

As an extension of the above, suppose that the value of y if the factor is required is not a fixed number but is drawn with probability p from a probability density function $f(x)$, which has mean and standard deviation μ_x, σ_x. Then the above expressions become:

$$\mu_y = p\mu_x$$
$$\sigma_y^2 = p\sigma_x^2 + p(1-p)\mu_x^2$$

(These reduce to the previous expressions if $\mu_x = Y$ and $\sigma_x = 0$).

These two moments, μ_y, σ_y^2 may be added along with the moments for continuous variables, as discussed earlier in this chapter. Note however that, if there is a large proportion of these discrete variables, the actual probability distribution of the sum may deviate from the Normal distribution. The problem is not in summing the moments; it lies in identifying what the distribution of the resulting sum should be.

3.5   Practice Problems

Problem 3.5.1 You are in charge of a small project comprised of six sequential activities. VP for Engineering asked you to prepare a presentation updating the project review board on the uncertainty about the project cost outcomes. Project engineers have prepared for you project schedule and cost estimates shown in Table 3.6. Determine the mean and standard deviation of the cost distribution and

Table 3.6 Problems data

Activity	Schedule		Labor/equip cost per period		Total material cost	
	Mean	Variance	Mean	Variance	Mean	Variance
A	17	9	10	2	500	40
B	9	5.44	5	1	50	0
C	6	1.78	4	0	40	0
D	5	1.78	5	0	150	20
E	4	0.44	5	2	50	0
F	9	5.44	10	3	300	20

calculate one-sided upper 95% confidence interval for the total project cost. Assume that the random variables are independent.

Problem 3.5.2 Consider a project situation defined in Problem 3.5.1; how would your answer change if you consider that the activities are correlated? (see the correlation matrix below).

$$
\begin{bmatrix}
 & A & B & C & D & E & F \\
A & 1 & 0 & 0 & 0 & 0 & 0 \\
B & 0 & 1 & 0.8 & 0.8 & 0.8 & 0.8 \\
C & 0 & 0.8 & 1 & 0 & 0 & 0 \\
D & 0 & 0.8 & 0 & 1 & 0 & 0 \\
E & 0 & 0.8 & 0 & 0 & 1 & 0 \\
F & 0 & 0.8 & 0 & 0 & 0 & 1
\end{bmatrix}
$$

Problem 3.5.3 Now consider four discrete risk factors that if materialize can result in an increase in activity duration (see Problem 3.5.1). These risk factors are listed in Table 3.7.

Table 3.7 Problem data

Risk factor	Affects activity	Probability	Consequence
I	A, D	0.05	C(A) = 4; C(D) = 1
II	B	0.10	C(B) = 2
III	C	0.01	C(C) = 6
IV	E	0.05	C(E) = 1

(A) Find one-sided upper 95% confidence interval on project completion time that includes both discrete risk factors and general variances in duration of the activities defined in Problem 3.5.1. What are the key assumptions?
(B) How would your answer change if you consider that the risk factors above are correlated? (see the correlation matrix below)
(C) How would your answer change if you consider that the consequences of the risk factors are uncertain with coefficient of variation = 0.5
(D) Can you define the distribution of the extra time added due to risk factor for project completion time?

$$
\begin{bmatrix}
 & I & II & III & IV \\
I & 1 & 0 & 0 & 0 \\
II & 0 & 1 & 0.4 & 0.4 \\
III & 0 & 0.4 & 1 & 0.4 \\
IV & 0 & 0.4 & 0.4 & 1
\end{bmatrix}
$$

Table 3.8 Problem data

Variable	Mean (μ)	Standard deviation (σ)
X1	150	20
X2	60	10
X3	20	10
X4	50	5
X5	0.1	0.05

Problem 3.5.4 Formulate a cost function for an activity A given the following information: X_1 = Daily cost of the crew and tools/equipment required to complete a task A; X_2 = Unit production rate of the crew per day of the scope of work defined by task A; X_3 = The total cope of work defined by task A; X_4 = Total cost of materials required to complete the scope of work defined by task A; and X_5 = Profit and overhead percentage. Assuming these are random variables with parameters shown in Table 3.8, determine a two-sided 95% confidence interval for the cost of activity A.

Problem 3.5.5 Determine the mean and the variance of the future project cost X [μ_x, σ_x] using the following parametric scaling cost model $X = Y \left(\dfrac{C_x}{C_0} \right)^m$ where, X is the cost of project X; C_x is the unit capacity of project X = 80; C_0 is the unit capacity of a similar project Y = 100; Y is the cost of project Y = \$10,000,000; and m is the scaling parameter considered to be a random variable [$\mu_m = 0.6$; $\sigma_m = 0.1$].

References

Denrell J (2004) Random walks and sustained competitive advantage. Manag Sci 50(7):922–934

Howard RA (1971) Proximal decision analysis. Manag Sci 17(9):507–541

Keefer DL, Bodily SL (1983) Three-point approximations for continuous random variables. Manag Sci 29(5):595–609

Chapter 4
Monte Carlo Simulation Approach for the Probability Distribution of Project Performance Functions

Abstract In this chapter we discuss the implementation of Monte Carlo simulation evaluating of project performance functions such as the total project cost and the total project duration. We focus on the key considerations that are often ignored when Monte Carlo simulation is implemented in project risk analysis – the effect of correlation and the sample size selection. Further, we provide the methods to determine if the correlation matrix is positive-semi definite, if not, how to fix it. Finally we show the method to evaluate the effect of sample size on the confidence intervals of decision variables.

Keywords Monte Carlo simulation · Correlated random variates · Sample size · Confidence intervals

4.1 Introduction

The availability of inexpensive and fast computing systems has brought Monte Carlo simulation within reach of practically everyone. This is a great advance in the ability of project teams to develop models and to perform analyses, as Monte Carlo simulation is a very flexible and versatile tool for project engineers and managers. Unfortunately, in a hurry to get answers, many engineers make mistakes. Two most common types are: (a) ignoring the effect of correlation among simulated random variables and (b) underestimating the number of trials needed to achieve reasonable confidence limits on the results of a Monte Carlo simulation.

In many ways, a Monte Carlo simulation is like a physical experiment. The simulator makes a number of random draws from specified probability distributions over the population. However, unlike physical experiments, the population probability distributions are known in advance, as the analyst specifies them therefore the results become very sensitive to the underlying assumptions. Therefore selecting the type of probability distributions and determining the correlation among the variables must be done in a rigorous manner. Unfortunately many engineers and risk analysts skip this step all together and proceed with the simulation experiment assuming the variables are normal and uncorrelated.

© Springer Nature Switzerland AG 2020
I. Damnjanovic, K. Reinschmidt, *Data Analytics for Engineering and Construction Project Risk Management*, Risk, Systems and Decisions,
https://doi.org/10.1007/978-3-030-14251-3_4

Like physical measurements, Monte Carlo simulations are also subject to errors due to small samples. That is, the observed relative frequencies in a Monte Carlo simulation may differ considerably from the population probability distribution due to small sample size. To overcome small sample size errors, the analysis must perform enough Monte Carlo runs to achieve acceptable confidence bounds on the derived parameters. Therefore, determination of the number of runs depends on the confidence bounds, which must be explicitly stated. Many analysts do not understand confidence bounds, and do not bother to state them, which means that the results of Monte Carlo simulations are virtually meaningless, just as the results of physical experiments would be if there were no attempt to quantify the measurement errors and confidence limits. Managers who use (or pay for) Monte Carlo simulations should insist that explicit confidence limits be provided for all derived decision variables.

4.2 Generating Independent Random Variates from a Normal Distribution

There are a number of methods for generating normal random variates. Here we present a method based on independent Uniform variates, just to provide context. Any textbook on simulation and/or probability and statistics should provide a more detailed description of the methods for generating independent random variates.

Let z be a standard unit Normal variate [$N(0, 1)$], which means that it has zero mean and variance 1. Let K be a number of repetitions of the random number generator, yielding successive independent Uniform variates $r_1, r_2, \ldots, r_j, \ldots, r_K$. on the interval [0, 1]. Then we can approximate the unit Normal variate z by:

$$z = \frac{\sum_{j=1}^{K} r_j - K/2}{\sqrt{K/12}}$$

An obvious choice of K to simplify this expression is $K = 12$, in which case:

$$z = \sum_{j=1}^{12} r_j - 6$$

That is, we generate an approximately Normal variate by taking the sum of 12 instances of the variate and subtracting 6. It can be seen that $K = 12$ truncates the Normal distribution to within the interval [−6, +6], which is to say, six standard deviations above or below the mean.

Another common approach to generating independent Normally distributed random variates is to use z variates and then generate Normal random variates from the inverse of cumulative Normal distribution function.

4.3   Consistency in Covariance Matrices

If the variances and covariances are subjective estimates, perhaps elicited from different experts, is it guaranteed that the covariance matrix is consistent and valid? Unfortunately, no. Consider a simple 3-by-3 example:

$$\mathbf{V} = \begin{pmatrix} \sigma_1^2 & \rho_{12}\sigma_1\sigma_2 & \rho_{13}\sigma_1\sigma_3 \\ \rho_{12}\sigma_1\sigma_2 & \sigma_2^2 & \rho_{23}\sigma_2\sigma_3 \\ \rho_{13}\sigma_1\sigma_3 & \rho_{23}\sigma_2\sigma_3 & \sigma_3^2 \end{pmatrix} = \begin{pmatrix} \sigma_1^2 & 0.9\sigma_1\sigma_2 & 0.8\sigma_1\sigma_3 \\ 0.9\sigma_1\sigma_2 & \sigma_2^2 & -0.6\sigma_2\sigma_3 \\ 0.8\sigma_1\sigma_3 & -0.6\sigma_2\sigma_3 & \sigma_3^2 \end{pmatrix}$$

In which the correlation estimates are $\rho_{12} = 0.9$, $\rho_{13} = 0.8$, and $\rho_{23} = -0.6$.

Each individually may be valid, but the set is inconsistent. This set of correlations states that the cost of work package 1 is highly positively correlated with the cost of work package 2 and is highly correlated with the cost of work package 3. This implies that, if the cost of work package 1 is higher than average, the costs of work packages 2 and 3 are likely to be higher than average as well; the costs of work packages 2 and 3 tend to move in the same direction as the cost of work package 1. But this matrix also says that the costs of work packages 2 and 3 are rather highly negatively correlated; that is, the cost of work package 3 would tend to be higher than average when the cost of work package 2 is lower than average, and vice versa, so the costs of work packages 2 and 3 move in the opposite direction. These statements are inconsistent. Therefore, the correlation coefficients must be mis-specified.

The variances are always consistent; so any positive number suffices; but the off-diagonal covariances may be inconsistent and this inconsistency shows up in computing the *determinant* of the covariance matrix. This determinant must be positive. We will not discuss determinants here in any depth; they are covered in any elementary textbook on linear algebra. We can, however, write out explicitly the determinant for a general 3-by-3 matrix $\mathbf{V}$. Let

$$\mathbf{V} = \begin{pmatrix} v_{11} & v_{12} & v_{13} \\ v_{21} & v_{22} & v_{23} \\ v_{31} & v_{32} & v_{33} \end{pmatrix} = \begin{pmatrix} \sigma_1^2 & \rho_{12}\sigma_1\sigma_2 & \rho_{13}\sigma_1\sigma_3 \\ \rho_{12}\sigma_1\sigma_2 & \sigma_2^2 & \rho_{23}\sigma_2\sigma_3 \\ \rho_{13}\sigma_1\sigma_3 & \rho_{23}\sigma_2\sigma_3 & \sigma_3^2 \end{pmatrix}$$

The determinant of $\mathbf{V}$ is:

$$|\mathbf{V}| = v_{11}v_{22}v_{33} - v_{11}v_{23}v_{32} - v_{12}v_{21}v_{33} + v_{12}v_{23}v_{31} + v_{13}v_{21}v_{32} - v_{13}v_{22}v_{31}$$

If $\mathbf{V}$ is a covariance matrix, then substitution of the covariances as in the above matrix gives:

$$|\mathbf{V}| = \sigma_1^2\sigma_2^2\sigma_3^2\left[1 - \left(\rho_{12}^2 + \rho_{13}^2 + \rho_{23}^2\right) + 2\rho_{12}\rho_{13}\rho_{23}\right]$$

This value can be positive only if

$$1-\left(\rho_{12}^{\ 2}+\rho_{13}^{\ 2}+\rho_{23}^{\ 2}\right)+2\rho_{12}\rho_{13}\rho_{23}>0$$

In the specific example above, this becomes

$$1.0-\left(0.81+0.64+0.36\right)+2\left(0.9\right)\left(0.8\right)\left(-0.6\right)=1.0-1.81-0.432=-1.242<0$$

which clearly violates the condition. Hence, the determinant is negative and the specified correlation coefficients are not consistent.

4.4   Generating Correlated Random Variates

Suppose we have some function or process $y=y(\underline{x})$ in which $\underline{x}$ is a vector of input variables and y is the output variable of interest. Note that $y(\underline{x})$ may be a closed form algebraic function or a whole computer program. If all the inputs $\underline{x}$ are assumed to be deterministic (that is, we have absolutely $\underline{x}$ no uncertainty about any of them), then we simply use the equation (or the computer program) to compute $y=y(\underline{x})$.

Suppose, however, that we have considerable uncertainty about the values of the inputs. These values are not deterministic, but are drawn from some histogram (if we have lots of data) or some assumed probability distribution. Therefore, what we are seeking is the uncertainty, or probability distribution, on the output y given the uncertainties about all the values for $\underline{x}$.

If $y(\underline{x})$ happens to have the special linear form:

$$y=y(\underline{x})=\sum_{j=1}^{m}a_{j}x_{j}$$

then we know from the previous work that we can use the sum of the means of the $\underline{x}$ and the sum of all the terms in the covariance matrix for $\underline{x}$ to compute the means and variance of y. In addition, we know that, for this special case of linear addition, y will be distributed approximately as the Normal. Therefore, we can determine the risk for any desired value of y.

However, if $y(\underline{x})$ does not have this special form, we cannot use these simplifications. We use Monte Carlo simulation when we desire to find the uncertainty (probability distribution) of y for any type of $y(\underline{x})$. We do this by, in effect, performing numerical experiments: we generate a random set of values for the input variables $\underline{x}$ using the assumed probability distributions and the correlation coefficients, and then compute the value of y for this particular random set of $\underline{x}$. Then we repeat the process with a new set of random inputs. Each instantiation of y is input to a histogram, in which we count the number of times we get a value of y within a given interval. When we have done a great number of these numerical experiments, we have the histogram for the outputs y, from which we can compute the risks for any particular value.

To use Monte Carlo simulation to generate random variates that are not statistically independent, we need to satisfy a covariance matrix $\mathbf{V}$. Suppose that one wishes variates that are Normally distributed. Any programming language will provide random numbers drawn from a Uniform distribution on the interval [0, 1]. From these Uniform random variates one can generate unit Normal variates by one of several methods, and from these generate variates that are Normally distributed with any desired mean and variance and are statistically independent. To generate correlated random variates, an additional step is required. If there are m correlated variates, they must be generated all at once, in order to accommodate all the cross-correlations.

Let:

$\mathbf{z}$ be a vector of m standard unit Normal variates (generated independently by any algorithm)
$\mathbf{x}$ be a vector of the desired m dependent Normal variates
$\boldsymbol{\mu}$ be a m-vector of the mean values of x
$\mathbf{C}$ be a $m*m$ lower triangular matrix, such that $\mathbf{x} = \mathbf{Cz} + \boldsymbol{\mu}$

Then it can be shown that the vector $\mathbf{x} - \boldsymbol{\mu}$ has the m-by-m variance-covariance matrix $\mathbf{V}$, where:

$$\mathbf{V} = \mathbf{CC}^{\mathrm{T}}$$

The objective is to factor the covariance matrix $\mathbf{V}$ into the lower triangular matrix $\mathbf{C}$; the values of $\mathbf{C}$ are generated by the following algorithm, Cholesky decomposition:

$$c_{i,1} = \frac{\sigma_{i,1}}{\sqrt{\sigma_{1,1}}} \quad \forall i, \ 1 \leq i \leq m$$

$$c_{i,i} = \sqrt{\sigma_{i,i} - \sum_{k=1}^{i-1} c_{i,k}^2} \quad \forall i, \ 1 \leq i \leq m$$

$$c_{i,j} = \frac{\left(\sigma_{i,j} - \sum_{k=1}^{j-1} c_{i,k} c_{jk}\right)}{c_{j,j}} \quad \text{for } 1 < j < i \leq m$$

Because $\mathbf{C}$ is a lower triangular matrix, $c_{i,j} = 0$ for all values of $j > i$. Having computed the elements of the matrix $\mathbf{C}$, from the above, then one obtains the m Normal random variates $x_n \sim N[\mu_{xn}, \sigma_{xn}]$, $1 \leq n \leq m$, from the matrix multiplication:

$$\mathbf{x} = \mathbf{Cz} + \boldsymbol{\mu}$$

Caveat: note the square root function in the calculation for $c_{i,i}$. If at any step the term inside the square root is negative, the process fails. That is, it is impossible to find a set of dependent random variates satisfying the given covariance matrix. Hence, Monte Carlo simulation will *fail* if the covariance matrix is not consistent.

It turns out that, for the covariance matrix to be consistent, and the algorithm given above to succeed, the matrix **V** must be *positive semidefinite*. If the covariance matrix is *indefinite*, the algorithm fails.

A method to determine if a square matrix is positive semidefinite is to compute its *eigenvalues*, or *characteristic values*. Then, If all the eigenvalues are strictly positive (nonzero), the matrix is *positive definite*. If all of the eigenvalues are non-negative, then the matrix is *positive semidefinite;* otherwise the matrix is *indefinite*.

Suppose than that a covariance matrix is generated based on subjective estimates of various experts and computation of the eigenvalues of this matrix shows that the matrix is indefinite; hence inconsistent. What is to be done?

The first step is to examine the matrix for any obvious inconsistencies, such as shown in the 3-by-3 example above. If there are apparent inconsistencies, these should be fixed, and the eigenvalue computation repeated for the revised covariance matrix.

In many cases, however, the inconsistencies may not be apparent, especially if the covariance matrix is large. A *heuristic* that often helps in this case is based on computing the eigenvalues.

If there are any negative eigenvalues, set all the negative eigenvalues equal to zero (or some small positive number). Then, using the computed *eigenvector matrix*, back-compute a revised covariance matrix **V′**. This matrix will probably be positive definite; to make sure, recompute the eigenvalues for the revised covariance matrix. If all are positive or zero, the revised covariance matrix is consistent and can be used; if some of the eigenvalues for the revised matrix **V′** are negative, set them to zero, back-compute a new revised matrix **V″**, and continue if necessary until a positive definite or positive semidefinite matrix is obtained. Of course, there are other methods to fixing inconsistent correlation matrix based on minimizing differences between original matrix values and the corrected semi-definite version (Higham 2002).

4.5   Generating Correlated Lognormal Variables

Log-normal probability distribution is of particular interest to project data analysts. This is because it is always positive and skewed to the right.

Let Y_j and Y_k be two out of a set of Normally–distributed random variates, normalized to zero means, with variances σ_j^2, σ_k^2 and correlation coefficient $\rho_{Yj,k}$. Now, let $X_j = e^{Y_j}, X_k = e^{Y_k}$. Then X_j and X_k are lognormally distributed (Law and Kelton 1991). The correlation between X_j and X_k is given by:

$$\rho_{Xj,k} = \frac{e^{\rho_{Yj,k}\sigma_j\sigma_k} - 1}{\sqrt{\left[e^{\sigma_j^2} - 1\right]}\sqrt{\left[e^{\sigma_k^2} - 1\right]}}$$

Or, inverting this last expression, we have

$$\rho_{Xj,k}\sqrt{\left[e^{\sigma_j^2} - 1\right]}\sqrt{\left[e^{\sigma_k^2} - 1\right]} = e^{\rho_{Yj,k}\sigma_j\sigma_k} - 1$$

$$e^{\rho_{Yj,k}\sigma_j\sigma_k} = 1 + \rho_{Xj,k}\sqrt{\left[e^{\sigma_j^2} - 1\right]}\sqrt{\left[e^{\sigma_k^2} - 1\right]}$$

$$\rho_{Yj,k}\sigma_j\sigma_k = \ln\left[1 + \rho_{Xj,k}\sqrt{\left[e^{\sigma_j^2} - 1\right]}\sqrt{\left[e^{\sigma_k^2} - 1\right]}\right]$$

$$\rho_{Yj,k} = \left(\frac{1}{\sigma_j\sigma_k}\right)\ln\left[1 + \rho_{Xj,k}\sqrt{\left[e^{\sigma_j^2} - 1\right]}\sqrt{\left[e^{\sigma_k^2} - 1\right]}\right]$$

Note that there may be some values of $\rho_{Xj,k}$ that are excluded (that is, cannot be valid) by the requirement that the argument of the natural logarithm must be positive. That is,

$$\left[1 + \rho_{Xj,k}\sqrt{\left[e^{\sigma_j^2} - 1\right]}\sqrt{\left[e^{\sigma_k^2} - 1\right]}\right] > 0$$

$$\rho_{Xj,k}\sqrt{\left[e^{\sigma_j^2} - 1\right]}\sqrt{\left[e^{\sigma_k^2} - 1\right]} > -1$$

$$\rho_{Xj,k} > \frac{-1}{\sqrt{\left[e^{\sigma_j^2} - 1\right]}\sqrt{\left[e^{\sigma_k^2} - 1\right]}}$$

Therefore, to generate random variates X_j and X_k that are lognormally distributed, with correlation $\rho_{Xj,k}$:

- First, determine the desired parameters (mean, shift factor, variance, and correlation $\rho_{Xj,k}$) for the original lognormal variables X_j and X_k.
- From these, find the parameters of the associated Normally distributed variables, Y_j and Y_k, and the correlation between them, $\rho_{Yj,Yk}$, from the above expression.
- From these, generate the full correlation matrix with correlations $\rho_{Yj,k}$ for all j and k. Equivalently, generate the full covariance matrix $\rho_{Yj,k}\sigma_j\sigma_k$ for all j and k. Then, generate the set of Normally distributed variables Y_k for all k.
- Then convert to the desired lognormal variates by $X_j = e^{Y_j}, X_k = e^{Y_k}$

4.6   On the Number of Trials in Monte Carlo Simulation

Many engineers who use Monte Carlo simulation specify the number of runs in the order of several hundreds. However, the actual number required for reasonable confidence limits may be more in the order of tens of thousands.

As an example, suppose that a sponsor of a prospective project is interested in estimating the risk before deciding whether or not to proceed with it. One view of risk assessment with Monte Carlo is to compute in each simulated run an attribute of the simulated project that depends on whether the project meets or does not meet some criterion. For example, the attribute might be whether the project costs overrun the budget, or risk-adjusted cost estimate, or that the project will be late, compared to the risk-adjusted schedule. It might be based on whether the present worth of the project, considering all future costs and benefits discounted to the present, exceeds some specified value. It might whether the rate of return on the project investment (that is, the discount rate for which the net discounted present worth is zero) exceeds some value. It might be some weighted combination of many factors.

One might ask, how can one determine the risk-adjusted budget or schedule? By repeated Monte Carlo simulations. That is, one runs the Monte Carlo simulator for various values for the risk adjustment (or contingency) until one determines the budget (or risk-adjusted cost estimate) such that both the budget and the probability of overrunning it are acceptable to the project sponsor. Or, until one determines that there is no budget acceptable to the sponsor that meets the sponsor's risk requirements.

Suppose that the Monte Carlo simulation computes the criterion or attribute for each of a number of projects using random numbers as inputs. Then the output attribute is a random variable. To be specific, suppose that the Monte Carlo simulator computes the rates of return (ROI) for some number, n, of simulated projects. The sponsor would naturally prefer projects that maximize his ROI (if the criterion is cost or duration, he would prefer to minimize the critical attribute). For the discussion here, we will assume that the critical attribute is to be maximized, and that there is some critical value of this attribute, set by the sponsor, such that a project with a ROI greater than this value is considered a success, and one with a ROI less than this valued is considered a failure. Thus, each random simulation has two possible attribute outcomes:

- Success, that is, the simulated ROI is greater than or equal to the sponsor's minimum rate of return (often called the *hurdle rate*); or
- Failure, the simulated ROI is less than the minimum rate of return.

For brevity, call a project in the first class a *good project* and a simulated outcome in the second class a *bad project*. If the probability of failure (that is, the probability of a rate of return less than the specified hurdle rate) is greater than some value, the sponsor will not proceed with the project. Suppose the number of simulation trials is n, a number to be determined. The probability of financial failure, that is, the probability that the ROI will be less than the minimum, is then estimated from the Monte Carlo results as the ratio of the number of bad projects to the total number of good and bad simulated projects. Let p_{sim} be this ratio and let p_{crit} be the sponsor's critical probability. Then,

If $p_{sim} \leq p_{crit}$ then the risk is acceptable to the sponsor and the project goes forward, but

If $p_{sim} > p_{crit}$ then the risk is too great, and the project is terminated.

This decision process may be regarded as an instance of hypothesis testing. Because sponsors typically feel that the burden of proof is on the project to justify proceeding, we may say that the *null hypothesis* is:

H_o: the project is not acceptable by the critical decision attributes and should not be done.

The *alternate hypothesis* is:

H_a: the project meets the financial or other criteria and should proceed.

Then to proceed with the project, the sponsor must *reject* the null hypothesis, which is equivalent to accepting the alternate hypothesis. The sponsor is assumed to make a decision between H_o and H_a based on the Monte Carlo ratio p_{sim}, the frequency of failing projects out of the number of random trials n. However, p_{sim} is itself a random variable and therefore it has a probability distribution of its own. We will consider how to determine the number of Monte Carlo trials needed to achieve some specified confidence bounds with respect to the hypothesis test, considering the probability distribution of p_{sim}.

Note that this error analysis does *not* consider modeling error. That is, the discussion here deals only with the random errors in the Monte Carlo simulation due to small sample size. This analysis does *not* address the issue of the degree to which the simulation model actually represents the performance of the real project. This does not mean that modeling error is not important; it only means that modeling error cannot be addressed by statistical means. Modeling error is not a random error, it is a bias, and no amount of random testing can reduce it. In fact, modeling error may be much more serious than statistical error, and modeling error may actually vitiate the Monte Carlo simulation, but this is a separate problem, considered elsewhere.

The binary definition of the outcomes (simulated projects are good or bad, by the specified criteria) means that the results can be described by the Binomial distribution. Suppose that x is the number of bad projects in n simulation trials. Then, by the Binomial distribution,

$$P\{X = x\} = f(x) = \binom{n}{x} p^x q^{n-x}$$

in which p is the true population probability of a failure on any trial, $q = 1 - p$, and $\binom{n}{x}$ means the number of combinations of n things taken x at a time. Obviously, the population considered here is the population of all Monte Carlo simulation runs, not the population of real-world projects. Of course, the value of p is unknown, being the value that is being estimated by the simulation. For any x, let $\hat{p} = x / n$ be an estimate of p (is the value that the simulation would produce if it did an infinite number of trials). In the Binomial distribution,

$$E[x] = \mu = np$$

so the expected value of the estimate of p is $E[\hat{p}] = E\left[\dfrac{x}{n}\right] = \dfrac{E[x]}{n}$.

Also, for the Binomial distribution,

$$Var[x] = \sigma^2 = npq$$

By the definition of variance, then,

$$\sigma_p^2 = Var[\hat{p}] = Var\left[\frac{x}{n}\right] = \frac{1}{n^2} Var[x] = \frac{pq}{n}$$

Hence, the standard deviation of $\dfrac{x}{n}$ is

$$\text{Std. Dev.}\left(\frac{x}{n}\right) = \text{Std. Dev.}[\hat{p}] = \sqrt{pq / n}$$

Then the mean plus k standard deviations is $\hat{p} + k\sqrt{pq / n}$

The Binomial distribution can be approximated by the Normal distribution with good agreement if $np > 5$ and $n(1 - p) > 5$. Then, the values for mean and variance of the estimate $\hat{p}$ can be used in a Normal distribution if n is chosen large enough to meet these conditions. Suppose that this is true (we will check these conditions later). To define confidence intervals, we will consider several alternate approaches.

In the first example, we will define a simple symmetric confidence interval around the value of the population proportion of bad projects, p. Then,

$$\hat{p} - \Delta\hat{p} \leq \text{population } p \leq \hat{p} + \Delta\hat{p}$$

Using the equation above, we can write:

$$\hat{p} - k\sqrt{\frac{\hat{p}\hat{q}}{n}} \leq p \leq \hat{p} + k\sqrt{\frac{\hat{p}\hat{q}}{n}}$$

Then,

$$\Delta\hat{p} = k\sqrt{\frac{\hat{p}(1 - \hat{p})}{n}}$$

If the Monte Carlo simulation has been performed, n and $\hat{p}$ are known, and so the confidence band $\Delta\hat{p}$ can be computed for any significance level defined by k. This confidence band is taken to be symmetric, so we use a two-tailed test based on the Normal approximation, discussed above. For example, if we want a 95% confidence band, this corresponds to probabilities of 0.025 in each tail in a two-tailed test, and $k = 1.96$ from tables of the Normal distribution.

To compute a required value of n before performing the Monte Carlo simulation, we take the square of both sides of the above equation, to give:

$$n = \left(\frac{k}{\Delta p}\right)^2 p(1-p)$$

To compute this equation before the simulation, we must guess a value for p. We might assume that the value of p of interest will be of the order of p_{crit}, the critical decision value. As an example, suppose that the sponsor wants a probability of 1% or less that the project will be bad (fail to meet the financial and other criteria). Then set $p = 0.01$. Suppose that the sponsor will accept a confidence band that is 10% of this value, above and below. Then, set $\Delta p = 0.001$. That is, the confidence interval will be

$$0.009 \le p \le 0.011$$

If the sponsor wants 95% confidence that this confidence interval includes the true population value p, then set $k = 1.96$. Using these numerical values in the equation for n, above, gives:

$$n = \left(\frac{1.96}{0.001}\right)^2 (0.01)(0.99) = 38{,}032 \text{ iterations}$$

Earlier, it was noted that the Normal approximation to the Binomial distribution is approximately accurate if $np > 5$ and $n(1 - p) > 5$. From this solution we see that $np = 380 \gg 5$ and $n(1 - p) = 37,652 \gg 5$. Therefore, the Normal approximation is valid.

Although the numerical values used above are invented, they are not necessarily unrealistic. If we widen the confidence band by a factor of five, by setting $\Delta p = 0.0025$, so the confidence interval is

$$0.0075 \le p \le 0.0125$$

then the equation for the required n gives:

$$n = \left(\frac{1.96}{0.0025}\right)^2 (0.01)(0.99) = 6085 = \text{iterations}$$

which shows that it takes a substantial number of Monte Carlo trials to obtain narrow confidence bands on the computed attributes.

A second example takes a more detailed approach. Here, the sponsor sets two statistical criteria:

1. The probability of rejecting a project that would really be a good project, with a true population failure rate less than $p_\alpha < p_{crit}$, should not be greater than α. This is the probability that the project will be terminated (that is, the sponsor does not reject the null hypothesis) for being too risky, based on the limited number of Monte Carlo simulations, when it is not too risky, by the sponsor's criteria.
2. Conversely, the probability of accepting a project that will really turn out to be a bad project, with a true population failure rate $p_\beta < p_{crit}$, should not be greater than β This is the risk that the project will be given the go-ahead (that is, the sponsor rejects the null hypothesis), based on the limited Monte Carlo simulations, when it is actually too risky, by the sponsor's criteria.

Note that α and β do not have to be identical. Also, many sponsors would say that it is better to miss an opportunity than to take on a bad situation, and therefore would require that the probability of mistakenly going ahead with a project that is too risky should be much less than the probability of mistakenly rejecting a project that is not risky, or $\beta < \alpha$.

Let k_α be a coefficient taken from the tables for the Normal distribution corresponding to the error α and let k_β be a coefficient taken from the tables corresponding to the error β. Then we have the confidence band around the (unknown) value of the true risk factor p given by:

$$P_\alpha - k_\alpha \sqrt{P_\alpha q_\alpha / n} \le \text{ true population } p \le P_\beta + k_\beta \sqrt{P_\beta q_\beta / n}$$

Hence, the width of the confidence band is given by:

$$k_\alpha \sqrt{P_\alpha q_\alpha / n} + k_\beta \sqrt{P_\beta q_\beta / n} = \left(\frac{1}{\sqrt{n}} \right) \left(k_\alpha \sqrt{P_\alpha q_\alpha} + k_\beta \sqrt{P_\beta q_\beta} \right) = P_\beta - P_\alpha$$

By rearranging this equation we get:

$$\sqrt{n} = \frac{k_\alpha \sqrt{P_\alpha q_\alpha} + k_\beta \sqrt{P_\beta q_\beta}}{P_\beta - P_\alpha}$$

$$n = \left(\frac{k_\alpha \sqrt{P_\alpha q_\alpha} + k_\beta \sqrt{P_\beta q_\beta}}{P_\beta - P_\alpha} \right)^2$$

As a numerical example, suppose the sponsor is somewhat risk averse and will accept only a 5% chance of going forward, based on the Monte Carlo simulations, with a project for which the failure rate would really be more than 1%. Then, say, $p_\beta = 0.011$ and $\beta = 0.05$ Suppose that, in addition, the sponsor wants only a 10% chance of rejecting a good project as being too risky, based on the Monte Carlo simulations, for which the failure rate would really be less than 0.01. Then, say, $p_a = 0.009$ and $\alpha = 0.10$. For these calculations, we use one-sided values in the Normal distribution tables. Therefore, $k_\beta = 1.645$ and $k_\alpha = 1.282$. Substituting in the above equation for n:

$$n = \left(\frac{1.282\sqrt{0.009(0.991)} + 1.645\sqrt{0.011(0.989)}}{0.002} \right)^2$$

which evaluates to $n = 21{,}411$ Monte Carlo iterations required. Suppose that the sponsor is now willing to accept a 20% chance of mistakenly rejecting a good project. Then $p_a = 0.009$, $\alpha = 0.20$ and $k_\alpha = 0.842$. Then, substituting these values with the previous:

$$n = \left(\frac{0.842\sqrt{0.009(0.991)} + 1.645\sqrt{0.011(0.989)}}{0.002} \right)$$

which evaluates to $n = 15{,}762$ Monte Carlo iterations. Thus, if the sponsor is willing to accept a larger risk of missing good projects, the number of random trials may be reduced. Other results for n will be obtained for different assumptions about acceptable risks. Nevertheless, the number of iterations is likely to remain above 10,000.

There are other approaches to determining the stopping criteria for sampling including a double sampling plan. In this approach, for Monte Carlo simulation,

1. A random sample of size n_1 is simulated.
2. If the sample contains c_1 or fewer failures (bad projects), then the project is accepted (the null hypothesis is rejected).
3. If the sample contains more than c_2 failures, then the project is rejected (the null hypothesis is not rejected).
4. If there are x_1 failures, where $c_1 \leq x_1 \leq c_2$, then a second run of n_2 simulation trials is made.
5. If the number of failures in both runs, $x_1 + x_2 \leq c_3$, then the project is accepted (the null hypothesis is rejected); if not, the project is rejected.

This method often allows the project to be accepted or rejected at a smaller number of trials than a single-step sampling plan. One could extend this approach to a multiple-step sampling plan, in which the number of failures would be tested after every iteration, with three outcomes: reject the project, accept the project, or continue simulating. The details of how to do this are left to the reader.

4.7   Practice Problems

Problem 4.7.1   Consider correlation matrix from Practice Problem 3.5.2 from Chap. 3 (i.e. a small project comprised of six sequential activities). Is the correlation matrix consistent? If not, why not?

Problem 4.7.2 Consider a small project comprised of four activities A, B, C, and D. The correlation matrix, developed by a project manager, of the activity duration for such project is shown below:

$$
\begin{bmatrix}
 & A & B & C & D \\
A & 1 & 0.9 & 0.7 & 0.1 \\
B & 0.9 & 1 & 0.3 & 0.3 \\
C & 0.7 & 0.3 & 1 & 0.3 \\
D & 01 & 0.3 & 0.3 & 1
\end{bmatrix}
$$

Is this correlation matrix positive semi-definite? [explain your answer] If it is not, how would you make it positive semi-definite without significantly affecting the results? What would be the new correlation matrix?

Problem 4.7.3 A project engineer employed by an asphalt paving contractor collected data one summer's day on the arrival and processing of asphalt trucks at a paving job (see Table 4.1). For each of the following observed or derived quantities, determine the frequency diagrams (histograms) of the actual data:

- The inter-arrival time between trucks arriving at the site.
- The processing time for the paving machine (time actually paving).
- The time spent by trucks waiting to go on line.
- The number of trucks waiting in the queue at any time.
- Turnaround time for individual trucks.

For each of the above quantities, recommend a type of probability distribution that seems to give a good fit to the histogram, if there is one. For example, does it appear that the inter-arrival time for trucks has the same frequency distribution for the entire job? Does it appear that the frequency distribution for the paving machine processing time is the same over the period of the job? Determine the parameters of each of these distributions from the dataset. Compare the functional form of the probability mass function or probability density function to the histogram. (That is, if you believe the underlying probability function for truck inter-arrival times is exponential [trucks arrive by a Poisson process], determine the best value for the single parameter of the exponential and plot the function along with the histogram. Similarly, for the other quantities.)

Develop a Monte Carlo simulation and simulate the job above assuming that the critical probability is defined when the total job time exceeds 700 minutes; in other words the operation will need to be redesigned (i.e. project fails) if the total job time exceeds 700 minutes. Assume that the job size is always 58 truckloads. From the Monte Carlo results, show frequency diagrams for:

- Truck delays
- Paving machine delays (time waiting for loaded trucks and not paving)
- Total job time

Table 4.1 Problem data

I	II	III	IV	V	VI	VII	VIII	IX	X	XI
1	0	6:25	6:27	0:02		6:32	0:05	CT1		0:03
2	1	6:28	6:35	0:07	0:03	6:38	0:03	W365		0:00
3	2	6:28	6:38	0:10	0:00	6:40	0:02	DL1		0:00
4	2	6:33	6:40	0:07	0:05	6:41	0:01	RT1		0:02
6	2	6:40	6:43	0:03	0:07	6:49	0:04	JH12		0:03
5	1	6:44	6:50	0:06	0:04	6:52	0:02	SCH1		0:24
7	0	7:14	7:16	0:02	0:30	7:20	0:04	JS2K		0:00
8	1	7:14	7:20	0:06	0:00	7:22	0:02	BH1		0:02
9	2	7:17	7:24	0:07	0:03	7:26	0:02	CT1	0:45	0:01
10	1	7:25	7:27	0:02	0:08	7:29	0:02	W365	0:47	0:01
11	1	7:24	7:30	0:06	0:01	7:32	0:02	DL1	0:44	0:05
12	0	7:35	7:37	0:02	0:10	7:44	0:03	AE6		0:05
13	1	7:35	7:50	0:15	0:11	7:53	0:03	DJ2		0:06
14	1	7:35	7:45	0:10	0:00	7:47	0:02	RT1	0:54	0:01
15	1	7:37	7:48	0:11	0:02	7:50	0:02	JH12	0:48	0:00
16	2	7:45	7:59	0:14	0:08	8:19	0:20	SCH1	0:53	0:00
17	1	7:58	7:59	0:01	0:13	8:00	0:01	RG23		0:16
18	0	8:15	8:16	0:01	0:17	8:19	0:03	CT1	0:49	0:07
19	0	8:25	8:26	0:01	0:10	8:50	0:24	JS2K	1:05	0:05
20	2	8:25	8:27	0:02	0:00	8:29	0:02	BH1	1:03	0:04
21	0	8:30	8:33	0:03	0:05	8:40	0:07	W365	1:01	0:02
22	1	8:31	8:40	0:09	0:01	8:41	0:01	DL1	0:59	0:01
23	2	8:35	8:42	0:07	0:04	8:45	0:03	RT1	0:48	0:01
24	3	8:40	8:46	0:06	0:05	8:48	0:02	AE6	0:56	0:00
25	2	8:45	8:55	0:10	0:05	8:58	0:03	DJ2	0:52	0:02
26	2	8:47	9:00	0:13	0:02	9:02	0:02	JH12	0:57	0:01
27	1	9:00	9:03	0:03	0:13	9:05	0:02	TD4801		0:01
28	2	9:02	9:06	0:04	0:02	9:12	0:06	TD4802		0:02
29	2	9:03	9:13	0:10	0:01	9:16	0:03	TD4806		0:04
30	3	9:03	9:20	0:17	0:00	9:21	0:01	TD4800		0:01

Repeat the Monte Carlo simulation for a situation in which you change one or more of the parameters. For example, what if you increase the frequency of truck arrivals? Decrease the frequency of truck arrivals? If the paver processing time is exponential, the standard deviation is the same as the mean time. Can you think of a way to reduce the standard deviation without changing the mean time? What would be the effect of this on the production efficiency?

Table 4.2 shows definition of the terms used in Table 4.1.

Warning: Check the data. The information was recorded under field conditions. The construction engineer who collected the data is no longer employed by this firm. You are responsible for detecting and resolving any omissions, discrepancies, or other issues about the data.

Table 4.2 Description of terms

Table term	Description
I	Load number
II	Number of trucks in queue when a loaded truck arrives
III	Truck arrival time – clock time when a truck arrives on site
IV	Time on line – clock time when truck reaches paving machine
V	Truck waiting time = elapsed time from arrival to paver
VI	Inter-arrival time = elapsed time between truck arrivals
VII	Truck time out – clock time when a truck leaves site
VIII	Process time = elapsed time when truck is at paving machine
IX	Truck number
X	Truck turnaround = elapsed time from empty truck leaving to full truck arriving back
XI	Elapsed time, paving crew waiting for full truck

References

Higham NJ (2002) Computing the nearest correlation matrix—a problem from finance. IMA J Numer Anal 22(3):329–343

Law AM, Kelton WD (1991) Simulation modeling and analysis, vol 2. McGraw-Hill, New York

Chapter 5
Correlation in Projects

Abstract In this chapter we extend our discussion on the assumption of statistical independence among the work packages and provide the theoretical justification why such assumption can lead to poor results. We also provide an overview of autoregressive models and the examples of how such models can be applied to forecasting of project outcomes and ultimately risk assessment and management.

Keywords Statistical independence · Correlation · Autoregressive models

5.1 Introduction

In all of the discussion of risks so far, the issue of statistical dependence and independence of the variates has been raised. Dependence (correlation) does make a difference, and we shall see here, and later, that neglect of correlation can lead to poor results. So, in order to handle variables that are not statistically independent or perfectly dependent, we need to define independence and dependence more precisely. Suppose there are two events, X and Y, and that $P\{X\}$ is the *marginal probability* of event X happening, $P\{X\}$ is the marginal probability of event Y happening, and $P\{X \cap Y\}$ is the *joint probability* of both events X and Y happening together. The general expression for the joint probability is

$$P\{X \cap Y\} = P\{X|Y\} P\{Y\} = P\{Y|X\} P\{X\}$$

Where, $P\{X|Y\}$ is the *conditional probability* of event X, *given that* event Y occurs. In fact, we may consider that most probabilities of interest are conditional, even though they may not be written that way. For example, we may say that X is the event that a project overruns its schedule, and $P\{X\}$ is the probability that the project overruns its schedule, but what we may really mean is $P\{X|Y\}$, the probability that the project overruns, conditional on (given that) event Y, the project manager takes no action to stop it from overrunning. All predictions of project costs and durations are really of this conditional nature.

I. Damnjanovic, K. Reinschmidt, *Data Analytics for Engineering and Construction Project Risk Management*, Risk, Systems and Decisions, https://doi.org/10.1007/978-3-030-14251-3_5

The necessary and enough condition for X and Y to be *statistically independent* is that the joint probability of the two events be the product of the two marginal probabilities:

$$P\{X \cap Y\} = P\{X\}P\{Y\}$$

This is the same as requiring that the conditional probabilities be equal to the marginal probabilities, that is,

$$P\{X|Y\} = P\{X\} \quad \text{and} \quad P\{Y|X\} = P\{Y\}$$

In the independent case, the probability that event X occurs, given that event Y also occurs, is just the probability that X occurs regardless of what happens with Y. That is, event X is *independent* of event Y if $P\{X|Y\} = P\{X\}$.

This is the condition for complete statistical independence, but of more concern are the degrees of dependence, as defined by the *correlation coefficient* ρ, where $-1 \leq \rho \leq +1$. Consider the case of the joint Normal distribution of two continuous variates x and y, defined over $\forall x$, $-\infty \leq x \leq +\infty$ and $\forall y$, $-\infty \leq y \leq +\infty$. The joint probability density function for the bivariate Normal is:

$$f(x,y) = \frac{\exp\left(-\frac{1}{2}Q\right)}{2\pi\sigma_x\sigma_y\sqrt{1-\rho^2}}, \text{ in which}$$

$$Q = \left[\frac{1}{1-\rho^2}\right]\left[\frac{(x-\mu_x)^2}{\sigma_x^2} - 2\rho\frac{(x-\mu_x)(y-\mu_y)}{\sigma_x\sigma_y} + \frac{(y-\mu_y)^2}{\sigma_y^2}\right]$$

Where μ_x is mean value of x, μ_y is mean value of y, σ_x^2 is variance of x, σ_y^2 is variance of y, and ρ is correlation between x and y.

If $\rho = 0$, then it can be easily seen the equation above separates into two parts, such that $f(x,y) = f(x)f(y)$, where $f(x)$ is the marginal univariate Normal probability density function for x, and therefore x and y are statistically independent. This not a sufficient condition in general, however; it is possible that two variates with zero correlation are not independent. If $\rho = 0$ and $\sigma_x = \sigma_y$, the contours of equal probability plot in the $x - y$ plane as circles. If $\rho > 0$, the contours of equal probability are ellipses, inclined upward to the right, and if $\rho < 0$, the ellipses of equal probability are inclined upward to the left. If $\rho \to +1$, the ellipses contract toward a straight line with positive slope; as $\rho \to -1$, the ellipses flatten out to a straight line with negative slope.

Note that one of the features of the Normal distribution that is not present in other probability distributions is that the correlation coefficient (sometimes called the *Pearson* correlation coefficient) appears *explicitly* as a parameter in the mathematical expression for the joint probability density function. This fact determines how the correlations are to be determined; correlation can be computed only one way (see below) when one is using the Normal distribution. However, there are other possible definitions for correlations. One may define higher order correlations (third, fourth, fifth, etc.), or one may use different definitions entirely (e.g., Spearman's rank order correlation coefficient). Random variables may be correlated even if their marginal probability distributions are not Normal. This is because correlation (or covariance) is a property of the data, and if data are available, one can determine estimates for any definition of correlation. If data are not available, which is true much of the time; one can still estimate subjective values for correlations, for any form of probability distribution.

Recall from elementary probability (Walpole et al. 1993) that the *Expectation Operator*, written usually as $E[\cdot]$, computes the *mean value* of any function inside the brackets. In the discrete case, with data z_j

$$E[z] = \frac{1}{n}\sum_{j=1}^{n} z_j$$

In the continuous case, if $f(z)$ is the known probability density function on z, then

$$E[z] = \int_{z=-\infty}^{z=+\infty} zf(z)\,dz$$

Also, for any function of z, say $g(z)$,

$$E[g(z)] = \int_{z=-\infty}^{z=+\infty} g(z)f(z)\,dz$$

Some obvious but useful results of the definition are as follows:

- $E[a] = a$, where a is a constant
- $E[ax] = aEx = a\mu_x = a\bar{x}$ where $\mu_x = \bar{x} = E[x]$ and x is a random variate
- $E[ax + by] = aE[x] + bE[y] = a\mu_x + b\mu_y$, where b is a constant and y is a random variate

The expectation $E[z]$ is also known as the *first moment* of z. The *second moment* of z is $E[z^2]$. The second *central* moments of x and y are the variances of x and y, σ_x^2 and σ_y^2, defined by the expectations taken around the means:

$$\sigma_x^2 \equiv E\left[(x-\mu_x)^2\right] \text{ and } \sigma_y^2 \equiv E\left[(y-\mu_y)^2\right]$$

That is, the variance of x is defined to be the average of the squares of the deviations of x taken about the mean of x. Expanding the expression for the variance, by simple algebra,

$$\sigma_x^2 \equiv E\left[\left(x-\mu_x\right)^2\right] = E\left[x^2 - 2\mu_x x + \mu_x^2\right] = E\left[x^2\right] - E\left[2\mu_x x\right] + E\left[\mu_x^2\right]$$

$$\sigma_x^2 \equiv E\left[x^2\right] - 2\mu_x E\left[x\right] + \mu_x^2 = E\left[x^2\right] - \mu_x^2$$

$$\sigma_x^2 \equiv E\left[x^2\right] - \left(E\left[x\right]\right)^2$$

This last result is a useful expression.

More generally, the k-th central moment of any probability distribution is defined by the equation:

$$v_k = E\left[\left(x-\mu_x\right)^k\right], \forall k \geq 1$$

Note that the first moment of the probability distribution $f(x)$ is μ_x but the first *central* moment of $f(x)$ is 0, because:

$$E\left[x-\mu_x\right] = E\left[x\right] - E\left[\mu_x\right] = \mu_x - \mu_x = 0$$

The *variances* of x and y, σ_x^2 and σ_y^2, were defined above by the expectations of the squares of the deviations from the mean:

$$\sigma_x^2 = E\left[\left(x-\mu_x\right)^2\right] \text{and} \sigma_y^2 = E\left[\left(y-\mu_y\right)^2\right]$$

and the *covariance* of x and y, σ_{xy}, is then defined by the expectation of the cross-product of the deviations from the means in x and y:

$$\sigma_{xy} = E\left[\left(x-\mu_x\right)\left(y-\mu_y\right)\right]$$

The linear *correlation coefficient* between x and y is defined as

$$\rho_{x,y} = \frac{\sigma_{xy}}{\sigma_x \sigma_y}$$

If there is a set of *statistically independent* variates $x^{(1)}, x^{(2)}, \ldots, x^{(j)}, \ldots, x^{(n)}$, with means $\mu^{(1)}, \mu^{(2)}, \ldots, \mu^{(j)}, \ldots, \mu^{(n)}$; variances $\sigma^{2(1)}, \sigma^{2(2)}, \ldots, \sigma^{2(j)}, \ldots, \sigma^{2(n)}$; and k-th central moments $v_k^{(1)}, v_k^{(2)}, \ldots, v_k^{(j)}, \ldots, v_k^{(n)}, k \geq 2$; then the moments of the probability distribution of the sum of all the $x^{(1)}, x^{(2)}, \ldots, x^{(j)}, \ldots, x^{(n)}$ variates are[1]:

---

[1] Conventionally, the symbol for the third central moment is μ_3, but μ is already used here for the mean.

$$\mu = \sum_{j=1}^{n} \mu^{(j)}$$

$$\sigma^2 = \sum_{j=1}^{n} \sigma^{2(j)}$$

$$v_k = \sum_{j=1}^{n} v_k^{(j)}$$

The third central moment is associated with skewness, which is a departure from symmetry; the fourth central moment with kurtosis, or the flatness or pointedness of the distribution. Note that the *odd* central moments, v_1, v_3, v_5, v_7, etc., are identically zero for the Normal distribution, and for any other distribution that is symmetric about its mean. For the specific case of the Normal distribution, the central moments are given by:

$$v_1 = 0$$
$$v_2 = \sigma^2$$
$$v_3 = 0$$
$$v_4 = 3\sigma^4$$
$$v_5 = 0$$
$$v_{2k-1} = 0$$
$$v_{2k} = 1(3)(5)(7)...(2k-1)\sigma^{2k}$$

The Normal distribution, conventionally abbreviated as $N[\mu, \sigma^2]$, has many moments but only two parameters, μ and σ, as shown in the general equation. The unit standard Normal distribution, $N[0, 1]$, has zero mean and variance equal to one.

The *second moment approach* is a method for approximating probability distributions using, as the name implies, only the first two moments, the mean and variance.

Consider the linear combination of random variates x_j given by:

$$y = a_0 + a_1 x_1 + a_2 x_2 + ... + a_n x_n$$

in which the value of a_j are constants. Then,

$$\mu_y = a_0 + E\left[\sum_{j=1}^{n} a_j x_j\right] = a_0 + \sum_{j=1}^{n} a_j E\left[x_j\right] = a_0 + \sum_{j=1}^{n} a_j \mu_j$$

$$\sigma_y^2 = E\left[(y - \mu_y)^2\right] = E\left[\left\{\sum_{j=1}^{n} a_j (x_j - \mu_j)\right\}^2\right]$$

Expanding the last equation by multiplying out all the terms in the square of the summation gives:

$$\sigma_y^2 = \sum_{j=1}^{n}\sum_{k=1}^{n} a_j a_k \sigma_{jk}$$

Here σ_{jk} is the *covariance* between x_y and x_k, and, by convention, $\sigma_{kk} \equiv \sigma_k^2$.

If all the x_j are statistically independent, then the cross-product terms in the covariance vanish for $\forall j \neq k$, and the variance of y reduces to

$$\sigma_y^2 = \sum_{j=1}^{n} a_j^2 \sigma_{jj} = \sum_{j=1}^{n} a_j^2 \sigma_j^2$$

But *only* if all the variates are statistically independent.

In matrix notation, the *variance-covariance matrix* (often written as simply *covariance matrix*), V, consists of all the variance and covariance terms, with σ_{jk} the entry in row j, column k:

$$V = \left[\sigma_{jk} \right]$$

The terms on the main diagonal are the variances $\sigma_{kk} = \sigma_k^2$ and the off-diagonal terms are the covariances. If all the covariances (all the correlation coefficients) are zero, then the matrix is *diagonal* and all the variables are independent.

Thus, there is substantial mathematical simplification if the variates are statistically independent, but there is also potential for substantial error if they are assumed to be independent when they are not. To get a simple bound on the possible error, rewrite the equation for the variance of y as

$$\sigma_y^2 = \sum_{j=1}^{n}\sum_{k=1}^{n} a_j a_k \sigma_{jk}$$

and assume that all the a_j, ρ_{jk} and σ_k are the same; i.e.,

$$a_j = a_k = a, \quad \text{for } \forall j, 1 \leq j \leq n; \quad \text{for } \forall k, 1 \leq k \leq n$$

$$\sigma_{jk} = \sigma^2 \quad \text{for } k = j$$

$$\sigma_{jk} = \rho_{jk}\sigma_j\sigma_k = \rho\sigma^2 \quad \text{for } k \neq j$$

Then the variance of y reduces to:

$$\sigma_y^2 = \sum_{j=1}^{n} a_j^2 \sigma_j^2 + \sum_{\substack{j=1 \\ k \neq j}}^{n}\sum_{k=1}^{n} a_j a_k \rho_{jk}\sigma_j\sigma_k$$

$$\sigma_y^2 = \sum_{j=1}^{n} a^2\sigma^2 + \sum_{\substack{j=1 \\ k \neq j}}^{n}\sum_{k=1}^{n} a^2 \rho\sigma^2$$

$$\sigma_y^2 = na^2\sigma^2 + n(n-1)\rho a^2\sigma^2$$

When $\rho = 0$ this reduces further to $\sigma_y^2 = na^2\sigma^2$ but as $\rho \to +1$, $\sigma_y^2 \to n^2a^2\sigma^2$. This indicates that, for $\rho > 0$, an assumption that the variates are statistically independent can greatly underestimate the variance, by as much as a factor of n. The standard deviation is an indicator of uncertainty or risk, so assuming statistical independence, $\rho = 0$, may underestimate the risk, and therefore lead to unconservative results if $\rho > 0$ (A corresponding bound cannot be stated for $\rho < 0$, as it is not possible for all the variates to be negatively correlated simultaneously).

5.2   Project Time Series and Autocorrelation

Another indication of the effects of statistical dependence can be obtained by considering a project as a time series, and applying some simple results from time series analysis (Hamilton 1994). Consider the simplest case, the first order stationary autoregressive process AR(1) (note that this is a Markov process):

$$z_t = \rho z_{t-1} + \mu(1-\rho) + u_t$$

in which,

- z_t is a random variate representing the value of the time series at time t; and
- u_t is a random disturbance, usually assumed to be Normal, with zero mean and variance σ_u^2, which is uncorrelated with any previous disturbance, u_{t-k}, or with any value of the series z_t.

The equation just above may also be written, by subtracting μ from both sides, as:

$$\{z_t - \mu\} = \rho\{z_{t-1} - \mu\} + u_t$$

This discrete time series is *first-order* because it considers only the first-order difference $z_t - z_{t-1}$; it does not consider the second-order term $z_t - z_{t-2}$; the third-order term $z_t - z_{t-3}$; etc. One needs higher-order processes to cover these situations.

As this process is assumed to be *stationary*, the mean of the process is constant, or

$$E[z_t] = E[z_{t-1}] = E[z_{t-k}] \quad \text{for } \forall k < t$$

Then, taking expectations of all the terms in the AR(1) equation given above,

$$E[z_t] = \rho E[z_{t-1}] + \mu(1-\rho) + E[u_t] = \rho E[z_t] + \mu(1-\rho) + 0$$

$$(1-\rho)E[z_t] = \mu(1-\rho) \quad \text{or} \quad E[z_t] = \mu$$

Here μ us the (constant) mean of the process, which can be determined by computing the average of the process. Similarly, it is possible to find a simple expression for the variance of the process: First write the defining equation for the variance of z_t:

$$E\left[\left(z_t - \mu\right)^2\right]$$

Then substitute the AR(1) equation for z_t, expand the terms in the squares, take advantage of the terms assumed to be zero, and clean up the algebra:

$$E\left[\left(z_t - \mu\right)^2\right] = E\left[\left(\rho z_{t-1} - \rho\mu + u_t\right)^2\right]$$

$$E\left[\left(z_t - \mu\right)^2\right] = E\left[\rho^2\left(z_{t-1} - \mu\right)^2 + 2\rho\left(z_{t-1} - \mu\right)u_t + u_t^2\right]$$

$$E\left[\left(z_t - \mu\right)^2\right] = \rho^2 E\left[\left(z_{t-1} - \mu\right)^2\right] + 2\rho E\left[\left(z_{t-1} - \mu\right)u_t\right] + E\left[u_t^2\right]$$

$$\left(1 - \rho^2\right) E\left[\left(z_t - \mu\right)^2\right] = \sigma_u^2 ; \text{as } \rho E\left[\left(z_{t-1} - \mu\right)u_t\right] = 0$$

$$\sigma_z^2 = E\left[\left(z_t - \mu\right)^2\right] = \frac{\sigma_u^2}{1 - \rho^2}$$

The result is a relationship between the variance of z_t (the signal), the variance of u_t (the noise), and ρ, the correlation between z_t and z_{t-1}.

Similarly, for the autocovariance,

$$E\left[\left(z_t - \mu\right)\left(z_{t-1} - \mu\right)\right] = E\left[\left\{\rho z_{t-1} - \rho\mu + u_t\right\}\left(z_{t-1} - \mu\right)^2\right]$$

$$E\left[\left(z_t - \mu\right)\left(z_{t-1} - \mu\right)\right] = E\left[\rho\left(z_{t-1} - \mu\right)^2\right] + E\left[u_t\left(z_{t-1} - \mu\right)\right]$$

$$E\left[\left(z_t - \mu\right)\left(z_{t-1} - \mu\right)\right] = \rho E\left[\left(z_{t-1} - \mu\right)^2\right] + 0 = \rho\sigma_z^2$$

The first order autocorrelation coefficient between z_t and z_{t-1} is ρ, the second order autocorrelation coefficient between z_t and z_{t-2} is ρ^2, and the kth order autocorrelation coefficient between z_t and z_{t-k} is ρ^k. Hence the variates in the first order stationary time series are correlated with each other in an exponentially decreasing pattern.

Thus, one may be able to model a stationary process as a first-order autoregressive process that predicts the one-step-ahead value of the variable z_t from the observed value z_{t-1} using the equation:

$$z_t = \rho z_{t-1} + \mu\left(1 - \rho\right) + u_t$$

where,

z_{t-1} is the observation at the previous time

μ is the observed mean value (computed from the historical data)

ρ is the observed value for the autocorrelation between z_t and z_{t-1} (computed from the previous data).

σ_z^2 is the observed value of the variance of z_t (computed from the previous data).

μ_t is a Gaussian random number drawn from a Normal distribution with mean 0 and variance $\sigma_u^2 = \sigma_z^2 \left(1 - \rho^2\right)$

Note that in the equation $z_t = \rho\, z_{t-1} + \mu(1 - \rho) + u_t$, u_t is a residual term and hence is a random variable does not have any deterministic value. In Monte Carlo simulation, one often plots the histogram of these residuals to determine the goodness of fit of the model to the data.

By substituting

$$z_{t-1} = \rho z_{t-2} + \mu\left(1-\rho\right) + u_{t-1}$$

the first order autoregressive AR(1) series can be written as the equivalent infinite moving average [$MA(\infty)$] series $z_t = \mu + u_t + \rho u_{t-1} + \rho^2 u_{t-2} + \ \dots\ + \rho^k u_{t-k} + \dots$.

Note that this last equation is in the same form as the finite linear combination of variates examined earlier,

$$y = a_0 + a_1 x_1 + a_2 x_2 + \dots + a_n x_n$$

and all the disturbances or noise terms are statistically independent. Then, by the Central Limit Theorem, z_t, the sum of many independent random variates, should, in the limit, be Normally distributed. Actually, recourse to the Central Limit Theorem is not necessary if all the disturbances are approximately Normally distributed; the weighted sum z_t will then be Normal by the reproductive property of the Normal distribution.

Now to put this in a project context, assume that y_t represents the time at which the job reaches milestone t, and z_t represents the time a job takes in the $t - th$ state; that is, the time elapsed from milestone $t - 1$ to milestone t. Then z_t is the change in y_t, the incremental time between milestone $t - 1$ and milestone t:

$$z_t = y_t - y_{t-1} = \Delta y_t$$
$$z_1 = y_1 - y_0 = \Delta y_1$$

The total time to reach milestone t is y_t, where

$$y_t = \sum_{j=1}^{t} \Delta y_j + y_0 = \sum_{j=1}^{t} z_j + y_0$$
$$y_t = y_0 + \left(y_1 - y_0\right) + \left(y_2 - y_1\right) + \left(y_3 - y_1\right) + \dots + \left(y_t - y_{t-1}\right)$$

where y_0 is the start time.

The problem is to find the mean and variance of y_t, the predicted time to reach any milestone t, including the completion time, $t = n$, given that the time spent in any state is represented by the autocorrelated AR(1) process. To do this, we use the moving average representation because the variates in it, the u_{t-k} terms, are uncorrelated. However, the series is not infinitely long, as the actual job has a definite starting point and an ending point. The finite $MA(t)$ series is not equivalent to the original AR(1) series. Hence, using the finite $MA(t)$ series instead of the $[MA(\infty)]$ series introduces some error, especially for small t and large ρ, but this error diminishes for large t (say, $t > 20$).

To predict the future state of the project, set $y_0 = 0$ for convenience, and substitute z_t into the equation for y_t. To show the pattern, we do this explicitly for $t = 5$, and then generically for any t:

$$y_5 = \sum_{j=1}^{5} z_j = z_5 + z_4 + z_3 + z_2 + z_1$$

$$y_5 = 5\mu + u_5 + (1+\rho)u_4 + (1+\rho+\rho^2)u_3 + (1+\rho+\rho^2+\rho^3)u_2$$
$$+ (1+\rho+\rho^2+\rho^3+\rho^4)u_1$$

$$y_t = \sum_{j=1}^{t} z_j = \mu t + \Psi_0 u_t + \Psi_1 u_{t-1} + \Psi_2 u_{t-2} + \ldots + \Psi_k u_{t-k} + \ldots + \Psi_{t-1} u_1$$

where:

$$\Psi_0 = 1$$
$$\Psi_1 = 1 + \rho$$
$$\Psi_2 = 1 + \rho + \rho^2$$
$$\Psi_3 = 1 + \rho + \rho^2 + \rho^3$$
$$\ldots$$
$$\Psi_k = \sum_{j=0}^{k} \rho^j$$

Now, the expression above for y_t is also a linear combination of statistically independent variates, u_{t-k}, each with zero mean. We can then see that the mean of y_t is:

$$E[y_t] = \mu t$$

The variance of y_t is, using the previous results,

$$\text{var}(y_t) = E\left[(y_t - \mu t)^2\right] = \sum_{k=0}^{t-1} \Psi_k^2 \sigma_u^2$$

$$\text{var}(y_t) = \sum_{k=0}^{t-1} \sigma_u^2 \left[\sum_{j=0}^{k} \rho_j\right]^2$$

This expression can be simplified by using the well-known identity for the sum of an infinite series:

$$\sum_{j=0}^{\infty} x^j = \frac{1}{1-x} \quad \text{for } |x| < 0$$

$$\sum_{j=0}^{\infty} x^j = \sum_{j=0}^{n} x^j + \sum_{j=n+1}^{\infty} x^j$$

$$\sum_{j=0}^{n} x^j = \sum_{j=0}^{\infty} x^j - \sum_{j=n+1}^{\infty} x^j = \sum_{j=0}^{\infty} x^j - x^{n+1} \sum_{j=0}^{\infty} x^j = \frac{1-x^{n+1}}{1-x}$$

Using this result, with $x = \rho$, gives:

$$\Psi_k = \sum_{j=0}^{k} \rho^j = \frac{1-\rho^{k+1}}{1-\rho}$$

Hence,

$$\text{var}(y_t) = \sum_{k=0}^{t-1} \sigma_u^2 \left[\sum_{j=0}^{k} \rho^j \right]^2 = \sigma_u^2 \sum_{k=0}^{t-1} \left[\frac{1-\rho^{k+1}}{1-\rho} \right]^2$$

Squaring the term inside the summation and then applying the same identity for a partial series gives

$$\text{var}(y_t) = \frac{\sigma_u^2}{(1-\rho)^2} \sum_{k=0}^{t-1} \left(1 - 2\rho^{k+1} + \rho^{2k+2} \right)$$

$$\text{var}(y_t) = \frac{\sigma_u^2}{(1-\rho)^2} \left[t - 2\rho \sum_{k=0}^{t-1} \rho^k + \rho^2 \sum_{k=0}^{t-1} \left(\rho^2 \right)^k \right]$$

$$\text{var}(y_t) = \frac{\sigma_u^2}{(1-\rho)^2} \left[t - \frac{2\rho(1-\rho^t)}{1-\rho} + \frac{\rho^2(1-\rho^{2t})}{1-\rho^2} \right]$$

To recap, what was done here was:

1. The milestones, or states of a job or work package are defined such that the expected or average time to successfully complete each state is a constant, μ. Then the time to accomplish a given state t of the job, $z_t = y_t - y_{t-1} = \Delta y_t$, is expressed as a stationary autoregressive time series with constant mean μ, first order autocorrelation ρ, and variance $\sigma_z^2 = \dfrac{\sigma_u^2}{1-\rho^2}$

2. This one stage autoregressive series is converted to an infinite moving average series in the uncorrelated noise disturbances u_{t-k}.
3. The infinite moving average series is approximated by a finite moving average series of length t.
4. The series in the increments z_j is summed to give the time at which the t-th milestone is reached $y_t = \sum_{j=1}^{t} \Delta y_j + y_0 = \sum_{j=1}^{t} z_j + y_0$ expressed in terms of the random disturbances u_{t-k}
5. Taking expectations gives the mean and variance of the completion time y_t. From these parameters we can forecast the completion time at any level of confidence, using the assumption that the resulting probability distribution on the time at any milestone is Normal.

Note that the assumptions here are consistent with the Markov process, with the transition probability λ constant for all states. One could also interpret z_j as the progress achieved in reporting period t, and y_t as the total cumulative progress reported up to time t. That is, in one approach we establish certain levels of progress (milestones or states) and report the time it takes to reach each of these milestones; in the other approach we set certain fixed reporting intervals and report the progress at the end of each interval. However, the assumption that the expected value of the progress in each interval, $E[z_j] = \mu = a$ constant would not be consistent with the assumptions used in deriving the logistic curve, in which it was assumed that $\Delta y_t = by_t(S - y_t)$. Which of these assumptions, or others, may be more realistic is up to the user to determine; the only recommendation that can be given here is that the assumptions, whatever they are, be made explicit and open, and checked against reality whenever possible.

Using the equation above, we can estimate the variance of the completion time for $Var[y_t]$ if we assume some disturbance variance σ_u^2 and correlation ρ between successive states. As an illustration, suppose that we instead neglected ρ and the complication of this equation and computed $Var[y_t]$ simply as the sum of the variances of each state, $Var[y_t] = t\sigma_z^2$. Call the variance computed in this way $\underline{Var}[y_t]$, which as we have seen before will underestimate the variance if $\rho > 0$, and call the variance computed from the equation in ρ derived above $Var^*[y_t]$. Then, at $t = 30$, which is sufficient to virtually eliminate the error due to truncating the infinite moving average series, the ratio $\dfrac{Var^*[y_t]}{\underline{Var}[y_t]}$ is well over a factor of three as $\rho \to +1$. On the other hand, $\underline{Var}[y_t]$ greatly overestimates the variance if $\rho \to -1$.

Positive ρ indicates that, if z_{t-1} is greater than the mean, μ, then z_t will tend to be greater than the mean, because $(z_t - \mu) = \rho(z_{t-1} - \mu) + u_t$. Conversely, if z_{t-1} is less than the mean, then z_t will tend to be less than the mean, or successive deviations from the mean will tend to be of the same sign. Negative ρ indicates the opposite: successive deviations from the mean tend to be of the opposite sign. There seems to be no theoretical reason why negative correlations should not be as prevalent as positive correlations, but in reality, just as with matter and antimatter, positive correlations seem to be very much more common than negative correlations. This is perhaps because the observed variables z_t, z_{t-1}, etc., are in fact conditioned on common underlying or hidden variables that are not or cannot be observed. Recalling an earlier discussion, we have here Type 2 uncertainty, which is attributable to our ignorance of the values of the real underlying variables. Whatever the reason, in our experience, positive correlations are almost universal and negative correlations are rare.

Example 5.1

As an illustration of a very simple project process model, let: $\mu = 1$; $n = 100$; $\sigma_z = 0.4$; $\sigma_u = 0.4$; and $\rho = 0$. That is, the example is a simple linear process with 100 steps, each with expected time 1 week. Then using the equations developed above, the Autoregressive (1) process predicts the future job progress as in the following figure. The central line is the expected time to reach each milestone, and the other lines represent probability contours representing confidence bands on the predictions: Mean ± 1 standard deviation; Mean ± 2 standard deviations; and Mean ± 3 standard deviations.

There is about one chance in a thousand of a random data point lying outside the 3σ line, so we might estimate a probability of about 0.001 that the project would take longer to finish than 112 weeks. Therefore, as shown in Fig. 5.1, we would be very confident in committing to a job completion in 112 weeks.

However, if there is correlation, the uncertainty becomes greater. Figure 5.2 shows the same process but now with correlation coefficient $\rho = 0.50$. Note that the confidence bands have gotten wider, due entirely to the increased correlation.

Figure 5.3 shows the predictions for project completion if the time series is highly correlated.

Note that the prediction bands are now very wide, indicating the greatly heightened degree of uncertainty with high values of the correlation coefficient. There is now a significant probability that the project will take longer than 112 weeks to complete.

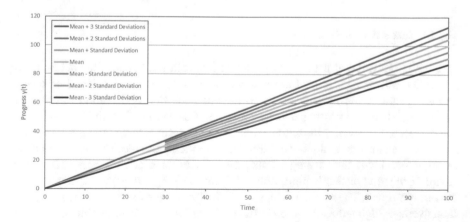

Fig. 5.1 Time series forecast $\rho = 0$

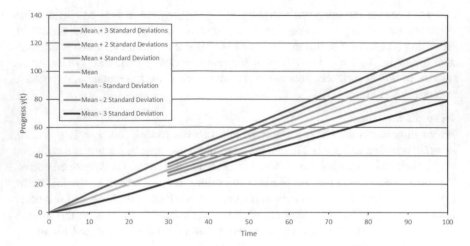

Fig. 5.2 Time series forecast $\rho = 0.50$

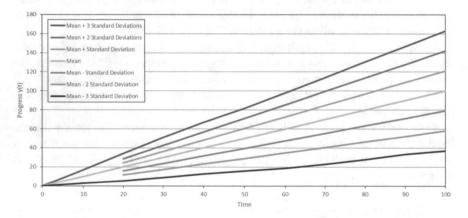

Fig. 5.3 Time series forecast $\rho = 0.95$

Qualitatively, the effect may be explained as follows: suppose the actual time to the first milestone is much longer than the mean. If the values in the time series are independent, the incremental time between the first and second milestones could be higher or lower than average; if lower, this would offset the high value in the first step. However, if the autocorrelation coefficient is large (and positive), the time between the first and second milestones is much more likely to be higher than average than lower than average. This process continues, and the project is likely to fall further and further behind. Runs of durations longer than the expected value become common. The same logic applies if the time to the first milestone is shorter than average; high autocorrelation is likely to lead to a project that finishes earlier than scheduled. Either way, the variance (uncertainty) increases.

5.3   Statistical Dependence and Independence in Project Costs

The discussion above of statistical dependence has mainly considered examples of job progress. In looking a project costs, consider the straightforward case in which the total job cost is the sum of a number of work package costs (or estimates). Then, if T is the total cost of n work packages, each of which has cost x_j, using the linear expression above with all coefficients equal to 1 gives the total cost as:

$$T = x_1 + x_2 + \ldots + x_n = \sum_{j=1}^{n} x_j$$

Then T has the expected value μ_T and variance σ_T^2

$$E[T] = \mu_T = \sum_{j=1}^{n} E[x_j] = \sum_{j=1}^{n} \mu_j$$

$$\sigma_T^2 = E\left[(T - \mu_T)^2\right] = E\left[\sum_{j=1}^{n}(x_j - \mu_j)^2\right]$$

giving the variance of the total cost as:

$$\sigma_T^2 = \sum_{j=1}^{n}\sum_{k=1}^{n} \sigma_{j,k}$$

Where $\sigma_{j,k}$ is the covariance between x_j and x_k. If all the x_j are statistically independent, then the cross product terms in the covariance vanish for $j \neq k$, and the variance of T reduces to $\sigma_T^2 = \sum_{j=1}^{n} \sigma_{j,j} = \sum_{j=1}^{n} \sigma_j^2$ if all variates are statistically independent.

Thus, there is substantial simplification if the variates are truly statistically independent, but there is also the potential for substantial error if they are assumed to be independent when they are not. To get a simple bound on the possible error, rewrite the equation for the variance of T as

$$\sigma_T^2 = \sum_{j=1}^{n} \sigma_j^2 + \sum_{j=1}^{n}\sum_{\substack{k=1 \\ k \neq j}}^{n} \rho_{j,k}\sigma_j\sigma_k$$

$$\sigma_T^2 = \sum_{j=1}^{n} \sigma_j^2 + 2\sum_{j=1}^{n}\sum_{k>j}^{n} \rho_{j,k}\sigma_j\sigma_k$$

The previous discussion has pointed out that the assumption that the $\rho_{j,k}$ are zero when they are not zero can lead to a significant underestimation of the variance and hence of the risk. However, why should it not be the case that all the $\rho_{j,k}$ are zero or close to it? After all, just earlier in discussing the Moving Average $MA(\infty)$ time series, we postulated the existence of an infinite number of disturbances, the white noise variates u_j, all of which were supposed to be uncorrelated. If there can be an infinite number of uncorrelated u_j, why shouldn't there be a finite number, n, of cost variables x_j? In fact, there is no prohibition to all the x_j being statistically independent, but the difference between the u_j and the x_j is that the former are mathematical constructs generated specifically to be statistically independent, whereas the latter are variables that arise in the real world. As an analogy from linear algebra: it is certainly possible for n vectors to be orthogonal in n-space, but n vectors picked at random would almost certainly not be orthogonal.

Therefore, one reason to question whether x_j are statistically independent is their number: in a complex project, there may be hundreds of work packages or cost elements. Is it reasonable to believer that none of these is correlated with another, or with some unspecified underlying variates? Let us consider some of the implications of this assumption.

To simplify the discussion for a moment, let all the x_j have the same mean and the same variance; that is,

$$\mu_j = \mu_x = E\left[x_j\right]$$

$$\sigma_{j,j} = \sigma_j^2 = \sigma_x^2 = E\left[\left(x_j - E\left[x_j\right]\right)^2\right], \ \forall j, \ 1 \le j \le n$$

Then $E[T] = \mu_T = nE[x_j] = n\mu_x$ and, if the x_j values are assumed all statistically independent, then

$$\sigma_T^2 = \sum_{j=1}^{n}\sigma_j^2 = n\sigma_x^2$$

$$\sigma_T = \sigma_x\sqrt{n}$$

As an aside, note that the mean and standard deviation have dimensions (dollars, tons, man-hours, whatever). A non-dimensional measure of dispersion is the coefficient of variation, COV, which is defined by COV = standard deviation/mean. Then

$$COV\left(x_j\right) = \frac{\sigma_x}{\mu_x}$$

$$COV\left(T\right) = \frac{\sigma_T}{\mu_T} = \frac{\sigma_x\sqrt{n}}{n\mu_x} = \left(\frac{1}{\sqrt{n}}\right)COV\left(x_j\right)$$

That is, the relative dispersion or coefficient of variation of the sum of statistically independent variables is not only smaller than the COV of one of its elements, but it becomes vanishingly small as n becomes large, unless we were to assume that all the $COV(x_j)$ simultaneously become very large (as $\sqrt{n}$) as n becomes large. So, if one were to subdivide a project into more and more cost elements by increasing n, either the relative dispersion of the total would approach zero, or the relative dispersion of each cost element would have to approach infinity as $n \rightarrow \infty$. As it is very difficult to accept that either of these could be true, then it becomes very difficult to accept the assumption of statistical independence.

Now consider the situation when determining the budget, or total estimated cost, for a project. Let ξ_j be the estimate prepared for work package or cost element j. Now suppose that ξ_j is not necessarily the expected value of the cost x_j, but is something more, to provide a safety factor, just to be on the safe side. Let us assume that the estimator or some manager adds some percentage of the estimate as a safety factor or contingency. That is, if η is the fraction of the expected value added to the expected value to produce the estimate, then

$$\xi_j = E\left[x_j\right] + \eta E\left[x_j\right] = \mu_j + \eta\mu_j = \mu_x\left(1 + \eta\right)$$

The Coefficient of Variation was defined above by:

$$COV = \frac{\sigma_x}{\mu_x}$$

Then the equation above may be rewritten as:

$$\xi_j = \mu_x + \eta\mu_x = \mu_x + \frac{\eta\sigma_x}{COV} = \mu_x + \left(\frac{\eta}{COV}\right)\sigma_x = \mu_x + \delta\sigma_x$$

$$\delta = \frac{\eta}{COV}$$

In this expression, it may be seen that the safety factor or buffer can written as a fraction δ of the standard deviation. Then we can write the estimate as

$$\xi_j = \mu_x + \delta\sigma_x$$

With this expression, we can get some idea of what this contingency factor might be. Suppose that the cost estimate is adjusted upward from the expected value by 10%, so $\eta = 0.1$. Suppose the work package in question has a substantial amount of variability, so the standard deviation is, say, 40% of the mean value, so COV = 0.4. Then

$$\delta = \left(\frac{\eta}{COV}\right) = \frac{0.1}{0.4} = 0.25$$

Using the Normal distribution, this corresponds to a probability of about 40% that this cost would overrun the estimate.

If the cost estimate is adjusted upward by 20%, so $\eta = 0.2$, and the work package in question has the same amount of variability, COV = 0.4, then

$$\delta = \frac{\eta}{COV} = \frac{0.2}{0.4} = 0.5$$

Using the Normal distribution, this corresponds to a probability of about 31% that this cost would overrun.

If the adjustment factor on the cost estimate is 10% ($\eta = 0.1$) and the work package in question has relatively little variability, say 10%, so COV = 0.1, then

$$\delta = \frac{\eta}{COV} = \frac{0.1}{0.1} = 1.0.$$

Using the Normal distribution, this corresponds to a probability of about 16% that this cost would overrun.

Notice that, from a risk viewpoint, the practice of adding some percentage of the expected value to get the estimate, without regard for variance, provides the largest safety factor where it is least needed – for work packages with the least variability.

Does this happen? Do cost estimates include a cushion? Some estimators assert that their estimates represent the expected values, but conversely many project managers admit openly that contingency has been pushed down into the individual estimates, that conservatism has been built into each work package estimates. That is, in our statistical terms, $\delta > 0$. Although it is difficult to establish what value may be used for δ, because for one thing there are no estimates of σ_x, it is highly likely that people at several levels respond to perceived variability or risk by trying to increase their margin for error. This behavior is to be expected, and it is certain that project managers and corporate managers expect it, as they often cut estimates to reduce the perceived safety factors in order to make projects happen when the total estimate is larger than the available budget.

Using the expression just above, we can see that the estimate ξ_j corresponds to some confidence factor, the probability that the estimate will not be exceeded. (Note that ξ_j is not a random variable, it is a number; x_j is the corresponding random variable). For example, if we assume that x_j is Normally distributed, and apply some modest safety factor, $\delta = 0.25$, say, then $\xi_j = \mu_x + 0.25\sigma_x$ corresponds to a confidence factor of 60% or a probability of 0.60 that ξ_j will not be exceeded and a probability of 0.40 that ξ_j will be exceeded. This, or even more, might seem to be a reasonable adjustment. Now consider the distribution of the total estimate, say τ, under the assumption of statistical independence of all variables:

$$\tau = \sum_{j=1}^{n} \xi_j = \sum_{j=1}^{n} \left(\mu_j + \delta\sigma_j \right) = \sum_{j=1}^{n} \left(\mu_x + \delta\sigma_x \right) = n\mu_x + n\delta\sigma_x$$

Using the previous results,

$$\tau = n\mu_x + \delta n\sigma_x = \mu_T + \sigma_T \delta \sqrt{n} \text{ because } \sigma_T = \sigma_x \sqrt{n}$$

Let $\eta = \delta\sqrt{n}$ then $\tau = \mu_T + \eta\sigma_T$ corresponds to some probability that the project estimate τ will be exceeded. Table 5.1 illustrates this probability for various values of n, using $\delta = 0.25$, and the Normal distribution:

According to this table, the probability of a cost overrun decreases greatly with increasing n, so that large projects with many work packages are much less likely to have cost overruns than small projects, according to this approach. In fact, the probability of a cost overrun for large n is so low that, using these assumptions, one would have to say that no large project has ever overrun in the history of projects since the pyramids.

Obviously, this result is untenable. Large projects are known to overrun, at least as often as small projects. We must therefore reject one or more of the assumptions; either

- Cost estimates do not ever have adjustments for conservatism, so $\delta = 0$ and $\tau = n\mu_x$;
- The variates cannot be statistically independent.

The subjective arguments as to why one expects $\delta > 0$, even $\delta \gg 0$, were given earlier. We now repeat the above simplified analysis by relaxing the assumption of statistical independence. We have, from the results above,

$$\sigma_T^2 = \sum_{j=1}^{n}\sigma_j^2 + \sum_{j=1}^{n}\sum_{\substack{k=1 \\ k \neq j}}^{n}\rho_{j,k}\sigma_j\sigma_k$$

To simplify the expressions, assume that all the cost variables x_j are correlated, with the same correlation coefficient $\rho_{j,k} = \rho$. Then the equation for the variance of the total cost reduces to:

$$\sigma_T^2 = \sum_{j=1}^{n}\sigma_j^2 + \sum_{j=1}^{n}\sum_{\substack{k=1 \\ k \neq j}}^{n}\rho\sigma_j\sigma_k = n\sigma_x^2 + n(n-1)\rho\sigma_x^2$$

$$\sigma_T^2 = \sigma_x\sqrt{\{n[1+(n-1)\rho]\}}$$

Table 5.1 Probability of not exceeding the estimates given n work packages

n	η	Confidence factor $= P\{\tau = \mu_T + \eta\sigma_T\}$ will not be exceeded
1	0.25	0.60
10	0.80	0.79
100	2.53	0.9943
1,000	8.01	0.99999999999...
10,000	25.3	0.9999999999999999999....

Modifying an earlier expression, we have, for the total project cost estimate:

$$\tau = n\mu_x + \delta n\sigma_x = \mu_T + \frac{n\sigma_T\delta}{\sqrt{\{n[1+(n-1)\rho]\}}}$$

$$\tau = \mu_T + \eta\sigma_T, \text{where } \eta = \frac{n\delta}{\sqrt{\{n[1+(n-1)\rho]\}}}$$

Suppose we would like the confidence level in the total estimate τ to be the same as the confidence level in the estimates for each work package. Then we set $\eta = \delta$ and solve for ρ; the solution is $\rho = 1$, for any value of n. That is, if the total project estimate has the same probability of being exceeded as each of the individual cost elements, then the work package costs must be perfectly correlated. This is consistent; if all the variables are completely dependent, then there is really only one variable. All the rest can be determined exactly from any one.

With these assumptions, there is no way that the confidence level in the total estimate τ can be less than the confidence level in the individual elements separately, as this would require $\eta < \delta$ which in turn would require $\rho > 1$, which is impossible by the definition of ρ.

This result cannot be accepted either, as it is not possible that all the cost elements are perfectly correlated. Actually, an assumption that would better explain the observed facts, that

- The probability of a cost overrun on a project is greater than the probability of an overrun on any of its individual work packages; and
- The probability of a cost overrun on a project increases with the size and complexity of the project, that is, as n increases

would be to assume that each individual work package estimate is slightly *less* than the expected value; that is

$$\xi_j = E[x_j] - \delta\sigma_j = \mu_j - \delta\sigma_j = \mu_x - \delta\sigma_x; \ \delta > 0$$

Under what circumstances might this occur? It might be true if some cost elements are omitted from the estimate or if some manager reduces the cost estimates in order to win a contract, for example. Another possibility might be that the estimate ξ_j corresponds to the median of an asymmetric probability distribution, in which, mode < median < mean. Suppose that each work package cost x_j is distributed as an Erlang distribution with $k = 2$, $\lambda = 4$. Then the mode is 0.5000, the median is 0.6685, and the mean is 0.7500 (all in millions of dollars). Although there is no simple expression for the median, it can easily be calculated numerically. It can be seen that the median is below the mean. Suppose then that the estimate is the median, the value that would be exceeded half the time, as might seem to be reasonable. That is,

$$\xi_j = \text{median} = \hat{\xi}$$

and, assuming for simplicity of presentation that all the work packages are identically distributed, then the total estimate is

$$\tau = \sum_{j=1}^{n} \xi_j = n\hat{\xi}$$

The Erlang distribution has the property of being locking or self-reproducing; That is, if n *statistically independent* variates are Erlang, each with parameters (λ, k_j), the sum is Erlang with parameters $\left(\lambda, \sum_{j=1}^{n} k_j \right)$. If the n variates are identically distributed, $\sum_{j=1}^{n} k_j = nk$. We can then easily compute the probability, using the Erlang distribution with parameters (λ, nk), that the total cost estimate τ will not be exceeded, for various values of n (Table 5.2).

Here we see the opposite behavior from that in the previous table; because the individual work package cost estimates are the medians, which are less than the means, the probability of a cost overrun (1 − Confidence Factor) for the whole contract is greater than that for any single work package, and increases with the number of work packages. Using the sample numbers here, a modest sized project with 125 work packages has only about a 1% chance of meeting the total cost estimate, if these are based on the medians. If the estimates are the modes, which are even less than the medians, the decrease in the confidence factor with n will be even more pronounced. Of course, if the probability distribution is symmetric, then the mode = median = mean, and no bias is introduced either up or down.

The self-reproducing property of the Gamma or Erlang would be useful in quickly estimating the probability distribution of the sum of variables, as this distribution represents the skewness that is likely in the probability distributions of cost estimates, if it were not for the fact that these variates must be statistically independent. As we have seen, statistical dependence may have a significant effect on the distribution of the total, unless there is a clear justification for believing that all the variates are uncorrelated.

Table 5.2 Probability of cost overrun given n work packages

n	Confidence factor (CF) = $P\{\tau = n\hat{\xi}\}$ will not be exceeded	$P\{\text{cost overrun}\} = 1 - \text{CF}$
1	0.5000	0.5000
5	0.3631	0.6369
25	0.1740	0.8260
125	0.0151	0.9849

The situation we have illustrated, in which the confidence level in meeting the total project estimate is very sensitive to the number of work packages, is unsatisfactory, and the reason for this problem is the use of point estimates for the costs. Point estimates, being single numbers, cannot convey information about both the mean and variance (let alone any other parameters) of the individual work package costs. If one is determined to use point estimates, then the best estimates are the means,

$$\xi_j = E\left[x_j\right] = \mu_j; \ \forall j, \ 1 \le j \le n$$

The means don't convey any information about the variances, obviously, but they are at least unbiased. Estimates that are either above or below the mean, as in

$$\xi_j = E\left[x_j\right] + \delta\sigma_j = \mu_j + \delta\sigma_j; \ \delta \ne 0$$

contaminate information about the means with information about the variances. This is a case in which individual work package or subcontract estimators may think they are doing the right thing by adding a little to the estimate so that their work package is less likely to overrun, or their subcontract less likely to lose money. But when every work package does the same, for the same reasons, the result can be seriously in error. The position here is that contingency is appropriate, but the amount of contingency should be based on the determination of the total project uncertainty, not buried and hidden in individual cost estimates. Therefore, some good advice is: If you have to give a point estimate, then use the expected value. Much better advice is: Don't use point estimates. Notice that, to determine that a point cost estimate is the mean, we would have to know something about the probability distribution on the cost. And if we knew something about the probability distribution, we would know something about the variance.

This chapter has spent a lot of time trying to illustrate the consequences of statistical dependence or independence. All the examples have been made to demonstrate the point that the assumption of statistical independence can be highly erroneous and may be seriously unconservative. From the simple models above, it seems that it would be good practice not to assume that the cost elements are independent unless there is definite reason to believe that this is the case. Therefore, a much better approach to estimating total project costs (material quantities, resource requirements, and other factors) is to provide information on the probability distributions of all the x_j, plus the correlation coefficients $\rho_{j,k}$.

Given this information, one can perform Monte Carlo simulation to estimate the probability distribution of the total project cost. Monte Carlo simulation is widely used, and even spreadsheet programs now provide this capability. However, one must always be sure to execute enough random trials in order to get reasonably tight confidence bounds on the results. The number of iterations required can be quite high (often greater than 10,000), and often to save time users do not do enough, and do not specify correlations between variables.

5.4 Practice Problems

Problem 5.4.1 An offshore platform project was built to provide natural gas, 55 million SCF per day, to a Power Plant. The project was hampered by weather delays and mechanical failures. The pile driving activity (including the weld-splice activities considered here) finished 52 days behind schedule. Each day over the drill hole cost the contractor over $100,000. Moreover, these delays pushed the project duration into seasonally bad weather, which magnified the effect.

The man-hours recorded on the platform for various pipe welding activities are given in Table 5.3. Welding activities are:

- Prepare and preheat
- Preheat
- Weld splice
- Cool and clean.
- Ultrasonic test

Consider each welding-related operation given above as a time series. Compute the first few lagged autocorrelation coefficients for each process. The autocorrelation coefficients describe how much of the cost of one activity can be explained (predicted) by the cost of the preceding activity (lag 1), etc. Could each separate process be adequately modeled as a first-order autoregressive process? Explain why or why not. What conditions have to be met for a process to be first-order autoregressive? If a process is not first-order autoregressive, can you devise another appropriate model to fit the data? Are there any time trends in the data? Is it better to generate one model to predict total time for each splice or to fit five separate models and then add them up?

Examine the cross-correlations. For example, is Weld Splice time correlated with the Prep time or Preheat time for each splice? Is Cool and Clean correlated with Preheat time? Are the separate processes independent of each other or dependent? Is this what you would have expected? Explain your results.

Problem 5.4.2 Suppose a project manager is involved in a commercial speculative building development. For economic viability, the cost estimate for the structure, after engineering and design are complete, cannot exceed $Budget. The project manager can retain a structural engineer to design the building in steel or in concrete, or both, and to make the necessary engineer's cost estimates, $Steel and $Concrete, when the designs are complete. If one or both of the designs is less than the budget, the project manager will choose the smaller and the project will proceed; otherwise it will be terminated. Therefore, using the identity between joint probability distributions and conditional probability distributions, the probability of termination of the project before the engineering work has started, is:

$$
\begin{aligned}
P[\text{Termination}] &= P\big[(\$Steel > \$Budget) \cap (\$Concrete > \$Budget)\big] \\
&= P\big[(\$Concrete > \$Budget)\big|(\$Steel > \$Budget)\big]P\big[(\$Steel > \$Budget)\big]
\end{aligned}
$$

Table 5.3 Problem data

Welding	Prepare and preheat (hours)	Preheat (hours)	Weld splice (hours)	Cool and clean (hours)	Ultrasonic test (hours)	Total welding (hours)
1	18.2	2.5	7.2	6.3	1	35.2
2	18.2	2.5	7	6.5	1	35.2
3	11.5	1.8	7.1	3.3	2.5	26.2
4	13.2	1.8	7.3	5.6	3	30.9
5	3.5	2.8	11.7	7.1	1.5	26.6
6	5	2.9	10.4	7.3	1.5	27.1
7	9.5	6.8	11.3	5.3	1.2	34.1
8	10	4.3	9.8	5.8	1.3	31.2
9	1.5	3.8	8.8	13	0.8	27.9
10	2.5	2.3	8.4	8.7	1.1	23
11	3.5	3	8.1	7.2	1.2	23
12	5.7	2.8	7.8	6.5	1.2	24
13	3.8	2.4	8.7	3.3	0.8	19
14	2	3.8	7.7	3.3	1.5	18.3
15	2.2	2.2	9	3.8	0.7	17.9
16	1.7	2.3	8	4.5	1	17.5
17	1.4	2.8	6.5	3.9	1.1	15.7
18	2	2.5	8.3	4.3	1.2	18.3
19	1.5	2.3	6	4.8	0.9	15.5
20	1.5	1.5	6.5	5.3	0.7	15.5
21	2.3	4	7	4	1.1	18.4
22	1.3	2.8	7.5	4.3	1.3	17.2
23	2	2.5	6.7	5.2	0.7	17.1
24	1.5	2.5	7.6	5.3	0.6	17.5
25	2.2	2.5	6.5	4.9	0.8	16.9
26	2.2	2.3	6.8	5.5	0.7	17.5
27	1.5	4	8.2	3.5	0.8	18
28	1.3	3.8	7	4.5	1.1	17.7
29	3.2	3.6	11	4.1	0.5	22.4
30	5	3	12.8	5.8	0.8	27.4
31	1.7	3.3	10.5	5	0.7	21.2
32	1.5	3	11.3	4.2	0.8	20.8
33	2.1	2.9	7.6	5	1	18.6
34	2.6	2.2	11.5	6.6	0.7	23.6
35	1.8	2.2	10.8	5	1	20.8
36	2	2.1	11.6	5	1	21.7
37	2	4.3	6	4.3	1.2	17.8
38	2.5	3	6.7	5.9	2.8	20.9
39	1.7	3.4	9.4	4.2	0.4	19.1
40	2.6	3.8	8.6	4.2	0.8	20

(continued)

Table 5.3 (continued)

Welding	Prepare and preheat (hours)	Preheat (hours)	Weld splice (hours)	Cool and clean (hours)	Ultrasonic test (hours)	Total welding (hours)
41	3.6	2.4	8.8	4.7	0.8	20.3
42	3.7	5.1	11.1	4.9	0.6	25.4
43	2.2	3.5	6.5	4.8	1	18
44	2	3.6	7.1	4.8	1.2	18.7
45	7	4	11.8	5.5	1	29.3
46	12.1	3	11.2	5.5	0.8	32.6
47	10.8	4.2	14.1	5.5	0.5	35.1
48	11.7	1.9	12.8	5.5	1.2	33.1
49	4.5	4	14.8	5.5	0.8	29.6
50	3	3.1	14.9	5	0.6	26.6
51	3.8	2.4	14	6	0.6	26.8
52	7.5	3.7	15.1	5	1.8	33.1
53	3	3.5	15.3	5.5	1	28.3
54	2.8	3.3	15.8	6.5	1	29.4
55	2.8	2.3	14.8	6.5	0.6	27
56	2.5	2.6	14.3	6.2	1	26.6
57	2.6	2.8	9.5	4	1	19.9
58	4	2.8	12.7	6.5	1	27
59	4.5	3	15.5	6.7	1.2	30.9
60	5.7	2.8	12.8	5.8	1.5	28.6
61	1.9	3.5	16.1	5.8	1	28.3
62	1.4	19	18	18	0.8	57.2
63	4.8	10	19	10	0.8	44.6
64	3.8	4.5	14.2	7.3	0.8	30.6
65	3.2	3.8	19.2	4.9	0.7	31.8
66	1.7	3.5	18.5	5.3	1	30
67	2	3.5	17.8	4.2	1.3	28.8
68	2	3.5	14.8	4.5	1.3	26.1
69	1.8	4.5	13.8	5.1	0.7	25.9
70	1.8	2.4	15.1	3.4	0.8	23.5
71	1.3	2	14.2	3.5	0.7	21.7
72	3	5.7	14.1	4	0.9	27.7
73	3	2.5	19	3.9	1	29.4
74	1.9	3.5	17.5	3.5	0.8	27.2
75	1.7	3.3	19.2	4	0.8	29
76	3.8	3.5	19.9	4.1	1	32.3
77	2.5	2.3	14.5	4	0.8	24.1
78	1.6	3.8	15.8	3.3	0.8	25.3
79	3.2	3.6	14.6	3.9	0.9	26.2
80	3.8	3.4	15.7	4	1	27.9

(continued)

Table 5.3   (continued)

Welding	Prepare and preheat (hours)	Preheat (hours)	Weld splice (hours)	Cool and clean (hours)	Ultrasonic test (hours)	Total welding (hours)
81	2.1	4.6	14.5	3.5	0.8	25.5
82	2	4.6	13.7	3	0.8	24.1
83	3	3.7	14.2	3.1	0.9	24.9
84	2.1	3.4	15	3.2	0.7	24.4
85	2.3	2.4	19.3	3.5	1	28.5
86	1.4	2.8	13.9	4.8	0.8	23.7
87	3.2	3.4	19.2	3.3	0.8	29.9
88	3.5	2.5	13.5	5.1	0.9	25.5
89	2	2.5	14	2.8	0.8	22.1
90	1.3	3.2	14.7	3.2	1	23.4
91	1.3	2.2	14.7	3.8	0.8	22.8
92	3	1.8	15.2	3.6	1.6	25.2

The question is, Are the two cost estimates, $Steel and $Concrete, independent or correlated, and how does correlation affect the project manager's decision whether to contract for one of them or both of them?

If the two cost estimates are independent, then:

$$P\left[\left(\$Concrete > \$Budget\right)|\left(\$Steel > \$Budget\right)\right] = P\left[\left(\$Concrete > \$Budget\right)\right]$$
$$\therefore P\left[\text{Termination}\right] = \left(\$Concrete > \$Budget\right)\left(\$Steel > \$Budget\right)$$

If the two cost estimates are positively correlated, then:

$$P\left[\left(\$Concrete > \$Budget\right)|\left(\$Steel > \$Budget\right)\right] > P\left[\left(\$Concrete > \$Budget\right)\right]$$
$$\therefore P\left[\text{Termination}\right] > \left(\$Concrete > \$Budget\right)\left(\$Steel > \$Budget\right)$$
$$\therefore P\left[\text{Termination|Positive Dependence}\right] > P\left[\text{Termination|Independence}\right]$$

In the limit, as the correlation between the two estimates approaches 1.0.

$$P\left[\left(\$Concrete > \$Budget\right)|\left(\$Steel > \$Budget\right)\right] \to 1.0$$
$$\therefore P\left[\text{Termination|Positive Dependence}\right] \to \left(\$Steel > \$Budget\right)$$

Conversely, if the two cost estimates are positively correlated, then:

$$P\left[\left(\$Concrete > \$Budget\right)|\left(\$Steel > \$Budget\right)\right] < P\left[\left(\$Concrete > \$Budget\right)\right]$$
$$\therefore P\left[\text{Termination}\right] < \left(\$Concrete > \$Budget\right)\left(\$Steel > \$Budget\right)$$
$$\therefore P\left[\text{Termination}\middle| \text{Negative Dependence}\right] < P\left[\text{Termination}\middle| \text{Independence}\right]$$

In the limit, as the correlation between the two estimates approaches -1.0

$$P\left[\left(\$Concrete > \$Budget\right) | \left(\$Steel > \$Budget\right)\right] \to 0$$
$$\therefore P\left[\text{Termination|Negative Dependence}\right] \to 0$$

Questions:

- What factors would cause the two cost estimates to be positively correlated?
- What factors would cause the two cost estimates to be negatively correlated?
- Do you think, on balance, that the two cost estimates would be positively correlated, negatively correlated, or uncorrelated?
- Given your answer to the question above, should the project manager contract for a steel design, a concrete design, or both?

NB: For the purposes of this exercise:

- A *steel building* is one in which the primary structural elements are made of structural steel, with concrete used for floor slabs, pile caps, piles, footings, shear walls, and other purposes.
- A *concrete building* is one in which the primary structural elements are made of reinforced concrete, with reinforcing steel used in the concrete, and steel used for piles, bar joists, and other purposes.

References

Hamilton JD (1994) Time series analysis. Princeton University Press, Princeton

Walpole RE, Myers RH, Myers SL, Ye K (1993) Probability and statistics for engineers and scientists. Macmillan, New York

Chapter 6
Estimating Means, Variances, and Correlations Based on Experts' Judgment

Abstract In this chapter we introduce the methods to estimate statistical moments and correlation coefficient based on expert judgements. We provide an overview of probability density functions that are suitable for integration with subjective data, and the elicitation procedures for estimating correlation coefficients.

Keywords Expert judgements · Statistical moments · Correlation · Probability distribution

6.1 Introduction

In the second moment approach or in stochastic simulation (i.e. Monte Carlo simulation), it is assumed that random variables (activity costs, durations, etc.) range across a continuum and the uncertainties about the true values of these variables can therefore be expressed by continuous probability distributions. The total risk, or uncertainty, in the project cost, for example, is the sum of all the uncertainties in the individual cost elements, assuming these elements are independent. Conversely, the total uncertainty in the project duration is the sum of all the activity uncertainties along the critical path, assuming these activities are independent. Of course, they are not independent, and so the correlation matrix is used to express how they are linked together.

As previously mentioned, bottoms-up risk assessment is based on the principle that the uncertainties in each individual element can be estimated more easily and more accurately than trying to estimate the uncertainty in the total project all at once (top-down risk assessment). That is, we break down the risk problem into individual elements (which are not necessarily independent) in order to estimate the uncertainties of each, and then we aggregate these risks by the second moment approximation or by stochastic simulation. Even in the case of simple summations, such as the sum of work package costs upward through the Work Breakdown Structure, or the sum of activity durations along a path through a network, we need the joint probability of all the random variables. Joint probability distributions are hard to come by so we are forced to deal with marginal distributions combined with correlation matrices.

© Springer Nature Switzerland AG 2020
I. Damnjanovic, K. Reinschmidt, *Data Analytics for Engineering and Construction Project Risk Management*, Risk, Systems and Decisions,
https://doi.org/10.1007/978-3-030-14251-3_6

And, even more, in the second moment approach we have only the mean and variance of each element. To apply stochastic simulation, we need the entire joint cumulative probability distribution, which again is approximated by the marginal probability distributions plus the correlations between all pairs of elements. These quantities can be estimated in various ways, as discussed below. Because project risk assessment is about the future, there are most often insufficient historical data from which to derive these probability distributions, and so the risk analyst is forced to use subjective estimates. This chapter is concerned with various methods for making subjective estimates of means, variances, and correlation coefficients.

6.2 Empirical Data

The obvious point, to use whatever historical data are available to estimate the expected value and uncertainty in future costs or durations, should not be overlooked. For example, if work package j has been performed under the same conditions in N previous projects, with reported values $x_{j,k}$ for $1 \le k \le N$, such that all the sample data can be considered to have been drawn from the same population, then the first and second moments (mean and variance) of these data are given by:

$$\overline{x}_j = \frac{1}{N}\sum_{k=1}^{N} x_{j,k}$$

$$s_j^{\,2} = \frac{1}{N-1}\sum_{k=1}^{N}\left(x_{j,k} - \overline{x}_j\right)^2$$

Of course, one problem with this method in practice is that the work packages or activities in the N previous jobs were most likely not performed under the same conditions. This assumption could be true in manufacturing, in which the environmental conditions are controlled to assure that the work stays within the specification limits. But projects differ, and because the external conditions cannot be controlled, the way to control the data is to use statistical methods, for example multivariate regression analysis, to correct for variable project conditions. However, to do this requires that all the conditions surrounding each historical project be recorded. This is, unfortunately, rarely done.

Therefore, even if one has some historical data about work package costs or activity durations, one may wish to adjust the computed sample means and variances to reflect the best judgment about the project at hand. For example, if there are N sample data for work package costs, all assumed to be from the same population, and x_{max} is the largest of these, then the probability that in the next occurrence (that is, the project being assessed) of this work package the cost will be greater than x_{max} is $\dfrac{1}{N+1}$. So, for example, if $N = 19$, then x_{max} is approximately the 95th percentile value, and the smallest value of the N data, x_{min}, is approximately the 5th percentile. One might use this empirical information in setting the 5th and 95th percentiles in the following methods.

6.3 Subjective Methods Depending on Expert Judgment

If there are historical data, the mean and variance of each element can be computed from them. If not, one might suppose that the solution is simply to ask experts to give their best judgments about the mean and variance (or mean and standard deviation), or the entire cumulative probability distribution, of each work package. Experience, and some experiments, have shown that this straightforward approach doesn't always work well, for the following reasons:

1. There is an effect known as *anchoring*, by which a person familiar with the (deterministic) estimated cost for a work package will cite this estimated value, or something very close to it, as the mean of the probability distribution. The deterministic estimate is not necessarily the mean, or even close to it. In many cases, the deterministic estimate is closer to the most optimistic value than to the mean. If one has substantial historical data on actual costs or durations, the frequency at which the deterministic estimates are exceeded provides an indication of the relation of the estimate to the mean – the mean and the estimate only coincide if approximately 50% of the actual costs are below the estimates.
2. Estimators tend to give consistently low values for uncertainty (or standard deviation) and therefore underestimate risk. This behavior is consistent with the position that the costs are deterministic (and hence have no uncertainty), but inconsistent with the objective of quantifying uncertainty in order to estimate project risks.

To overcome this bias due to anchoring, several methods for eliciting judgments about probability distributions have been developed. A few are discussed here. Many of these have the common approach, in which:

1. Two or three points on the cumulative probability distribution are estimated, typically with one point in the lower tail, another in the upper tail, and a third near the middle (median or mode), to avoid reliance on subjective estimates of the mean, which may be contaminated by anchoring.
2. These points are used to fit the two or three parameters (mean and variance, typically) of the probability distribution believed to be most suitable.

6.3.1 The Beta Distribution

The Beta distribution is widely known as the foundation for the PERT method. The developers of PERT (Malcolm et al. 1959; Clark 1962) chose the Beta distribution because it is unimodal, and because it can represent a whole family of distributions: symmetric, skewed to the left, or skewed to the right, by varying the two parameters, α and β. The equation for the Beta distribution in the standardized variable x ($0 \leq x \leq 1$) is:

$$f(x) = \frac{\Gamma(\alpha+\beta)}{\Gamma(\alpha)\Gamma(\beta)} x^{\alpha-1}(1-x)^{\beta-1}, \forall x, 0 \le x \le 1; \alpha, \beta > 0$$

$$f(x) = 0 \; elsewhere$$

Here, $\Gamma(\alpha)$ represents the Gamma function. In the special case that N is an integer, then $\Gamma(N) = (N - 1)!$ The mean and variance of the Beta distribution in the standardized variate x are given by:

$$\mu_x = \frac{\alpha}{\alpha+\beta}$$

$$\sigma_x^2 = \frac{\alpha\beta}{(\alpha+\beta+1)(\alpha+\beta)^2}$$

A function defined on the unit interval is not particularly useful, so the more general function for the variable y defined on the interval $[a, b]$, or $y = a + (b - a)x$, is:

$$f(y) = \frac{\Gamma(\alpha+\beta)}{\Gamma(\alpha)\Gamma(\beta)} \left[\frac{1}{(b-a)^{\alpha+\beta-1}} \right] (y-a)^{\alpha-1}(b-y)^{\beta-1}, a \le y \le b; \alpha, \beta > 0$$

$$f(y) = 0 \; elsewhere$$

From the expressions above for the mean and variance of x, the mean and variance of y are easily obtained from the transformation relations:

$$\mu_y = a+(b-a)\mu_x = a+(b-a)\left[\frac{\alpha}{\alpha+\beta} \right]$$

$$\sigma_y^2 = (b-a)^2 \sigma_x^2 = (b-a)^2 \left[\frac{\alpha\beta}{(\alpha+\beta+1)(\alpha+\beta)^2} \right]$$

The most likely value of y is the mode of the distribution. As the Beta distribution is unimodal, the mode is the value of y at which $f(y)$ is a maximum. Therefore, the mode, m, is the value of y at which the first derivative of $f(y)$ is zero:

$$\frac{df(y)}{dy} = 0 \, at \, y = m$$

After some manipulation, this evaluates to:

$$m = \frac{a(\beta-1)+b(\alpha-1)}{\alpha+\beta-2}$$

In PERT, the user estimates, for the duration of each activity, the three values for:

a, the most optimistic (shortest) value;
b, the most pessimistic (longest) value;
m, the most likely value (mode)

Using the Beta distribution, one can get just about any shape one wants (symmetric, skewed to the left, skewed to the right) by varying the two parameters. If one believes that the distribution is skewed to the right, then one should estimate $(b - m) > (m - a)$. On the other hand, the limits a and b are hard boundaries: the duration will never be less than a nor greater than b. Certainly $a = 0$ is a fixed lower limit for real variables such as cost and duration, and there may be values of $a > 0$ that delimit the absolute lowest value. However, one may be dissatisfied with the fixed upper bound b, and it is not appropriate to set $b = \infty$ in the Beta distribution. The difficulty with the hard upper bound b is not so much that one expects that costs could go to infinity or that project durations could last for eternity; the problem is that one doesn't know where the upper bound should be. In the general expression given earlier, if x_{max} is the largest value actually observed in N instances, there is always some probability $\dfrac{1}{N+1}$ that the next observation will be greater than x_{max} no matter how large x_{max} is, conflicting with the presumption that there is some known upper bound.

The Beta distribution can also be used for the probability distribution on cost for each work package: simply let a = the lower bound on cost; b = the upper bound on cost; and m = the most likely cost. The pros and cons of the use of the Beta in PERT of course apply here as well.

In PERT, the mean and variance of each activity duration are then derived from these three-point estimates, as follows:

$$\mu_{PERT} = \frac{a+4m+b}{6}$$

$$\sigma^2_{PERT} = \frac{(b-a)^2}{36}$$

Equating the PERT expression for the mean to the mean of the Beta distribution gives:

$$\mu_{PERT} = \frac{a+4m+b}{6} = \mu_y = a+(b-a)\mu_x = a+(b-a)\left[\frac{\alpha}{\alpha+\beta}\right]$$

After some straightforward algebraic manipulation, this reduces to:

$$\alpha = \left[\frac{5a-4m-b}{a+4m-5b}\right]\beta = \phi\beta$$

$$\phi = \left[\frac{5a-4m-b}{a+4m-5b}\right]$$

Equating the PERT expression for variance to the equation for the variance of the Beta distribution gives:

$$\frac{(b-a)^2}{36} = (b-a)^2 \left[\frac{\alpha\beta}{(\alpha+\beta+1)(\alpha+\beta)^2} \right]$$

Substituting $\alpha = \phi\beta$ into this expression gives:

$$36\phi\beta^2 = \beta^3 (1+\phi)^3 + \beta^2 (1+\phi)^2$$

This expression reduces to:

$$\beta = -\left[\frac{\varphi^2 - 34\varphi + 1}{(1+\varphi)^3} \right]$$

Then,

$$\alpha = \phi\beta = -\phi\left[\frac{\varphi^2 - 34\varphi + 1}{(1+\varphi)^3} \right]$$

However, the actual parameters of the Beta probability distribution are needed only if one is doing Monte Carlo simulation. If one is simply interested in obtaining the mean and variance, for an activity duration or for a work package cost, to be used in the second moment approach, using subjective three-point estimates (a, m, b), the process is:

1. Estimate the three values a, m, b, based on knowledgeable sources or the best expert advice.
2. Compute the mean and variance of each work package or activity for use in the second moment calculations from the following equations:

$$\mu = \frac{a + 4m + b}{6}$$

$$\sigma^2 = \frac{(b-a)^2}{36}$$

6.3.2   The Triangular Distribution

The Triangular distribution is similar in concept to the Beta distribution, although, of course, a different shape. Like the Beta distribution, it can be symmetric, skewed to the right, or skewed to the left. If, as before, a is the estimated lower limit, b is the estimated upper limit, and m is the estimated mode, the peak of the triangular density function, then the mean and variance are given exactly by:

$$\text{mean} = \overline{x} = \frac{1}{3}(a + m + b)$$

$$\text{var}[x] = \frac{1}{18}\left[a^2 + b^2 + m^2 - (am + ab + bm)\right]$$

The Beta distribution and the Triangular distribution give quite different results for the same estimates. If the distribution is skewed to the right, the mean of the Beta distribution will be less than the mean of the Triangular distribution. The Triangular distribution will always have much more variance than the Beta distribution. For illustration, assume that the estimates are: $a = 100$, $m = 125$, and $b = 200$ for both cases. Then the means and standard deviations by the two distributions are:

Beta: ($\mu_\beta = 133.333$; $\sigma_\beta = 16.667$) and Triangular: ($\mu_{Tri} = 141.667$; $\sigma_{Tri} = 90.139$)

In short, the Triangular distribution conveys a much higher level of uncertainty than the Beta distribution with the same bounds. Figure 6.1 plots the probability density functions for the Beta and the Triangular for this set of values.

Figure 6.2 plots the cumulative probability functions for the Beta and Triangular distributions for the same set of values.

The PMBOK® Guide provides an example of the use of triangular distributions (PMI 2008, p. 297). Table 6.1 gives the data for a project of three work packages. In the notation used herein, a is the lower limit, b is the upper limit, and m is the mode (most likely value) of the triangular probability density function.

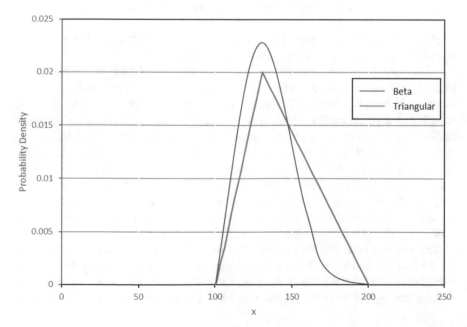

Fig. 6.1 Beta and triangular density distribution

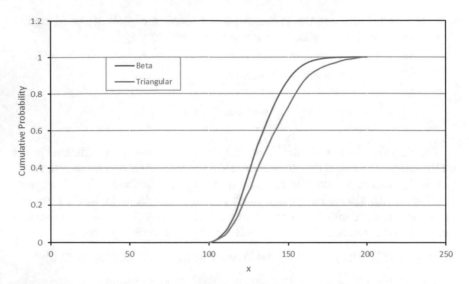

Fig. 6.2 Beta and triangular cumulative distribution

Table 6.1 Range of project cost estimates

WBS element	Low (a)	Most likely (m)	High (b)
Design	$4 M	$6 M	$10 M
Build	$16 M	$20 M	$35 M
Test	$11 M	$15 M	$23 M
Total project		$41 M	

Adapted from Project Management Institute (2008, p. 297)

Plots showing the three probability density functions are shown in Fig. 6.3. (Note that: This figure is not given in the PMBOK® Guide).

Also, note from the table above that the PMBOK® Guide gives the "Most Likely" total project cost estimate as $41M, the sum of the "most likely" values of all the work package probability density functions. This procedure is not founded in or justified by probability theory. As noted earlier, the n-th moment of a sum of random variates is the sum of the n-th moments of all the terms. That is, the first moment (the mean) of the sum is the sum of the means, and the second moment (the variance) of the sum is the sum of the second moments (variances and covariances). There is no rule about summing the most likely values (except in the special case that the density functions are symmetric and the modes are the same as the means). It is not generally true that the most likely value for the total project cost is the sum of the most likely values of all the work packages.

The probability distribution for the total project cost in this example is found, in the PMBOK® Guide, by Monte Carlo simulation. There is nothing wrong with using Monte Carlo simulation, but for this type of problem it is unnecessary. A solution by the second moment approach is indistinguishable from the Monte Carlo

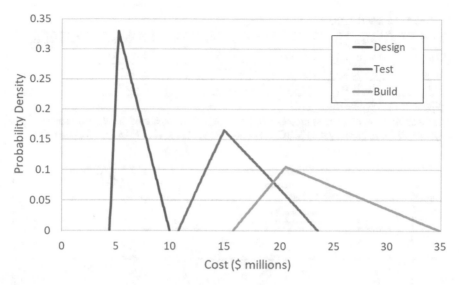

Fig. 6.3 Three triangle density functions

simulation, can be easily done by spreadsheet, and is faster. The PMBOK® Guide does not indicate how many Monte Carlo iterations were used to obtain the answer given. It also does not mention independence or dependence, so it is assumed that the PMBOK® Guide deals only with the independent case, without mentioning the assumptions used.

Table 6.2 shows the means and variances for the three work packages as determined by the equations given earlier and repeated here:

$$\mu = \bar{x} = \frac{1}{3}(a + m + b)$$

$$\sigma^2 = \frac{1}{18}\left[a^2 + b^2 + m^2 - (am + ab + bm)\right]$$

By the second moment approach, the variance of the total project cost is the sum of the variances and covariances in the variance-covariance matrix. Assuming independence of all work packages, the covariance matrix is:

$$\begin{bmatrix} & \text{Design} & \text{Build} & \text{Test} \\ \text{Design} & 1.55556 & 0 & 0 \\ \text{Build} & 0 & 16.72222 & 0 \\ \text{Test} & 0 & 0 & 6.22222 \end{bmatrix}$$

The matrix is symmetric as the correlations between different work packages are zero. The sum of the elements in the covariance matrix is 24.5, and the standard deviation is $4.95 M. The cumulative distribution function for the total project cost

Table 6.2 Moments of the three work packages

WBS element	Mean (μ)	Variance (σ²)	Standard deviation (σ)	Coefficient of Variation (COV)
Design	6.666667	1.555556	1.247219	0.187083
Build	23.66667	16.72222	4.089281	0.172787
Test	16.33333	6.222222	2.494438	0.152721
Total project	46.66667			

Note that the mean of the total project cost is the sum of the means of the work packages, in this case $46.67 M, as given by the PMBOK® Guide, Figs. 11, 12, 13, 14, 15, and 16, page 300

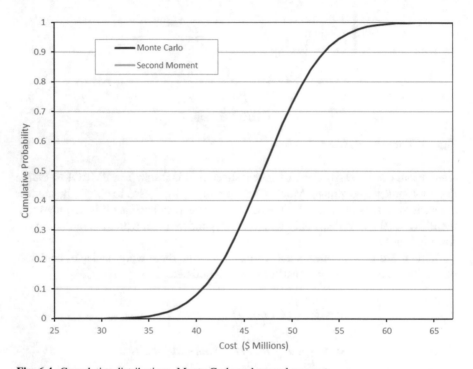

Fig. 6.4 Cumulative distributions, Monte Carlo and second moment

is shown graphically in Fig. 6.4. The Monte Carlo solution (the solid line) is that given in Figs. 11, 12, and 13 Cost Risk Simulation Results, on page 300 of the PMBOK® Guide. The second moment solution (the dashed line) is shown as a Normal distribution with mean $46.67 M and standard deviation $4.95 M, as computed above using the second moment method. The two solutions are identical. The probability of completing the project in $41 M or less (the so-called "Most Likely" value) is about 0.12. The "conservative organization" that wants a 75% chance of success (not overrunning the budget) should then have a budget of $50 M, that is, including a contingency of $9M above the "Most Likely" value.

However, the second moment approach allows for a straightforward analysis of the effects of dependencies. For example, suppose that one assumes a correlation of 0.9 between the work packages Design and Build. (For example, if the project turns out to be more complex that originally estimated, then it is likely that both the

Design costs and the Build costs will overrun). The other possible correlations are taken as zero. The covariance matrix for this case is:

$$
\begin{bmatrix}
 & \text{Design} & \text{Build} & \text{Test} \\
\text{Design} & 1.55556 & 4.590207 & 0 \\
\text{Build} & 4.590207 & 16.72222 & 0 \\
\text{Test} & 0 & 0 & 6.22222
\end{bmatrix}
$$

With this matrix, and everything else the same, the plot of the cumulative probability distribution on total cost is shown in Fig. 6.5. The plot for the Monte Carlo solution is the same as in the figure above, because the PMBOK® Guide provides no guidance on the condition of correlation between work packages (it does not even mention correlation). In this case, the effect of one correlation is not great, but it does indicate that the contingency allowance should be increased by perhaps $1 M, compared to the independent case.

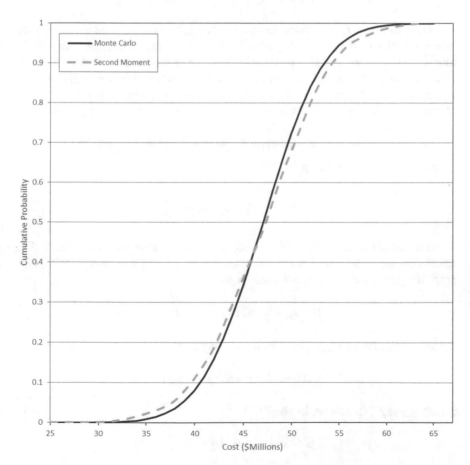

Fig. 6.5 Cumulative distribution, Monte Carlo and correlated second moment

6.3.3   Other Three Point Approximations

Many other approaches have been proposed for estimating the mean and variance of a probabilistic variate by eliciting three points on a subjective probability distribution. Many of these approaches are summarized and compared in Keefer and Bodily (1983), who ran a large set of computations to determine which approximations were the best. To do this, they assumed a set of Beta distributions with different parameters as the underlying probability functions. Then they compared the estimates made for the mean and variance, using various proposed approximation equations, with known true values for mean and variance.

In estimating the mean, the best approximation formulas were found to be the following.

Pearson-Tukey (Pearson and Tukey 1965):

$$\mu = 0.63x_{50} + 0.185(x_5 + x_{95})$$

In this notation, x_k represents the estimated value of the random variable at the k-th percentile of the probability distribution. Therefore, x_{50} is the value of the random variable at the 50th percentile, hence the median; x_5 is the estimated value of x at the 5th percentile; and x_{95} is the value of x at 95th percentile. Keefer and Bodily (1983) found that the maximum percentage error using this formula in their experiments as less than 0.1%, and the average percentage error was about 0.02%.

Other expressions giving good approximations to the mean were the following.

Perry-Greig (Perry and Greig 1975):

$$\mu = \frac{1}{2.95}(x_5 + 0.95m + x_{95})$$

Here, as in PERT, m is the mode, or most likely value. However, using the median rather than the mode gives better results, as shown by Perry and Greig (1975), who proposed "the extraordinarily accurate" equation:

$$\mu = x_{50} + 0.185(x_{95} + x_5 - 2x_{50})$$

Modified Davidson-Cooper (Davidson and Cooper 1976):

$$\mu = 0.16m + 0.42(x_{10} + x_{90})$$

Swanson-Megill (Swanson, in Megill 1977):

$$\mu = 0.40x_{50} + 0.30(x_{10} + x_{90})$$

Conversely, according to Keefer and Bodily (1983), "especially poor performances are provided by the original PERT approximation ... and the ... triangular model." "The Pearson-Tukey approximation for the mean outperforms the PERT and the ... triangular approximation by *more than three orders of magnitude* on average absolute error" (Emphasis added). Despite this "poor performance," the PERT approximation is still used, and the triangular distribution is very popular in Monte Carlo simulations using canned software packages.

In estimating the variance, Keefer and Bodily (1983) found the following approximation "to be the best":

Truncated Pearson-Tukey (1965):

$$\text{var} = \left[\frac{x_{95} - x_5}{3.29 - 0.1 \left(\dfrac{\Delta}{\sigma_0} \right)^2} \right]^2$$

$$\Delta = x_{95} + x_5 - 2x_{50}$$

$$\sigma_0^2 = \left[\frac{x_{95} - x_5}{3.25} \right]^2$$

Extended Pearson-Tukey (1965):

$$\sigma_0^2 = \left[\frac{x_{95} - x_5}{3.25} \right]^2$$

Perry and Greig (1975) favored the following expression for the standard deviation, which is identical to the Extended Pearson-Tukey equation:

$$\sigma = \frac{x_{95} - x_5}{3.25}$$

Again, for estimating the variance, "particularly poor approximations are the ... triangular model (unfortunately perhaps the most commonly used model), [and] the original PERT" (Keefer and Bodily 1983).

Other approximations were documented by Keefer and Bodily (1983) but are not reproduced here; for further information see their paper. The conclusion from this is that there are much better approximations for the means and variances than the original PERT formula or the results derived from the triangular distribution, and these better approximations should be preferred, inasmuch as "the differences in performance between the best and the worst are very large" (Keefer and Bodily 1983). If, for some reason, it is absolutely imperative to use a triangular distribution in a Monte Carlo simulation, Keefer and Bodily (1983) give three other forms of triangular distribution that give more accurate results for means and variances than the one in the earlier figure.

Triangular distributions are often used in Monte Carlo simulations to represent skewed distributions, which the Normal distribution cannot. If a probability distribution is thought to be skewed, Keefer and Bodily (1983) recommend the used of the lognormal distribution with parameters estimated by the Pearson-Tukey approximations given here.

6.3.4   The Normal Distribution

The Normal distribution can be used in a somewhat similar way, if one believes that the probability distribution should tail away and not come to abrupt stop, as with the Beta and the Triangular distributions. Of course, two limitations of the Normal distribution are:

1. It tails away on the low side as well as on the high side, and this may not seem realistic.
2. It is symmetric, and cannot represent skewness.

The primary justification for using the Normal is that it is so familiar. To use it, first define some probability level or percentile at which the points on the distribution will be estimated. Because only two values are to be computed, the mean and the variance, only two points are needed. To avoid anchoring, these points should be far from the mean.

Establish convenient, symmetric percentiles at the lower and upper ends of the range. Call the lower value $F_{low}(x)$ and the upper value $F_{high}(x) = 1.0 - F_{low}(x)$. The expert then estimates the values for the duration at these percentiles. That is, the expert estimates an optimistic value, x_{low}, such that the probability is $F_{low}(x)$ that the actual duration will be less than this. Conversely, the expert estimates the pessimistic value of the activity, x_{high}, such that the probability is $F_{high}(x)$ that it will not be exceeded. Some typical values are:

1. $F_{low}(x) = 0.10$; $F_{high}(x) = 0.90$. With these figures, it is assumed that the expert judgment on the most pessimistic value, x_{high}, has a likelihood of 10% of being conservative (that is, too low).
2. $F_{low}(x) = 0.05$; $F_{high}(x) = 0.95$. With these figures, it is assumed that the expert judgment on the most pessimistic value, x_{high}, has a likelihood of 5% of being conservative (too low). Note that this definition gives a smaller variance that that in case 1 above, for the same values of x_{low} and x_{high}.
3. Some people believe that experts can more easily deal with *odds* rather than *probabilities*, and therefore set the percentiles accordingly. For example, if the odds are believed to be 10 to 1 that the duration (or cost) x_{low} will be exceeded, then the probability of a duration less than x_{low} is $1/11$ and the probability of a duration greater than x_{low} is $10/11$. Then $F_{low}(x) = 0.09091$; $F_{high}(x) = 0.90909$.

The expert then adjusts his estimates of x_{low} and x_{high} to achieve the desired condition of odds of 10 to 1.

As the Normal distribution is symmetric, and assuming that symmetric values of $F_{low}(x)$; $F_{high}(x)$ are specified, the mean is determined from:

$$\mu = \bar{x} = x_{low} + \frac{1}{2}\left(x_{high} - x_{low}\right) = \frac{1}{2}\left(x_{low} + x_{high}\right)$$

The standard deviation is obtained from the published tables for the Normal distribution.

1. In case 1, the distance from the mean to the 10th percentile is 1.282σ, so the standard deviation is $\sigma = \dfrac{x_{high} - x_{low}}{2.564}$.

2. In case 2, the distance from the mean to the 5th percentile is 1.645σ, so the standard deviation is $\sigma = \dfrac{x_{high} - x_{low}}{3.29}$.

3. In case 3, the distance from the mean to the cumulative at 0.09090 is 1.3347σ, so the standard deviation is $\sigma = \dfrac{x_{high} - x_{low}}{2.6694}$.

6.3.5   The Gumbel Type II Extreme Value Distribution

This distribution is a limiting extreme value distribution. If there are a number of parallel paths to an activity; if all the paths are independent; if the number of independent paths is large; if the duration of each is exponentially distributed; and if the durations are unlimited to the right but not less than some specified minimum γ; in the limit the largest order statistic approaches the probability distribution is known as the Gumbel Type II Extreme Value Distribution, which is defined by the cumulative probability distribution $F_x(x)$ and the probability density function $f_x(x)$ below:

$$F_x\left(x\right) = \exp\left\{-\left[\frac{x-\gamma}{\alpha}\right]^{-\beta}\right\} \quad \text{for all } \gamma < x < \infty$$

$$f_x\left(x\right) = \frac{dF_x\left(x\right)}{dx} = \exp\left\{-\left[\frac{x-\gamma}{\alpha}\right]^{-\beta}\right\}\left(\frac{\beta}{\alpha}\right)\left[\frac{x-\gamma}{\alpha}\right]^{-(\beta+1)}$$

$$= \left(\frac{\beta}{\alpha}\right)\left[\frac{x-\gamma}{\alpha}\right]^{-(\beta+1)} F_x\left(x\right) \quad \text{for all } \gamma < x < \infty$$

Here, γ is a location or shift parameter that shifts the distribution to the right. (That is, the probability is zero for $x < \gamma$). The distribution has the properties:

$$\text{mode}(x) = \gamma + \alpha \left[\frac{\beta}{1+\beta} \right]^{1/\beta}$$

$$\text{median}(x) = \gamma + \alpha \left[\frac{-1}{\ln(0.5)} \right]^{1/\beta}$$

This function is bounded from below by γ. It is not bounded from above and the values of x can be indefinitely large. Of course, if one uses this distribution for real projects, the parallel paths for an activity cannot be all independent, so the conditions of the derivation cannot be met. However, this origin suggests that this distribution might be a reasonable approximation for actual activity durations, even with dependence. Generally speaking, it is highly skewed to the right.

We can fit this function to our best judgment by the following process:

1. Estimate the shift factor γ. This is the absolute lower limit for the variate x; x can never take on values less than γ.
2. Establish a convenient percentile at the lower end of the range, for example, 0.10. Call this value $F_{low}(x)$, with value 0.10 or 0.05, say. The expert then estimates the value for the duration at this percentile; that is, has probability $F_{low}(x)$ that this value will not be exceeded. Call this value x_{low}.
3. Establish a suitable percentile at the upper end of the range, for example, 0.90. Call this value $F_{high}(x)$, with value 0.90 or 0.95, say. Then estimate the value of the activity duration at this percentile; that is, the duration that has probability $F_{high}(x)$ that it will not be exceeded. Call this value x_{high}

Then the two parameters of the Type II distribution are computed as follows:

$$\beta = \ln \left[\frac{\ln(F_{low})}{\ln(F_{high})} \right] \frac{1}{\ln \left(\dfrac{x_{high} - \gamma}{x_{low} - \gamma} \right)}$$

$$\alpha = \frac{x_{high} - \gamma}{\left(\dfrac{-1}{\ln(F_{high})} \right)^{1/\beta}}$$

These three parameters, α, β, γ, define the Type II distribution. However, the values needed for the second moment process are the mean and variance. There are closed-form equations for the mode and median, but unfortunately there are no closed-form solutions for the mean and variance. This would seem to be an impediment to the use of this distribution in practice, but this is not necessarily so. The numerical integrations needed to find the values of the mean and variance are easily set up and performed by a spreadsheet program. This process is as follows:

1. Input γ, F_{low}, x_{low}, F_{high}, x_{high}
2. From these, compute α, β from the equations above.
3. Using α, β, γ, compute the probability density function $f_x(x_k)$ from the equation given above, for a number of values at regular intervals, δ. There should be enough intervals to cover the range from $x = \gamma$ to $x > x_{high}$, until the computed values of $f_x(x_k)$ become negligible.
4. Compute the mean by:

$$\mu = \overline{x} = \sum_{k=1}^{n} x_k f_x\left(x_k\right)\delta$$

5. Compute the variance by:

$$\sigma^2 = \sum_{k=1}^{n}\left(x_k - \overline{x}\right)^2 f_x\left(x_k\right)\delta$$

The integrations above essentially use the trapezoidal rule; more accurate results might be obtained with Simpson's rule or some other integration rule. The computed mean and variance are then used in the second moment process. The numerical integration is not quite as elegant as a closed form equation, but in practice the computation takes no more time on a spreadsheet. And, the spreadsheet can plot a picture of what the probability density looks like; if it seems too skewed, or not skewed enough, one can easily adjust the input parameter values until it looks right.

Figure 6.6 shows an example of a Gumbel Type II distribution with shift zero, 10th percentile 1100, and 90th percentile 2300.

6.3.6 Fitting the Gumbel Type II Distribution to a Histogram

If one has some data, sufficient to define a histogram, it is possible to fit a Gumbel Type II distribution to three points on the cumulative probability function, rather than two points and the shift factor, as above. To do this, consider three percentiles of the cumulative distribution: a low value, the median value, and a high value. Let these be denoted (F_{low}, x_{low}), (F_{50}, x_{50}), (F_{high}, x_{high}). Here x_{50} is the median, and so $F_{50} = 0.50$. The low and high values will be left unspecified for the moment. The cumulative distribution function for the Gumbel Type II was given above as:

$$F_x\left(x\right) = \exp\left\{-\left[\frac{x - \gamma}{\alpha}\right]^{-\beta}\right\} \quad \text{for all } \gamma < x < \infty$$

The shift factor can be eliminated by considering three points on the cumulative:

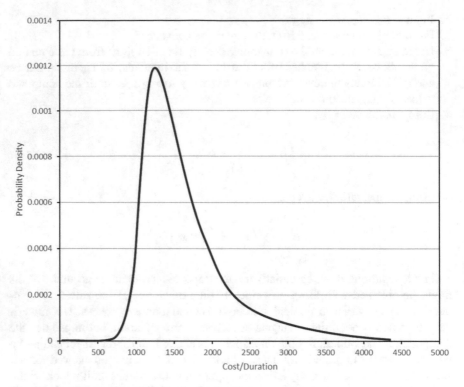

Fig. 6.6 Gumbel type II probability density function

$$F_{low} = \exp\left\{-\left[\frac{x_{low} - \gamma}{\alpha}\right]^{-\beta}\right\}$$

$$F_{50} = 0.50 = \exp\left\{-\left[\frac{x_{50} - \gamma}{\alpha}\right]^{-\beta}\right\}$$

$$F_{high} = \exp\left\{-\left[\frac{x_{high} - \gamma}{\alpha}\right]^{-\beta}\right\}$$

Now take the natural logarithms of all three expressions:

$$\ln[F_{low}] = -\left[\frac{x_{low} - \gamma}{\alpha}\right]^{-\beta}$$

$$\ln[0.50] = -\left[\frac{x_{50} - \gamma}{\alpha}\right]^{-\beta}$$

$$\ln[F_{high}] = -\left[\frac{x_{high} - \gamma}{\alpha}\right]^{-\beta}$$

Using these relations, form two ratios, as follows:

$$\frac{\ln[F_{low}]}{\ln[0.50]} = \left[\frac{x_{50} - \gamma}{x_{low} - \gamma}\right]^{\beta}$$

$$\frac{\ln[0.50]}{\ln[F_{high}]} = \left[\frac{x_{high} - \gamma}{x_{50} - \gamma}\right]^{\beta}$$

Taking logarithms of both sides of both expressions gives:

$$\ln\left[\frac{\ln[F_{low}]}{\ln[0.50]}\right] = \beta \ln\left[\frac{x_{50} - \gamma}{x_{low} - \gamma}\right]$$

$$\ln\left[\frac{\ln[0.50]}{\ln[F_{high}]}\right] = \beta \ln\left[\frac{x_{high} - \gamma}{x_{50} - \gamma}\right]$$

Solving both expressions for β gives:

$$\beta = \left\{\frac{\ln\left[\frac{\ln[F_{low}]}{\ln[0.50]}\right]}{\ln\left[\frac{x_{50} - \gamma}{x_{low} - \gamma}\right]}\right\}$$

$$\beta = \left\{\frac{\ln\left[\frac{\ln[0.50]}{\ln[F_{high}]}\right]}{\ln\left[\frac{x_{high} - \gamma}{x_{50} - \gamma}\right]}\right\}$$

Obviously, the values for β from both expressions must be identical, so the two expressions on the right can be equated:

$$\left\{\frac{\ln\left[\frac{\ln[F_{low}]}{\ln[0.50]}\right]}{\ln\left[\frac{x_{50} - \gamma}{x_{low} - \gamma}\right]}\right\} = \left\{\frac{\ln\left[\frac{\ln[0.50]}{\ln[F_{high}]}\right]}{\ln\left[\frac{x_{high} - \gamma}{x_{50} - \gamma}\right]}\right\}$$

Or, rearranging,

$$\ln\left[\frac{x_{high}-\gamma}{x_{50}-\gamma}\right]\ln\left[\frac{\ln\left[F_{low}\right]}{\ln\left[0.50\right]}\right]=\ln\left[\frac{x_{50}-\gamma}{x_{low}-\gamma}\right]\ln\left[\frac{\ln\left[0.50\right]}{\ln\left[F_{high}\right]}\right]$$

This equation must be solved for the shift factor, γ. The general solution of this equation would be difficult. However, the percentiles F_{low} and F_{high} have been left unspecified, so we may chose specific values of these parameters to make the solution easier. An obvious simplification would be to set F_{low} and F_{high} so that:

$$\ln\left[\frac{\ln\left[F_{low}\right]}{\ln\left[0.50\right]}\right]=1$$

$$\ln\left[\frac{\ln\left[0.50\right]}{\ln\left[F_{high}\right]}\right]=1$$

Thus,

$$\frac{\ln\left[F_{low}\right]}{\ln\left[0.50\right]}=e$$

$$\frac{\ln\left[0.50\right]}{\ln\left[F_{high}\right]}=e$$

Rearranging,

$$\ln\left[F_{low}\right]=e\ln\left[0.50\right]$$

$$\ln\left[0.50\right]=e\ln\left[F_{high}\right]$$

Hence,

$$F_{low}=0.50^{e}\cong0.15196$$

$$F_{high}=0.50^{1/e}\cong0.77492$$

That is, we choose the lower percentile, $F_{low}\equiv F_{15}$, to correspond to approximately the 15th percentile of the cumulative histogram, and the upper percentile, $F_{high}\equiv F_{77}$, to correspond to approximately the 77th percentile of the cumulative. With these specific values, the equation above,

$$\ln\left[\frac{x_{high}-\gamma}{x_{50}-\gamma}\right]\ln\left[\frac{\ln\left[F_{low}\right]}{\ln\left[0.50\right]}\right]=\ln\left[\frac{x_{50}-\gamma}{x_{low}-\gamma}\right]\ln\left[\frac{\ln\left[0.50\right]}{\ln\left[F_{high}\right]}\right]$$

becomes

$$\ln\left[\frac{x_{77}-\gamma}{x_{50}-\gamma}\right] = \ln\left[\frac{x_{50}-\gamma}{x_{15}-\gamma}\right]$$

This becomes

$$\frac{x_{77}-\gamma}{x_{50}-\gamma} = \frac{x_{50}-\gamma}{x_{15}-\gamma}$$

$$\left(x_{77}-\gamma\right)\left(x_{15}-\gamma\right) = \left(x_{50}-\gamma\right)\left(x_{50}-\gamma\right)$$

The solution to this is:

$$\gamma = \frac{x_{15}x_{77} - x_{50}^{2}}{x_{15} + x_{77} - 2x_{50}}$$

This value for the shift factor may then be used in the Gumbel Type II distribution, instead of estimating it directly. Note: this method of fitting three points is not the best fit or the least squares fit of the Gumbel Type II to observed data. An equation fitted by least squares will be that distribution with the minimum sum of squares of the deviations between the data and the function; the three-point fit matches the function to the data at exactly three points, and not necessarily elsewhere.

6.3.7 The "Binormal" Distribution

The "binormal" distribution was documented by King et al., in the aptly named paper "An Alternative to Monte Carlo Sampling in Stochastic Models" (1975). Their "… approach was to seek a single function that could be treated both as the underlying cdf of each x_i and as an approximation for the cdf of the sum, y. A promising candidate should be a mathematically tractable three-parameter function capable of approximating a wide variety of distribution shapes. In addition, if such a function were assumed to be the true cdf of each x_i, it then should provide a close approximation to the cdf of y for values on n in the practical range…."

"After screening a number of candidates, [the authors] selected a function which may be called "binormal" with the cumulative density function (cdf):

$$F\left(x; m, \sigma_1, \sigma_2\right) = \Phi\left(\frac{x-m}{\sigma_1}\right) \quad for\, x < m$$

$$F\left(x; m, \sigma_1, \sigma_2\right) = \Phi\left(\frac{x-m}{\sigma_2}\right) \quad for\, x \geq m$$

where $\Phi(x)$ is the cdf of a standardized unit normal distribution. The density function consists of the left half of one normal curve and the right half of another, both having the same mean, m. The parameters σ_1 and σ_2 are, in effect, the standard deviations of the two original Gaussian curves, and the parameter m becomes the median of the composite distribution. The special case of a normal distribution is obtained when $\sigma_1 = \sigma_2$.

The authors recommend the estimation of three values corresponding to the 10th, 50th, and 90th percentiles, x_{10}, x_{50}, x_{90}. However, other percentiles could be used by suitably modifying the expressions below. With these estimates, the three parameters of the binormal are computed from:

$$\sigma_1 = \frac{x_{50} - x_{10}}{1.282}$$

$$\sigma_2 = \frac{x_{90} - x_{50}}{1.282}$$

$$m = x_{50}$$

Of course, the factor 1.282 is obtained from tables of the Normal distribution, and should be changed if other percentiles are used. For example, if one prefers the 5th, 50th, and 95th percentiles, corresponding to the judgmental estimates x_{05}, x_{50}, and x_{95}, the factor would be 1.645.

With these three parameters estimated, the first three moments of the binormal can be computed. Only the first two moments are given here, because the second moment approach uses only the mean and the variance. They are:

$$\mu_1 = \bar{x} = m + \frac{\sigma_2 - \sigma_1}{\sqrt{2\pi}}$$

$$\mu_2 = \text{var}(x) = \frac{\sigma_2^2 + \sigma_1^2}{2} - \frac{(\sigma_2 - \sigma_1)^2}{2\pi}$$

King et al. used three estimates in order to fit a three-moment approach, but they assumed for ease of use that all the individual variables were independent. Therefore, their method as given cannot be applied when the variables are correlated, as is assumed in these notes. Here, we use a second moment method plus correlations.

Comparing the Gumbel Type II distribution and the King et al. binormal distribution, assume that the estimates are, with values at the 10th and 90th percentiles: shift = 500, x_{10} = 1200, x_{90} = 2000. Then the resulting means and variances are:

Type II distribution:

  mode = 1315, median = 1442, μ = 1519, σ = 350

Binormal distribution:

  mode = 1315, median = 1315, μ = 1492, σ = 340

Figure 6.7 plots the cumulative probability distributions for the Gumbel Type II and the binormal for the parameters given just above. As would be expected, the

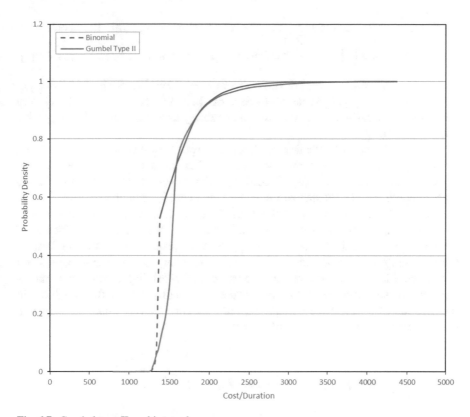

Fig. 6.7   Gumbel type II vs. binormal

binormal is not particularly suitable around the junction point (1315), but the binormal and the Gumbel give very similar results in both tails.

The Gumbel Type II distribution seems to be satisfactory, and perhaps more theoretically justified than the binormal, the origin of which, in combining two Normal distributions may seem somewhat arbitrary. Based on limited comparisons, however, the binormal distribution seems to give comparable results, and is easier to use, given than no numerical integration is required. Therefore, if one can accept the rather heuristic origin of the binormal, it seems sufficiently accurate to determine the mean and variance from three point estimates.

6.3.8   Reality Checks

Some students and practitioners object that they may be able to estimate single point values for some random variables, but they have no basis for estimating variances. Obviously, experience is valuable in making these estimates. Nevertheless, variances can be estimated by practically anyone. Instead of dealing with the variance directly, consider dealing with the Coefficient of variation, $COV[x] = \dfrac{\sqrt{\operatorname{var}[x]}}{E[x]} = \dfrac{\sigma_x}{\mu_x}$. If $E[x] \neq 0$, and we can estimate $COV[x]$, then we can get the standard deviation from $\sigma_x = COV[x]\mu_x$. So how can one get $COV[x]$? One thing people can do to develop their intuition for uncertainty is to determine the coefficients of variation on ordinary activities for which they have abundant data.

For example, consider the time it takes you to travel from your home to your first class, or your last class to home, or to run 2 miles every morning, or some other activity you do nearly every day. Record these times every day for a month or more, then compute the means, standard deviation, and coefficients of variation for each activity. Now use these values for the coefficients of variation to calibrate estimates of project times. This does not mean that everything has the same coefficient of variation. However, there have been many cases of project proponents estimating the variability for first of a kind, new technology, complex projects to have much smaller coefficients of variation than the variability in driving to work every morning. This doesn't imply that your daily experiments tell you what the variability in the duration of a new project might be, but they can tell you when the estimated coefficients of variation are absurdly small.

6.3.9 Expert Judgment for Correlations

Less work has been done on the best methods for eliciting expert judgment on correlation coefficients than has been done with means and variances. This is no doubt because most researchers assume all correlations to be zero. It appears that the best way found so far to elicit correlation coefficients from experts is simply to ask them to give their best estimates of the correlation coefficients. Some background discussion of the meaning of correlations is desirable. It may also be helpful to show scatter diagrams representing different values of the correlations between two variables.

The following method has been found to be workable in risk analyses for industrial and commercial engineering and construction projects:

1. Set up a small number of admissible correlations, associated with verbal descriptors that are comprehensible to the experts, such as, for example:

   High = 0.9
   Moderate = 0.6
   Low = 0.3
   Zero = 0.0

2. Ask each expert evaluator to identify all the pairs of variables (such as work packages or activities) that are related by each of the verbal descriptors.

3. Crosscut the above by asking each evaluator to identify potential common mode risks that would affect multiple work packages and lead to correlations between them.
4. Check for consistency by asking each evaluator to estimate the correlations for specific pairs of work packages.
5. Compare results across all evaluators and follow up on any apparent inconsistencies between experts.
6. Form the complete covariance matrix and determine if it is admissible; that is, it is positive definite and invertible. If it is not positive definite, adjust the correlation values until the matrix is positive definite.
7. Perform a sensitivity analysis of the risk analysis results to determine if the computed risks are sensitive to any particular correlation coefficients.

6.3.10 Other Methods for Estimating Correlation Coefficients

Assume that for every pair of work packages or cost accounts, j and k, there is a correlation coefficient ρ_{jk} such that

$$-1 \leq \rho_{jk} \leq +1 \ \text{ for all } j \text{ and } k$$

$$\rho_{jk} = \rho_{kj} \ (\text{symmetry})$$

$$\rho_{jk} = 1 \text{ if } \ k = j$$

The correlation coefficient describes the stochastic interaction between cost account j and cost account k. The correlation coefficient does not define causal relationships, such as $X_j = f(X_k)$, where f is some deterministic function. Without correlation, all the uncertain components would be independent entities; correlation links all these components together into a system.

- If, when the cost of WP_k is above its mean, the cost of WP_j also tends to be above its mean, then the two work packages are positively correlated, $\rho_{jk} > 0$
- If, when the cost of WP_k is above its mean, the cost of WP_j tends to be below its mean, then the two work packages are negatively correlated, $\rho_{j\,k} < 0$

The meaning of ρ_{jk} can be interpreted as follows: suppose that there are two work packages, j and k, with variances σ_j^2, σ_k^2 and correlation ρ_{jk}. Then the square of the correlation ρ_{jk}^2 is the fraction of the uncertainty (i.e., the variance) in the cost of Work Package j that is explained or removed by knowledge of the true cost of Work Package k.

That is, suppose the project manager initially does not know the true values of the Work Package costs X_j and X_k but attributes to them the uncertainties σ_j^2 and σ_k^2, respectively. Then suppose that work package k finishes, and the project manager now knows the true value of X_k. If the two work packages are correlated, this knowledge of the true value for Work Package k then tells the project manager something about the cost of the incomplete Work Package j. As the square of the correlation ρ_{jk}^2 is the fraction of the uncertainty in the cost of Work Package j that is explained or removed by knowledge of the true cost of Work Package k, then this knowledge reduces the variance of Work Package j by the amount $\rho_{jk}^2\sigma_j^2$. Then the remaining variance in Work Package j, the uncertainty remaining in this work package cost, is equal to the original uncertainty (before Work Package k was known) minus the reduction in variance due to the knowledge of Work Package k. Call the prior variance (before k) σ_j^2 and the posterior variance (after k) σ_j, where

$$\tilde{\sigma}_j^2 = \sigma_j^2 - \rho_{jk}^2\sigma_j^2 = \sigma_j^2\left(1 - \rho_{jk}^2\sigma_j^2\right)$$

The result states that, no matter what the actual value of X_k is, the project manager's uncertainty in the cost is always reduced by knowledge of X_k. The knowledge of X_k reduces the uncertainty in X_j due to the linkage ρ_{jk} between these two activities.

If the correlation is 0, then knowledge of the actual cost of the Work Package WP_k provides the project manager with no information about the cost of incomplete Work Package WP_j, as $\sigma_j^2 - \rho_{jk}^2\sigma_j^2 = \left(1 - \rho_{jk}^2\right)\sigma_j^2 = \sigma_j^2$.

If the correlation is $+1$ or -1, then knowledge of the actual cost of WP_k provides the project manager with complete information about the cost of uncompleted Work Package WP_j:

$$\tilde{\sigma}_j^2 = \sigma_j^2 - \rho_{jk}^2\sigma_j^2 = \left(1 - \rho_{jk}^2\right)\sigma_j^2 = 0$$

If the knowledge of the actual cost of WP_k would induce the project manager to consider revising the estimate of the cost of incomplete WP_j (either up or down), then the project manager believes that the two WP costs are correlated.

The amount of the project manager's revision depends on the value of the correlation coefficient. This does not imply that an overrun in WP_k necessarily *causes* an overrun in WP_j (there may be a common cause).

The most common method for eliciting subjective estimates of correlation coefficients from people with experience is simply to ask them to estimate the values. Experience has shown this method to be workable. However, some people may not relate to estimating correlation coefficients directly, but may be more able to estimate changes (ratios) in parameters such as variance due to additional knowledge about the values for some work packages. Let's define:

σ_j^2 = the variance or uncertainty in Work Package j when the costs of both Work Package j and Work Package k are unknown.

$\tilde{\sigma}_j^2$ = the variance or uncertainty in Work Package j if the true cost of Work Package k were known (Work Package j remains unknown)

Then the ratio of the variance (uncertainty) of WP_j *after* WP_k is known to the value before WP_k is known is:

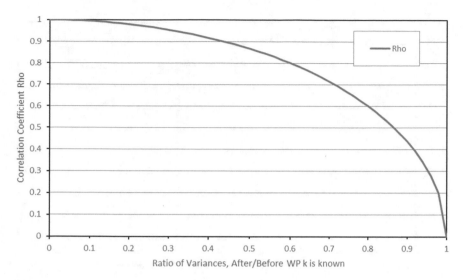

Fig. 6.8 Estimating correlation coefficient from the ratio of variances

$$\frac{\tilde{\sigma}_j^2}{\sigma_j^2} = \frac{\sigma_j^2 - \rho_{jk}^2 \sigma_j^2}{\sigma_j^2} = \frac{\left(1 - \rho_{jk}^2\right)\sigma_j^2}{\sigma_j^2} = \left(1 - \rho_{jk}^2\right)$$

$$\rho_{jk}^2 = 1 - \frac{\tilde{\sigma}_j^2}{\sigma_j^2}$$

$$\rho_{jk} = \sqrt{1 - \frac{\tilde{\sigma}_j^2}{\sigma_j^2}} = \sqrt{1 - \left(\frac{\tilde{\sigma}_j}{\sigma_j}\right)^2}$$

Figure 6.8 shows a plot of this relationship between variance ratio and correlation coefficient.

For example, if someone familiar with the project estimates that knowledge of Work Package k would reduce the estimated Coefficient of Variation of Work Package j to 60% of its value before this knowledge became available, then the estimator implicitly values the correlation between these two work packages at 0.90, as shown in Fig. 6.9.

6.3.11 Correlations Derived from Work Breakdown Structures

Assessment of correlations is not optional; it is necessary. If one defines a new model for a project, it is possible or at least conceivable that one can choose variables that are independent. However, when using common models, such as the Work Breakdown Structure (WBS) for project costs and the Critical Path Method

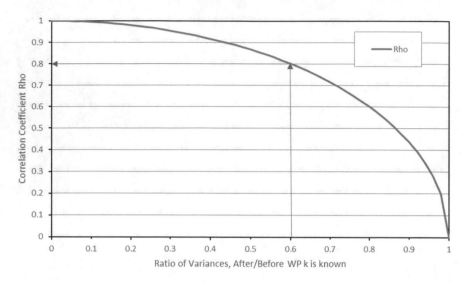

Fig. 6.9 Estimating correlation coefficient from the ratio of variances

(CPM) for project schedules, the usual variables (Work Package costs and Activity durations) are definitely not independent. Assuming that all variables are independent can be, as shown in various places in these notes, very unconservative, and so correlations must be assessed.

When dealing with means and variances, one has to estimate N values in each case, where N is the number of variables, but the obvious difficulty with correlations is that one must estimate all the correlation coefficients in the correlation matrix. Taking advantage of symmetry, this means estimating $\dfrac{N(N-1)}{2}$ values, but this is still a large number.

However, there may be a source of information about correlations readily at hand, in the form of the Work Breakdown Structure. The WBS is the standard method for dissecting projects into manageable parts for planning, estimating, scheduling, and reporting, and is required on virtually all projects that follow common project management principles. Although Work Breakdown Structures vary considerably from project to project, they all have the common factor that Work Packages are grouped by association. That is, the WBS is a subjective or qualitative expression of correlation between Work Packages.

For example, consider the common form of WBS, a tree structure rooted at the top level and branching at each level down to the Work Packages. Table 6.3 indicates one form of WBS:

Each branch at each level is given an identification number in the WBS dictionary, and the Work packages, in this arrangement, have five digit numbers. As an illustration, suppose that some project is the first project for the first program or client. Then the Project has the number 1.1. Suppose, for convenience in discussion, there are three branches from every node, so the project will have three tasks: 1.1.1,

Table 6.3 WBS level description

Level	Description
Level 1	Program or client
Level 2	Project
Level 3	Task
Level 4	Subtask
Level 5	Work package

Table 6.4 WBS levels

Level 1	Level 2	Level 3	Level 4	Level 5
Program	Project	Task	Subtask	WP
				1.1.1.1.1
			1.1.1.1	1.1.1.1.2
				1.1.1.1.3
				1.1.1.2.1
		1.1.1	1.1.1.2	1.1.1.2.2
				1.1.1.2.3
				1.1.1.3.1
			1.1.1.3	1.1.1.3.2
				1.1.1.3.3
				1.1.2.1.1
			1.1.2.1	1.1.2.1.2
				1.1.2.1.3
				1.1.2.2.1
1	1.1	1.1.2	1.1.2.2	1.1.2.2.2
				1.1.2.2.3
				1.1.2.3.1
			1.1.2.3	1.1.2.3.2
				1.1.2.3.3
				1.1.3.1.1
			1.1.3.1	1.1.3.1.2
				1.1.3.1.3
				1.1.3.2.1
		1.1.3	1.1.3.2	1.1.3.2.2
				1.1.3.2.3
				1.1.3.3.1
			1.1.3.3	1.1.3.3.2
				1.1.3.3.3

1.1.2, and 1.1.3. Assuming three subtasks per task, there are nine subtasks, labeled 1.1.1.1, 1.1.1.2, 1.1.1.3, etc. And assuming three Work Packages per subtask, there are 27 WPs. Table 6.4 shows the hierarchical tree, turned sideways to fit the paper.

Looking at the tree, one may say that work packages are associated if they have the same parent, that is, subtask. Therefore, work packages 1.1.1.1.1, 1.1.1.1.2, and 1.1.1.1.3 are associated because all have the same parent, subtask 1.1.1.1. This association was built into the Work Breakdown Structure when the WBS was made, and, because association is correlation, now we can take advantage of this effort to generate the correlation matrix.

Thus, we may say that work packages 1.1.1.1.1, 1.1.1.1.2, and 1.1.1.1.3 are correlated with a common correlation coefficient, say ρ_4, because they are associated through their common parent at level 4. Similarly, work packages 1.1.1.2.1, 1.1.1.2.2, and 1.1.1.2.3 are associated with their common parent, subtask 1.1.1.2. Then we may say that these WPs have a common correlation coefficient, ρ_4, because they are associated through their common parent at level 4. This logic continues for all WPs with common parents.

However, the set of WPs 1.1.1.1.1, 1.1.1.1.2, and 1.1.1.1.3 shares a common grandparent (task) with the set of WPs 1.1.1.2.1, 1.1.1.2.2, and 1.1.1.2.3. Therefore, we may say that these two sets are associated with correlation coefficient ρ_3, because they are related by a common grandparent at level 3. We would normally expect that $\rho_3 \leq \rho_4$. This logic continues for all WPs with common grandparents (tasks).

However, the set of WPs 1.1.1.1.1, 1.1.1.1.2, and 1.1.1.1.3 does not share a common grandparent (task) with the set of WPs 1.1.3.3.1, 1.1.3.3.2, and 1.1.3.3.3. These two sets do share a common great-grandparent (the project), and so are associated with correlation coefficient ρ_2, because they are related by a common great-grandparent at level 2. We would normally expect that $\rho_2 \leq \rho_3$. This logic continues for all WPs in the project. There may also be correlations between projects in programs, but multiple projects are not considered here.

Thus, by use of the project Work Breakdown Structure, the correlation issue has been reduced to the estimation of only three correlation coefficients: ρ_2, ρ_3, and ρ_4, for any size of project. Of course, this reduction may not suit all projects, but it may be used as a starting point, and values provided for specific correlations for any two WPs j and k, ρ_{jk}, by overriding the default values.

Figure 6.10 shows a portion of a 27×27 spreadsheet correlation matrix (truncated to fit the page) with three branches at every node, as discussed above, and $\rho_4 = 0.9$; $\rho_3 = 0.7$; and $\rho_2 = 0.5$.

6.4   Aggregating Expert Judgments

Expert judgments about the moments of probability distributions of random variables in project management are typically obtained in a group setting utilizing techniques such as Delphi or Kaplan method. Here the objective is to for the group of experts to converge to a single representation of the probability distribution. The key advantage of such a group-based exercise that comes typically in the form of 2-day workshop is to expose knowledge of many experts and get a broader backgrounds and more complete picture of the factors that may contribute to the project risks. However, often such exercises end up with a one-sided and skewed perspective due to a number of reasons such as organizational bias; for example, it is not too uncommon to have sideline conversations with participants that doubt the group logic and

WP	1.1.1.1.1	1.1.1.1.2	1.1.1.1.3	1.1.1.2.1	1.1.1.2.2	1.1.1.2.3	1.1.1.3.1	1.1.1.3.2	1.1.1.3.3	1.1.2.1.1	1.1.2.1.2	1.1.2.1.3
1.1.1.1.1	1	0.9	0.9	0.7	0.7	0.7	0.7	0.7	0.7	0.5	0.5	0.5
1.1.1.1.2	0.9	1	0.9	0.7	0.7	0.7	0.7	0.7	0.7	0.5	0.5	0.5
1.1.1.1.3	0.9	0.9	1	0.7	0.7	0.7	0.7	0.7	0.7	0.5	0.5	0.5
1.1.1.2.1	0.7	0.7	0.7	1	0.9	0.9	0.7	0.7	0.7	0.5	0.5	0.5
1.1.1.2.2	0.7	0.7	0.7	0.9	1	0.9	0.7	0.7	0.7	0.5	0.5	0.5
1.1.1.2.3	0.7	0.7	0.7	0.9	0.9	1	0.7	0.7	0.7	0.5	0.5	0.5
1.1.1.3.1	0.7	0.7	0.7	0.7	0.7	0.7	1	0.9	0.9	0.5	0.5	0.5
1.1.1.3.2	0.7	0.7	0.7	0.7	0.7	0.7	0.9	1	0.9	0.5	0.5	0.5
1.1.1.3.3	0.7	0.7	0.7	0.7	0.7	0.7	0.9	0.9	1	0.5	0.5	0.5
1.1.2.1.1	0.5	0.5	0.5	0.5	0.5	0.5	0.5	0.5	0.5	1	0.9	0.9
1.1.2.1.2	0.5	0.5	0.5	0.5	0.5	0.5	0.5	0.5	0.5	0.9	1	0.9
1.1.2.1.3	0.5	0.5	0.5	0.5	0.5	0.5	0.5	0.5	0.5	0.9	0.9	1
1.1.2.2.1	0.5	0.5	0.5	0.5	0.5	0.5	0.5	0.5	0.5	0.7	0.7	0.7
1.1.2.2.2	0.5	0.5	0.5	0.5	0.5	0.5	0.5	0.5	0.5	0.7	0.7	0.7
1.1.2.2.3	0.5	0.5	0.5	0.5	0.5	0.5	0.5	0.5	0.5	0.7	0.7	0.7
1.1.2.3.1	0.5	0.5	0.5	0.5	0.5	0.5	0.5	0.5	0.5	0.7	0.7	0.7
1.1.2.3.2	0.5	0.5	0.5	0.5	0.5	0.5	0.5	0.5	0.5	0.7	0.7	0.7
1.1.2.3.3	0.5	0.5	0.5	0.5	0.5	0.5	0.5	0.5	0.5	0.7	0.7	0.7
1.1.3.1.1	0.5	0.5	0.5	0.5	0.5	0.5	0.5	0.5	0.5	0.5	0.5	0.5
1.1.3.1.2	0.5	0.5	0.5	0.5	0.5	0.5	0.5	0.5	0.5	0.5	0.5	0.5
1.1.3.1.3	0.5	0.5	0.5	0.5	0.5	0.5	0.5	0.5	0.5	0.5	0.5	0.5
1.1.3.2.1	0.5	0.5	0.5	0.5	0.5	0.5	0.5	0.5	0.5	0.5	0.5	0.5
1.1.3.2.2	0.5	0.5	0.5	0.5	0.5	0.5	0.5	0.5	0.5	0.5	0.5	0.5
1.1.3.2.3	0.5	0.5	0.5	0.5	0.5	0.5	0.5	0.5	0.5	0.5	0.5	0.5
1.1.3.3.1	0.5	0.5	0.5	0.5	0.5	0.5	0.5	0.5	0.5	0.5	0.5	0.5
1.1.3.3.2	0.5	0.5	0.5	0.5	0.5	0.5	0.5	0.5	0.5	0.5	0.5	0.5
1.1.3.3.3	0.5	0.5	0.5	0.5	0.5	0.5	0.5	0.5	0.5	0.5	0.5	0.5

Fig. 6.10 Truncated correlation matrix

judgements but would like remain anonymous. Hence, in some instances is to our advantage to elicit experts' judgment on an individual basis which bring the question – How do we aggregate expert judgments of probability distributions?

Mathematical aggregation combines individual probability distributions into one single distribution. Opinion pooling is the most common method of aggregation. In linear opinion pool we consider a weighted average of the individual distributions with weights w_i summing to 1.

$$f(\theta) = \sum_{i=1}^{n} w_i f_i(\theta)$$

Simple average (equal-weighted) is $w_i = 1/n$ (for n experts). Generally, weights are chosen depending on the expertise of the experts.

In logarithmic option pool we take weighted geometric mean of the n individual distributions.

$$f(\theta) = k \prod_{i=1}^{n} f_i(\theta)^{w_i}$$

where k is a normalizing constant that ensures that $f(\theta)$ integrates to 1.

So when to use Linear Pool and When to use a Logarithmic Pool? Suppose there are two experts, then the logarithmic pool implies stronger information than that given by either expert separately, whereas the linear pool represents less knowledge than either expert alone.

There also other more complex methods of aggregating expert opinion including Cooke's method that takes advantage of information about experts' performance in a separate elicitation, as well as the methods that process aggregation based on monetary stake such as prediction markets. To choose the right approach in mathematical aggregation, one should consider the type of information available (full or partial probability distributions); the individuals performing the aggregation of probabilities; the degree of modeling to be undertaken (e.g. risk assessment team); the form of the combination rule (e.g. weighted average); the specification of parameters for the combination rule (e.g. weights); and the consideration of simple vs. complex rules (e.g. simple averages vs. complex models) .

6.5 Practice Problems

Problem 6.5.1 A construction company keeps records of cost and duration of certain standard work packages on previous projects, for example, the total cost and duration per cubic feet (cf) of reinforced concrete installed for standard foundation work. During planning and estimation phase of the project, the project manager needed to know what cost and schedule risks are associated with installation of 50,000 ft^3 of reinforced foundation on the project. The previous projects were deemed similar enough to provide a basis for the analysis. The data is shown in the Table 6.5.

- Assuming that distributions are normally distributed find the mean and variance of the distribution?
- What is the correlation coefficient between the two random variables (schedule and cost)? Explain your results and provide logic.

Table 6.5 Problem data

Project	Cost per cf ($)	Schedule per cf (days)
A	65	4
B	66	3
C	59	2
D	38	2
E	110	5
F	89	3
G	200	6
H	40	2
J	160	5

Table 6.6 Problem data

Work package	5-th percentile	50-th percentile	95-th percentile
1	950	1000	1050
2	900	1000	1100
3	600	1000	1400

Table 6.7 Problem data

Expert	New Orleans, LA – Stockton CA			Stockton CA – project site		
	5-th percentile	50-th percentile	95-th percentile	5-th percentile	50-th percentile	95-th percentile
Susan	15	22	35	1	2	3
Rodney	10	12	25	1	2	5
Maria	12	15	30	1	2	5

Problem 6.5.2 Archie is the project manager for a short-term retail sales facility development and Bernie is the construction manager. Archie gives Bernie the task of estimating the risk function of the construction cost. Bernie breaks down the construction phase into three work packages, all of which are performed in 1 year. Bernie makes three-point estimates of the optimistic cost (the 5th percentile, or x_{05}), the median (the 50th percentile, or x_{50}), and the pessimistic cost (the 95th percentile, or x_{95}). These three point estimates are shown in Table 6.6. All figures are in thousands of dollars. All costs are incurred in the same year. Use the Pearson-Tukey approximations to compute the mean and variance of each work package cost.

Bernie estimates the correlations between work packages at 0.95. Determine the probability function. What is the probability that the project will cost more than $3500?

Problem 6.5.3 Michael is in-charge of assessing risks associated the company logistics operations. More specifically, he was given a task to determine the risk that sensitive equipment will not be arriving on-time to be installed. As no empirical data for this is available for this job he has interviewed 3 "experts" to determine a probability density function including the estimates of mean time and the variance that would take the equipment to travel from New Orleans, LA to Stockton, CA and then from Stockton, CA to the project site. The experts' responses (days of travel) are shown in Table 6.7.

If the travel times are modeled using Beta distribution determine the total travel time by aggregating estimates using linear opinion pooling method (hint: use equal weights).

References

Clark CE (1962) The PERT model for the distribution of an activity time. Oper Res 10(3):405–416

Davidson LB, Cooper DO (1976) A simple way of developing a probability distribution of present value. J Pet Technol 28(9):1069–1078

Keefer DL, Bodily SL (1983) Three-point approximations for continuous random variables. Manag Sci 29(5):595–609

King EP, Sampson CB, Simms LL (1975) An alternative to Monte Carlo sampling in stochastic models. Manag Sci 21(6):649–657

Malcolm DG, Roseboom JH, Clark CE, Fazar W (1959) Application of a technique for research and development program evaluation. Oper Res 7(5):646–669

Megill RE (1977) An introduction to risk analysis. Petroleum Publishing Company, Tulsa

Pearson ES, Tukey JW (1965) Approximate means and standard deviations based on distances between percentage points of frequency curves. Biometrika 52(3–4):533–546

Perry C, Greig ID (1975) Estimating the mean and variance of subjective distributions in PERT and decision analysis. Manag Sci 21(12):1477–1480

Project Management Institute (2008) A guide to the project management body of knowledge, 4th edn. Newtown Square, Pennsylvania

Chapter 7
Sensitivity and Common Cause Effects

Abstract In this chapter we discuss the sensitivity of the project performance outcomes such as the total project cost to the uncertainty in work packages. We provide two approaches to this critical step for developing risk mitigation strategies, one based on calculating derivatives of the total variance with the respect to work package of interest, and the other one based on the correlation between the total cost and the work package. Further, this chapter introduces another important analysis for designing proper mitigation strategies – determining the effects of common cause events on the correlation and ultimately on the total cost.

Keywords Sensitivity · Common cause events · Correlation

7.1 Introduction

The total cost of a project is, according to the engineering or "bottom-up" model, the sum of the costs of all of its individual work packages. In a deterministic approach, it is obvious which work package makes the greatest contribution to the total cost: it is the largest work package. However, we are concerned here with the uncertainty in the total cost, which is, in some way, the combination of the uncertainties in all the individual work packages. Given that a project has many work packages, and that the project manager has limited time, it is of some importance to be able to assess priorities, in order to be able to determine which work packages are making the most contribution to the total uncertainty. In this chapter we discuss methods that can help determine which work packages should receive the most attention from the project manager. We focus on the sensitivity of the total cost to each work package as well as the impact of common cause factors on correlation.

© Springer Nature Switzerland AG 2020
I. Damnjanovic, K. Reinschmidt, *Data Analytics for Engineering and Construction Project Risk Management*, Risk, Systems and Decisions,
https://doi.org/10.1007/978-3-030-14251-3_7

7.2   Total Cost Sensitivity to Work Packages

Let T be the total cost of a project and let X_k be the cost of the k-th work package. All work package costs are assumed to be random variates, and so T is also a random variate. There are N work packages. Then the total cost is the sum,

$$T = \sum_{j=1}^{N} x_j$$

$$E\left[x_j\right] = \mu_j = \overline{x_j}$$

$$\sigma_j^2 = \sigma_{j,j} = E\left[\left(x_j - \mu_j\right)^2\right]$$

The expected value of the total cost is given by

$$E[T] = \mu_T = E\left[\sum_{j=1}^{N} x_j\right] = \sum_{j=1}^{N} E\left[x_j\right] = \sum_{j=1}^{N} \mu_j$$

That is, the mean value of the sum is the sum of the means for all the work packages.

In order to consider the risk associated with the total project cost, T, it is desirable to compute the uncertainty in the value of T. Consider the variance of T as a measure of this uncertainty or ignorance about the total cost of the project. Then,

$$\sigma_T^2 = \mathrm{var}[T] = \sum_{i=1}^{N}\sum_{j=1}^{N} \mathrm{cov}\left[x_i, x_j\right] = \sum_{i=1}^{N}\sum_{j=1}^{N} \sigma_{i,j} = \sum_{i=1}^{N}\sum_{j=1}^{N} \rho_{i,j}\sigma_i\sigma_j$$

In this expression, $\mathrm{cov}[x_i, x_j]$, $\sigma_{i,j}$ and $\rho_{i,j}\sigma_i\sigma_j$ are just different symbols standing for the same thing, namely the covariance of the variables x_i and x_j.

All these variances and covariances can be written compactly in the symmetric $N \times N$ Covariance matrix, $\mathbf{V}$, as,

$$\mathbf{V} \equiv \begin{pmatrix} v_{11} & v_{12} \cdots & v_{1N} \\ v_{21} & v_{22} \cdots & v_{2N} \\ v_{N1} & v_{N2} \cdots & v_{NN} \end{pmatrix} \equiv \begin{pmatrix} \sigma_{11} & \sigma_{12} \cdots & \sigma_{1N} \\ \sigma_{21} & \sigma_{22} \cdots & \sigma_{2N} \\ \sigma_{N1} & \sigma_{N2} \cdots & \sigma_{NN} \end{pmatrix} \equiv \begin{pmatrix} \sigma_1^2 & \rho_{12}\sigma_1\sigma_2 \cdots & \rho_{1N}\sigma_1\sigma_N \\ \rho_{12}\sigma_1\sigma_2 & \sigma_2^2 \cdots & \rho_{2N}\sigma_2\sigma_N \\ \rho_{1N}\sigma_1\sigma_N & \rho_{2N}\sigma_2\sigma_N \cdots & \sigma_N^2 \end{pmatrix}$$

In words, the variance of the sum of a number of random variates is the sum of all the terms in the covariance matrix; that is, the sum of the N^2 variances and covariances of the individual work package costs. This is true regardless of whether the individual costs are correlated or uncorrelated.

The summation can be compactly written in matrix notation. Define the N column vector consisting of all 1s:

$$1 = \begin{Bmatrix} 1 \\ 1 \\ 1 \\ 1 \end{Bmatrix}_N$$

Then the summation above may be written as:

$$\mathrm{var}[T] = 1^T \mathbf{V} 1$$

In order to concentrate our efforts on risk mitigation and management on the right areas, we will be interested in knowing which of the work package costs has the greatest impact on the total project cost. If the project had no uncertainty, and all values were deterministic, this question would be trivial: obviously a dollar increase in the cost of any work package would increase the total cost by 1 dollar. More formally, in the case of the mean values, we can write the sensitivity of the expected project cost to any of the work package costs as the rate of change of the expected value of the total cost with respect to the expected value of any work package cost, or, using the above expression for $E[T]$,

$$\frac{\partial E[T]}{\partial E[x_k]} = \frac{\partial \mu_T}{\partial \mu_k} = 1$$

This is trivial, but now we move on to the situation when there is uncertainty, which is not so trivial. We now look at the sensitivity of the *variance* in the total cost with respect to the uncertainty in any work package cost. This may be written as the rate of change in the variance of T with respect to the standard deviation of any work package cost. (We use the variance of T and the standard deviation of x_k to avoid square roots and to make the differentiation easy). Then, using the equation above for var[T],

$$\frac{\partial \mathrm{var}[T]}{\partial \sigma_k} = 2\sum_{j=1}^{N} \rho_{j,k}\sigma_j$$
$$(\rho_{k,k} \equiv 1)$$

Therefore, if we wish to know which work packages have the greatest impact on the uncertainty in the total project cost, we can compute the above expression for all values of k and then rank them from largest to smallest (that is, as a Pareto chart) (Wilkinson 2006) . The value of k that maximizes the quantity $\sum_{j=1}^{N}\rho_{j,k}\sigma_j$ has the greatest influence on the uncertainty in the total cost (the multiplier 2 doesn't affect the ranking and may be omitted). That is, if we want to know where to look in order to reduce the uncertainty (or risk) in the total project cost, a good place to start would be work package k, where k is the work package that maximizes the quantity $\sum_{j=1}^{N} \rho_{j,k} \sigma_j$.

Some of the correlation coefficients may be negative. However, the variance of T must be positive, in order that the total cost may have a non-negative variance and a real standard deviation. Therefore,

$$\sigma_T^2 = \mathrm{var}[T] = \sum_{i=1}^{N} \sum_{j=1}^{N} \rho_{i,j} \sigma_i \sigma_j \geq 0$$

This condition places a constraint on the values of the correlation coefficients, as it is impossible to have so many negative correlation coefficients that $\sigma_T^2 < 0$. We may also feel that the sensitivity of the uncertainty in the total cost should also be positive; that is, that $\dfrac{\partial \mathrm{var}[T]}{\partial \sigma_k} > 0$. If this were not true, then there would be some work package such that we could reduce the uncertainty in the total project cost by increasing the uncertainty in this work package cost. This would seem very unlikely, if not a contradiction, so we should have that

$$\frac{\partial \mathrm{var}[T]}{\partial \sigma_k} = 2 \sum_{j=1}^{N} \rho_{j,k} \sigma_j > 0$$

This can be rewritten as

$$\left(\frac{1}{2}\right) \frac{\partial \mathrm{var}[T]}{\partial \sigma_k} = \sum_{j=1}^{N} \rho_{j,k} \sigma_j = \sigma_k + \sum_{\substack{j=1 \\ j \neq k}}^{N} \rho_{j,k} \sigma_j > 0$$

Therefore, for consistency, the correlation coefficients should be specified such that

$$-\sum_{\substack{j=1 \\ j \neq k}}^{N} \rho_{j,k} \sigma_j < \sigma_k$$

In the special case in which it is established that all N of the work package costs are statistically independent, then $\rho_{j,k} = 0$, $\forall\, j \neq k$. Then the above expression for $\partial\,\mathrm{var}[T]/\partial \sigma_k$ reduces to

$$\frac{\partial \mathrm{var}[T]}{\partial \sigma_k} = 2 \sum_{j=1}^{N} \rho_{j,k} \sigma_j = 2\sigma_k$$

Therefore, in the special case of statistically independent work package costs, to identify the work package with the greatest impact on the uncertainty in the total project cost, look at the work package with the highest variance.

Suppose we do this for a project of, say, 20 work packages, sort them according to the largest values of the sensitivities, and plot the Pareto diagram as shown in Fig. 7.1. (The absolute values of the sensitivities have no importance; the only concern is the relative values).

7.3   Correlations Between Work Packages and Total Cost

The same question can be addressed by means of the correlation between the total project cost and each work package cost. To find the correlation coefficient between the total cost $T = \sum_{j=1}^{N} x_j$ and any work package cost x_k, first compute the covariance between T and x_k, which is given by

$$\text{cov}[T,x_k] = E\left[(T - \mu_T)(x_k - \mu_k)\right]$$

$$E[T] = \mu_T = \sum_{j=1}^{N} E[x_j] = \sum_{j=1}^{N} \mu_j$$

$$\therefore \text{cov}[T,x_k] = E\left[\left\{\left(\sum_{j=1}^{N} x_j\right) - \left(\sum_{j=1}^{N} \mu_j\right)\right\}(x_k - \mu_k)\right]$$

$$\text{cov}[T,x_k] = E\left[\left\{\sum_{j=1}^{N} (x_j - \mu_j)\right\}(x_k - \mu_k)\right]$$

$$\text{cov}[T,x_k] = E\left[\sum_{j=1}^{N} (x_j - \mu_j)(x_k - \mu_k)\right]$$

$$\text{cov}[T,x_k] = \sum_{j=1}^{N} E\left[(x_j - \mu_j)(x_k - \mu_k)\right]$$

By definition, the correlation coefficient between the work package costs x_j and x_k is

$$correlation[x_j,x_k] = \rho_{j,k} = \frac{\text{cov}[x_j,x_k]}{\sigma_j \sigma_k}$$

$$\therefore \text{cov}[x_j,x_k] = E\left[(x_j - \mu_j)(x_k - \mu_k)\right] = \rho_{j,k}\sigma_j\sigma_k$$

$$\therefore \text{cov}[T,x_k] = \sum_{j=1}^{N} \text{cov}[x_j,x_k] = \sum_{j=1}^{N} \rho_{j,k}\sigma_j\sigma_k$$

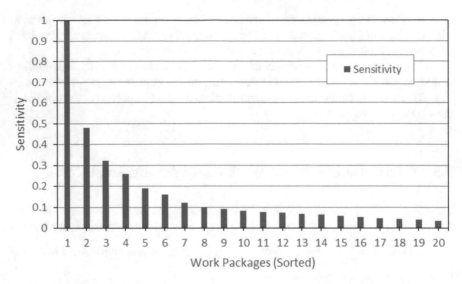

Fig. 7.1 Pareto sensitivity chart

Similarly,

$$\rho_{T,k} = \frac{\text{cov}[T,x_k]}{\sigma_T \sigma_k}$$

$$\sigma_T^2 = \text{var}[T] = \sum_{i=1}^{N}\sum_{j=1}^{N}\text{cov}[x_i,x_j] = \sum_{i=1}^{N}\sum_{j=1}^{N}\rho_{i,j}\sigma_i\sigma_j$$

Then, the correlation between the total cost T and the cost of any work package, x_k is

$$\rho_{T,k} = \frac{\displaystyle\sum_{j=1}^{N}\rho_{j,k}\sigma_j\sigma_k}{\sigma_k\sqrt{\text{var}[T]}} = \frac{\displaystyle\sum_{j=1}^{N}\rho_{j,k}\sigma_j}{\sqrt{\text{var}[T]}} = \frac{\displaystyle\sum_{j=1}^{N}\rho_{j,k}\sigma_j}{\sqrt{\displaystyle\sum_{i=1}^{N}\sum_{j=1}^{N}\rho_{i,j}\sigma_i\sigma_j}}$$

The work package with the largest correlation with the total cost is work package k, where k is the value that maximizes the numerator in the above expression. That is, to determine the work package with the greatest correlation with the total project cost, find the value of k for which $\displaystyle\sum_{j=1}^{N}\rho_{j,k}\sigma_j = \text{max}$.

This is the same expression obtained above by the use of derivatives.

If it is known for a fact that all the work package costs are statistically independent, then

$$\rho_{j,k} = 0 \text{ for } \forall j \neq k$$

$$\rho_{k,k} = 1$$

Then, in the case of complete statistical independence,

$$\text{cov}[T, x_k] = \sum_{j=1}^{N} \text{cov}[x_j, x_k] = E\left[(x_k - \mu_k)^2\right] = \sigma_k^2$$

$$\text{var}[T] = \sum_{j=1}^{N} \sigma_j^2$$

$$\therefore \rho_{T,k} = \frac{\sigma_k^2}{\sigma_k \sqrt{\text{var}[T]}} = \frac{\sigma_k}{\sqrt{\sum_{j=1}^{N} \sigma_j^2}}$$

In this case, the work package with the highest correlation with the total cost is the work package with the largest standard deviation (i.e., with the largest variance). Again, this is the same result as obtained before.

7.4   Finding the Correlation Due to Common Cause

Many people automatically assume that random variates such as work package costs or activity durations must be independent if there is no obvious direct causal relationship between them. However, this is an error. Although a causal relationship definitely means a dependency (although not necessarily a nonzero linear correlation coefficient), the converse is not true: correlation does not require or imply cause and effect. Variates can be and often are correlated because they all depend on some other variate, sometimes called the *common cause* (Wright 1921). This section is concerned with finding the correlation between work package variates that are affected by some common cause.

7.4.1   No Common Cause

First, for comparison, consider the case in which there is no common cause; the common factor will be added in the next section. Let x_j and x_k be two work package costs, for work package j and k, respectively. (Actually, they could just as well be activity durations, but they will be called costs here for simplicity). Now write x_j as the summation of a constant term plus a variable term:

$$x_j = m_j + u_j$$

in which m_j is a constant (that is, having no uncertainty) representing the point estimate for work package j; and u_j is a random variate representing the uncertainty in
the cost for work package j.

Some simplifying assumptions are made in order to make the model tractable.
The uncertainty term u_j is assumed to be additive (rather than, say, multiplicative),
and further, it is assumed to have zero mean. In other words, if the uncertainty term
had any nonzero mean value, that value would be incorporated into the constant m_j,
so that u_j would be unbiased (have zero mean).

The uncertainty term, being a random variate, also has an associated variance,
$\mathrm{var}\left[u_j\right] = \sigma_{u_j}^2$. In this derivation it is not assumed that the probability distribution
of the uncertainty u_j is necessarily Normal. Then define the following terms for the
mean, the variance, and the covariance of the uncertainty u_j:

$$E\left[u_j\right] = \bar{u}_j = 0$$

$$E\left[\left(u_j - \bar{u}_j\right)^2\right] = E\left[u_j^2\right] = \sigma_{u_j}^2$$

$$E\left[\left(u_j - \bar{u}_j\right)\left(u_k - \bar{u}_k\right)\right] = E\left[u_j u_k\right] = \sigma_{u_j u_k} = \rho_{u_j u_k} \sigma_{u_j} \sigma_{u_k}; \quad \forall k \neq j$$

Here, $\rho_{u_j u_k}$ is the linear correlation coefficient between the uncertainty in work
package j and the uncertainty in the cost of work package k.

Using the above definitions and identities, the mean value of the work package
cost x_j is determined by taking expectations of the terms in the expression given
earlier:

$$x_j = m_j + u_j$$

$$\bar{x}_j = E\left[x_j\right] = E\left[m_j + u_j\right] = E\left[m_j\right] + E\left[u_j\right] = m_j + 0 = m_j$$

This result shows that the point estimate m_j of the cost of work package j cannot
be any arbitrary number, it must be the mean or expected value of the cost x_j, so that
$m_j = E\left[x_j\right] = \bar{x}_j$.

The variance of the cost of work package j is determined as follows:

$$\sigma_{x_j}^2 = E\left[\left(x_j - \bar{x}_j\right)^2\right] = E\left[\left(m_j + u_j - m_j\right)^2\right]$$
$$= E\left[u_j^2\right] = \sigma_{u_j}^2$$

That is, all the variance or uncertainty in the work package cost x_j is contributed
by the estimation error term u_j.

The covariance between any two work packages, j and k, is given by:

$$E\left[\left(x_j - \bar{x}_j\right)\left(x_k - \bar{x}_k\right)\right] = E\left[\left(m_j + u_j - m_j\right)\left(m_k + u_k - m_k\right)\right]$$
$$= E\left[u_j u_k\right] = \rho_{u_j u_k} \sigma_{u_j} \sigma_{u_k}$$

This leads to the result:

$$\rho_{x_j x_k} = \rho_{u_j u_k}$$

That is, all the correlation between the costs for any two work packages j and k is contributed by the correlation between the estimation error terms, u_j and u_k.

7.4.2   With an Underlying Common Cause

Suppose that, in the model given just above, the estimation errors in two work package costs are uncorrelated, so that $\rho_{u_j u_k} = 0$ for $\forall j \neq k$. It will be shown here that work package costs x_j and x_k can nevertheless be correlated, if they both depend on some common factor.

Now, for all j, write x_j as an extended form of the model in the previous section:

$$x_j = m_j + u_j + b_j z$$

Here, x_j, u_j, and m_j are the same as before, and u_j is normalized so that:

$$E\left[u_j\right] = 0$$

and z is a random variate representing some external cause or risk factor; and b_j is a coefficient, constant for each work package j.

That is, the work package cost is represented as a constant (the point cost estimate m_j), an uncertainty term (the random variate u_j), an external random process (z), and a multiplier, b_j.

For example, the external process z might represent the weather. The cause must be quantified, so let z be the number of work days lost to rain during the period of activity of work package j. Then the coefficient b_j represents the additional cost for work package j for each work day lost to rain, in dollars per day.

As before, some assumptions are made in order to make the model tractable. As stated above, the uncertainty term u_j is assumed to be additive and to have zero mean. (In other words, any mean value of the uncertainty is incorporated into the constant m_j, so that u_j is unbiased). The uncertainty term also has an associated variance, $\operatorname{var}\left[u_j\right] = \sigma_{u_j}^2$. In this derivation it is not assumed that the probability distribution of the uncertainty u_j is necessarily Normal. The random variates u_j representing the estimation error in different work packages may be independent or correlated with other estimation error terms.

However, the uncertainty estimation error terms are assumed to be independent of (uncorrelated with) the external causal factor, z. Then:

$$E[z] = \overline{z}$$
$$E\left[(z-\overline{z})^2\right] = \sigma_z^2$$
$$E\left[(u_j - \overline{u}_j)(z-\overline{z})\right] = E\left[u_j(z-\overline{z})\right] = 0; \quad \forall j$$

That is, if z represents the number of rain days and u_j represents the uncertainty in the cost estimate made without regard to weather considerations (for example, if the cost estimate is based on the implicit assumption that the weather will be perfect), then u_j and z are independent.

Now, the mean value of the cost for work package j can be found using the expression $x_j = m_j + u_j + b_j z$:

$$E\left[x_j\right] = E\left[m_j + u_j + b_j z\right] = E\left[m_j\right] + E\left[u_j\right] + E\left[b_j z\right]$$
$$\overline{x}_j = m_j + 0 + b_j\overline{z} = m_j + b_j\overline{z}, \quad \forall j$$

Suppose now that, for some work package j, there is no effect of the external cause, z. Then the equation above becomes:

$$\overline{x}_j = m_j \text{ if } b_j = 0 \text{ or if } \overline{z} = 0$$

This shows again that m_j must be the mean or expected value of the work package cost when the estimation error is unbiased ($E[u_j] = 0$) and there is no effect of any external cause z.

Having found the expected value of the work package cost, the variance of that cost is obtained from the defining equation using the relations derived above. That is,

$$\text{var}\left[x_j\right] = E\left[(x_j - \overline{x}_j)^2\right] = E\left[(m_j + u_j + b_j z - m_j - b_j\overline{z})^2\right]$$
$$= E\left[(u_j + b_j\{z-\overline{z}\})^2\right]$$

Expanding the squared term and taking expectations gives:

$$\text{var}\left[x_j\right] = E\left[u_j^2 + 2b_j u_j(z-\overline{z}) + b_j^2(z-\overline{z})^2\right]$$
$$= E\left[u_j^2\right] + 2b_j E\left[u_j(z-\overline{z})\right] + b_j^2\left[(z-\overline{z})^2\right]$$
$$= \sigma_{u_j}^2 + 0 + b_j^2\sigma_z^2 = \sigma_{u_j}^2 + b_j^2\sigma_z^2$$

Note that the term $E\left[u_j\left(z-\bar{z}\right)\right]$ is zero in the last expression by use of the assumption that the variability in the external common cause is independent of the estimation errors in the individual work packages. Using this result gives an expression for the variance (and standard deviation) of the cost for work package j:

$$\sigma_{x_j}^2 = \sigma_{u_j}^2 + b_j^2 \sigma_z^2 \quad for \ \forall j$$

$$\sigma_{x_j} = \sqrt{\sigma_{u_j}^2 + b_j^2 \sigma_z^2}$$

That is, the variance in the work package cost is the estimation uncertainty increased by the product of the variance of the common cause and the square of the influence coefficient.

Having derived the expression for the variance of each work package cost, the next step is to determine the covariance between any pair of work packages. Proceeding as before from the defining equation and the results already obtained, some straightforward algebraic manipulation gives:

$$
\begin{aligned}
\operatorname{cov}\left[x_j, x_k\right] &= E\left[\left(x_j - \bar{x}_j\right)\left(x_k - \bar{x}_k\right)\right] = E\left[\left(u_j + b_j\{z-\bar{z}\}\right)\left(u_k + b_k\{z-\bar{z}\}\right)\right] \\
&= E\left[u_j u_k + b_j u_k\left(z-\bar{z}\right) + b_k u_j\left(z-\bar{z}\right) + b_j b_k\left(z-\bar{z}\right)^2\right] \\
&= E\left[u_j u_k\right] + b_j E\left[u_k\left(z-\bar{z}\right)\right] + b_k E\left[u_j\left(z-\bar{z}\right)\right] + b_j b_k E\left[\left(z-\bar{z}\right)^2\right] \\
&= E\left[u_j u_k\right] + 0 + 0 + b_j b_k E\left[\left(z-\bar{z}\right)^2\right] \\
\operatorname{cov}\left[x_j, x_k\right] &= \operatorname{cov}\left[u_j u_k\right] + b_j b_k E\left[\left(z-\bar{z}\right)^2\right] = \operatorname{cov}\left[u_j u_k\right] + b_j b_k \sigma_z^2
\end{aligned}
$$

Two terms in the expansion are zero by the assumption that the uncertainties in the cost estimates for different work packages are independent of the external cause. Using the linear correlation coefficient, the correlation between the costs of any two distinct work packages is:

$$\operatorname{cov}\left[x_j, x_k\right] = \operatorname{cov}\left[u_j u_k\right] + b_j b_k \sigma_z^2 = \rho_{u_j u_k}\sigma_{u_j}\sigma_{u_k} + b_j b_k \sigma_z^2 = \rho_{x_j x_k}\sigma_{x_j}\sigma_{x_k}$$

$$\rho_{x_j x_k} = \frac{\rho_{u_j u_k}\sigma_{u_j}\sigma_{u_k} + b_j b_k \sigma_z^2}{\sigma_{x_j}\sigma_{x_k}}$$

where, $\sigma_{x_j} = \sqrt{\sigma_{u_j}^2 + b_j^2 \sigma_z^2}$

Note that, if one sets $j = k$ to find the main diagonal elements, then:

$$\rho_{x_k x_k} = \frac{\rho_{u_k u_k}\sigma_{u_k}\sigma_{u_k} + b_k b_k \sigma_z^2}{\sigma_{x_k}\sigma_{x_k}} = \frac{1\sigma_{u_k}^2 + b_k^2 \sigma_z^2}{\sigma_{u_k}^2 + b_k^2 \sigma_z^2} = 1$$

as they should be.

The derivation shows that, if x_j and x_k are influenced by a common factor z, this hidden variable creates a correlation between x_j and x_k, even if the estimation errors are uncorrelated ($E[u_ju_k] = 0$). For example, assume that $E[u_ju_k] = 0$; the expression for the correlation then becomes (for $j = k$ distinct) :

$$\rho_{x_j, x_k} = \frac{b_j b_k \sigma_z^2}{\sigma_{x_j} \sigma_{x_k}}; \quad \forall j \neq k, \quad \text{where } \sigma_{x_j} = \sqrt{\sigma_{u_j}^2 + b_j^2 \sigma_z^2}$$

If the common cause has a large variance, so that

$$b_j \sigma_z \gg \sigma_{u_j} \text{ and } b_k \sigma_z \gg \sigma_{u_k}$$

then $\rho_{x_j, x_k} \rightarrow 1$ if b_j and b_k have the same signs, and $\rho_{x_j, x_k} \rightarrow -1$ if b_j and b_k have opposite signs.

This gives the perhaps unexpected result that, the larger the *variability* in the underlying common cause, the greater the *correlation* between the work package costs. As the variance σ_z^2 of the common cause increases, the probable values for z get larger, increasing the impact terms b_jz, and these dominate the expression for x_j i.e. $x_j = m_j + u_j + b_jz$, and as a result $x_j \rightarrow b_jz$ and $x_k \rightarrow b_kz$, which are perfectly correlated.

In summary, if the work package costs (or activity durations) x_j and x_k are to be considered statistically independent, then there must be no underlying common factor, either overt or latent, which affects both work packages. Or, if there is some common factor, it must have very small variance for the work package costs to be approximately uncorrelated. Whether or not this is true must be established by risk analysis for each project's conditions. An assumption that work packages are independent when in fact they are not, due to some common cause, may lead to a substantial underestimation of the project risk.

It is worth noting here that specification of all the work package correlations individually requires the estimation of $\frac{N(N-1)}{2}$ correlation coefficients (allowing for symmetry of the correlation matrix). If all the work packages are dependent to some degree on a single underlying factor, z, then the entire covariance matrix is determined by the N values for b_j plus one variance σ_z^2 for the underlying cause. Thus the number of parameters to be estimated falls from $\frac{N(N-1)}{2}$ to $N + 1$, which is about the same as the number of means and the number of variances to be estimated.

Table 7.1 Example 1 work package cost estimates

Fifth percentile	Fiftieth	Ninety-fifth	Mean	Standard deviation
$55,125	$60,000	$64,875	$60,000	$3000

Example 7.1

Assume that a simple project has $N = 10$ work packages, and each work package j has an unknown cost, x_j. The estimated 5th, 50th, and 95th percentiles are as follows (for simplicity here, it is assumed that all N work packages are identically distributed) (see Table 7.1).

The mean value given in the table is computed from the Pearson-Tukey formula (Pearson and Tukey 1965)

$$\mu = 0.63x_{50} + 0.185\left(x_5 + x_{95}\right)$$

and the standard deviation is computed by the extended Pearson-Tukey formula.

$$\sigma^2 = \left[\frac{x_{95} - x_5}{3.25}\right]^2$$

This particular three-point estimate gives a symmetrical distribution, but in a more general case the distribution might be skewed to the right to reflect the probability of costs in the upper tail.

Assuming that all the N work packages are independent and identically distributed; the expected value of the total cost is easily computed. The total cost is the random variate T, where

$$T = \sum_{j=1}^{N} x_j$$

Therefore the expected total job cost is:

$$E[T] = \bar{T} = \sum_{j=1}^{N} \bar{x}_j = 10(60,000) = \$600,000$$

Suppose that a contractor wishes to determine the cost for this job in preparation for making a bid. The contractor's problem is to determine the value that he/she should use as the cost (including contingency but not including overhead and profit) such that the probability of exceeding this cost is less than or equal to some number, say 0.10. Clearly the value that fits this requirement is not the median (which in the symmetric case is the same as the mean), because the probability that the cost would exceed the median is 0.50. The value needed must be the median (or mean) plus some contingency.

First, the contractor can estimate the variance of the total cost using the second moment equation:

$$\sigma_T^2 = \text{var}[T] = \sum_{j=1}^{N}\sum_{k=1}^{N} \text{cov}\left[x_j, x_k\right] = \sum_{j=1}^{N}\sum_{k=1}^{N} \rho_{x_j x_k} \sigma_{x_j} \sigma_{x_k}$$

As the work packages are considered independent, this reduces to:

$$\sigma_T^2 = \sum_{j=1}^{N} \sigma_{x_j}^2 = N(3000)^2 = 90,000,000$$

$$\sigma_T = \sqrt{90,000,000} = \$9487$$

Using the Central Limit Theorem, the probability distribution of the total cost should be approximately Normal, with mean \$600,000 and standard deviation \$9487. (Note that this is the only time in this discussion that the assumption of Normality is made). The 90th percentile of the Normal distribution is 1.282 standard deviations above the mean, or:

$$90\text{th } \% = 600,000 + 1.282(9487) = \$612,162$$

Thus, if the contractor wants to be 90% confident that his cost estimate will not be exceeded, he should add a contingency of $\$612, 162 - \$600, 000 = \$12, 162$ to the estimated mean value. This amounts to a contingency of $\dfrac{12,162}{600,000} = 2\%$ of the estimated mean, which is a very small amount. The coefficient of variation of each individual work package is $\dfrac{3,000}{600,000} = 5\%$, whereas the coefficient of variation of the total job is $\dfrac{9,487}{600,000} = 1.6\%$. Therefore, as measured by the coefficient of variation, the total job is less variable than any of the individual work packages. This is a consequence of the work packages being independent (or assumed independent). However, an experienced project manager may find it difficult to accept the proposition that increasing the number of work packages (N) makes the total project less uncertain. And so, in this exercise, the contractor may find it difficult to accept that a contingency of only 2% of the mean cost is adequate to provide a 90% confidence that the bid will be high enough to cover costs.

Example 7.2
Now suppose that, for the project in the previous case, it is recognized that all N work packages may be affected by the weather. Let z represent the total number of days of production lost to inclement weather on this job. Of course, z is, a priori, unknown. The contractor estimates that on the average he expects 12 days to be lost on the job, with a standard error of 1 day. (The estimates may be obtained by the three-point method as used above; here for brevity it is simply assumed that this has been done and the mean and standard deviation have been obtained). Using the notation developed above,

$$b_j = \$1000 \text{ per day lost due to weather}$$

$$E[z] = 12 \text{ days}$$

$$\sigma_z = 1 \text{ day}$$

$$m_j = \$48,000$$

Here the value of the point estimate for each work package, m_j, has been adjusted for comparison with the previous exercise, by factoring out the expected cost of the 12 days lost to bad weather. By this adjustment, the total cost in this exercise will be the same as in Exercise 1.

Again, all work packages are assumed to be identically distributed, for simplicity. Then the expected value of each work package cost is:

$$\bar{x}_j = m_j + b_j \bar{z} = \$48,000 + \$1000(12) = \$60,000$$

The expected total job cost is then, as in Exercise 1:

$$E[T] = \bar{T} = \sum_{j=1}^{N} \bar{x}_j = 10(60,000) = \$600,000$$

The standard deviation of each work package cost is:

$$\sigma_{x_j} = \sqrt{\sigma_{u_j}^2 + b_j^2 \sigma_z^2} = \sqrt{(3000)^2 + (1000)^2 (1)^2} = \sqrt{10,000,000} = \$3162$$

The covariance between any two work packages j and k, $j \neq k$ is:

$$\text{cov}\left[x_j, x_k\right] = \text{cov}\left[u_j u_k\right] + b_j b_k \sigma_z^2 = 0 + b_j b_k \sigma_z^2 = 1,000,000$$

The variances on the main diagonal of the covariance matrix are:

$$\sigma_{x_j}^2 = 3162^2 = 10,000,000$$

The sum of the terms in the covariance matrix may easily be determined in any case by a spreadsheet program. In this particular case, it is also easily done by pocket calculator. There are $N = 10$ main diagonal elements with variance 10,000,000 and 90 off-diagonal elements with covariance 1,000,000, which add up to a variance of the total cost of:

$$\sigma_T^2 = 90(1,000,000) + 10(10,000,000) = 190,000,000$$
$$\sigma_T = \$13,784$$

Addressing the same question as in the earlier exercise: if the contractor wishes to be 90% confident that the number will not be exceeded, the contractor should use as his estimate of the total cost,

$$90\text{th } \% = \$600,000 + 1.282(\$13,784) = \$617,671$$

In this case, the contingency of $1.282 \times 13,784 = \$17,671$ is added to the expected value to reach the 90% confidence value. This reflects a contingency of 2.9% of the expected cost, more than in the independent case, but still a very small contingency.

The correlation coefficients between work package j and work package k were, in the independent case, zero for all $j \neq k$. In this case, the common cause, weather, implies correlation coefficients for all $j \neq k$ given by:

$$\rho_{x_j x_k} = \frac{\rho_{u_j u_k} \sigma_{u_j} \sigma_{u_k} + b_j b_k \sigma_z^2}{\sigma_{x_j} \sigma_{x_k}} = \frac{b_j b_k \sigma_z^2}{\sigma_{x_j} \sigma_{x_k}} = \frac{(1000)^2}{(3162)^2} = 0.10$$

This slight amount of correlation has increased the coefficient of variation for each work package to $3162/600,000 = 5.3\%$, compared to 5% for the independent case. The coefficient of variation of the total job cost has increased to $13,784/600,000 = 2.3\%$ compared to 1.6% for the independent case. By all these metrics, the uncertainty in the weather has had very little effect on the dependence of the work packages and the uncertainty in the total job cost.

Example 7.3

Suppose, for the project in the previous case, that the contractor now recognizes that his meteorological forecasts are not very accurate. The contractor makes a three-point estimate of the probability distribution on the number of days lost to weather, as in the Table 7.2.

(Here, as in previous exercises, the values have been chosen to make the arithmetic easier. This probability distribution is highly skewed to the right, as one might expect).

If everything else remains the same as in the previous case, on the average the contractor expects 12 days to be lost on the job, except now the contractor believes that his standard error is 10 days.

Again, all work packages are assumed to be identically distributed, for simplicity. Then the expected value of each work package cost is the same as before (adding in the mean weather effect):

$$\bar{x}_j = m_j + b_j \bar{z} = 48,000 + 1000(12) = \$60,000$$

Table 7.2 Example 3 work package cost estimates

Fifth percentile	Fiftieth	Ninety-fifth	Mean	Standard deviation
2.5	8	35	12	10

The standard deviation of each work package cost is now:

$$\sigma_{x_j} = \sqrt{\sigma_{u_j}^2 + b_j^2 \sigma_z^2} = \sqrt{(3000)^2 + (1000)^2 (10)^2} = \sqrt{109,000,000} = \$10,440$$

The covariance between any two work packages j and k, $j \neq k$ is:

$$\text{cov}\left[x_j, x_k\right] = \text{cov}\left[u_j u_k\right] + b_j b_k \sigma_z^2 = 0 + b_j b_k \sigma_z^2 = 1000(1000)10^2 = 100,000,000 = 10^8$$

The variances on the main diagonal of the covariance matrix are:

$$\sigma_{x_j}^2 = 10,440^2 = 109,000,000 = 109\left(10^6\right)$$

There are $N = 10$ main diagonal elements with variance $109(10)^6$ and 90 off-diagonal elements with covariance 10^8, which add up to a variance for the total job cost of:

$$\sigma_T^2 = 90\left(10^8\right) + 10(109)\left(10^6\right) = 9000\left(10^6\right) + 1090\left(10^7\right) = 10,090\left(10^6\right)$$
$$\sigma_T = \$100,449$$

Addressing the same question as in the earlier exercises, the contractor, if he wishes to be 90% confident that the estimated cost will not be exceeded, should use as his estimate of the total cost:

$$90\text{th } \% = \$600,000 + 1.282\left(\$100,449\right) = \$728,776$$

In this case, the contingency of $1.282(100,449) = \$128,776$ is added to the expected value to reach the 90% confidence value. The contingency is now 21% of the estimated average cost.

In this example, the common cause, weather, implies correlation coefficients:

$$\rho_{x_j x_k} = \frac{\rho_{u_j u_k} \sigma_{u_j} \sigma_{u_k} + b_j b_k \sigma_z^2}{\sigma_{x_j} \sigma_{x_k}} = \frac{b_j b_k \sigma_z^2}{\sigma_{x_j} \sigma_{x_k}} = \frac{(1000)^2 (10)^2}{(10,440)^2} = 0.92$$

The correlation coefficients have increased due to the greater uncertainty in the weather forecasts, and in consequence the 90% confidence cost value has increased from \$612,162 to \$728,776. In another way of looking at this result, if the contractor had assumed independence, and estimated the total cost including contingency at \$612,162, when the work packages are actually highly correlated, the probability of the cost overrunning this amount would be not 10% but 45%. The contractor, by assuming independence, would have seriously underestimated the risk of the project.

Example 7.4

Figure 7.2 shows how the correlation coefficients $\rho_{x_j x_k}$, for $j \neq k$, vary with the standard error of the weather forecast, σ_z, with all other parameters the same as in the previous exercise. The correlation is, of course, asymptotic to 1.0. Figure 7.2 is a plot of the required contingency, as a percentage of the expected value of the total project cost, needed for the contractor to have 90% confidence that the estimate will not be overrun. (This is also known as the Value-at-Risk).

Figure 7.3 shows how the required contingency increases approximately linearly with the standard deviation of the error in the weather forecast. The figure below combines the two figures just above to show how the required contingency (as a percentage of the expected job cost) varies with the correlation coefficient. As can be easily seen, the required contingency increases rapidly if the correlation coefficient exceeds 0.80.

Example 7.5

In the previous exercises, for simplicity, all work packages were assumed to be equally affected by the weather. This may be the case, for certain types of projects in which all the work is outdoors, but in many projects the individual work packages are differentially affected by some common cause.

Consider a simplified version of Exercise 3, in which $N = 6$ for simplicity. Assume that everything is the same as in Exercise 3, except that only work packages 1, 2, and 3 are affected by the weather, and work packages 4, 5, and 6 are completely unaffected by the weather. Then

$$b_j = \$1000 \text{ per day} \quad \text{for } j = 1, 2, 3$$

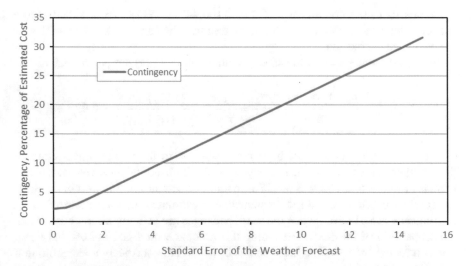

Fig. 7.2 Contingency vs. weather forecasting error

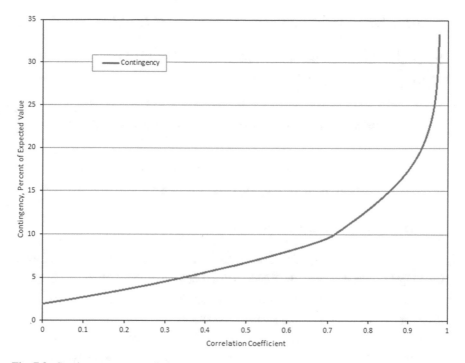

Fig. 7.3 Contingency vs. correlation

$$b_j = 0 \quad \text{for } j = 4,5,6$$

Then the expected value of each work package cost depends on the influence of the weather factor:

$$\bar{x}_j = m_j + b_j \bar{z} = 48,000 + 1000(12) = \$60,000 \quad \text{for } j = 1,2,3$$

$$\bar{x}_j = m_j + b_j \bar{z} = 48,000 + 0 = \$48,000 \quad \text{for } j = 4,5,6$$

The standard deviation of each work package cost is now:

$$\sigma_{x_j} = \sqrt{\sigma_{u_j}^2 + b_j^2 \sigma_z^2} = \sqrt{(3000)^2 + (1000)^2 (10)^2} = \sqrt{109,000,000}$$
$$= \$10,440 \quad \text{for } j = 1,2,3$$

$$\sigma_{x_j} = \sqrt{\sigma_{u_j}^2 + b_j^2 \sigma_z^2} = \sqrt{(3000)^2 + 0(10)^2} = \$3,000 \quad \text{for } j = 4,5,6$$

The covariance between any two work packages j and k, $j \neq k$ is:

$$\text{cov}\left[x_j, x_k\right] = \text{cov}\left[u_j u_k\right] + b_j b_k \sigma_z^2 = 0 + b_j b_k \sigma_z^2 = 1000(1000)10^2 = 10^8 \quad \text{if } j \leq 3 \text{ and } k \leq 3$$
$$\text{cov}\left[x_j, x_k\right] = \text{cov}\left[u_j u_k\right] + b_j b_k \sigma_z^2 = 0 \quad \text{if } j \geq 4 \text{ or } k \geq 4$$

The variances on the main diagonal of the covariance matrix are:

$$\sigma_{x_j}^2 = 10,440^2 = 109,000,000 \quad \text{for } j = 1,2,3$$
$$\sigma_{x_j}^2 = 3000^2 = 9,000,000 \quad \text{for } j = 4,5,6$$

The total 6 by 6 covariance matrix is, then:

	Landscaping	Pave Lot	Ext.Paint	Wiring	Conduit	Int.Paint
Landscaping	109000000	100000000	100000000	0	0	0
Pave Lot	100000000	109000000	100000000	0	0	0
Ext.Paint	100000000	100000000	109000000	0	0	0
Wiring	0	0	0	9000000	0	0
Conduit	0	0	0	0	9000000	0
Int.Paint	0	0	0	0	0	9000000

The sum of all the terms in the covariance matrix is the variance of the total cost, or:

$$\sigma_T^2 = 954000000$$
$$\sigma_T = \$30,887$$

The correlation matrix is then:

	Landscaping	Pave Lot	Ext.Paint	Wiring	Conduit	Int.Paint
Landscaping	1.00	0.92	0.92	0	0	0
Pave Lot	0.92	1.00	0.92	0	0	0
Ext.Paint	0.92	0.92	1.00	0	0	0
Wiring	0	0	0	1.00	0	0
Conduit	0	0	0	0	1.00	0
Int.Paint	0	0	0	0	0	1.00

The correlation matrix shows, for example, that the costs of Exterior Painting and Interior Painting are independent, although both are about painting. (Independent, at least, with regard to the weather; there might be other common causes, such as a shortage of painters, that would affect both Interior and Exterior Painting). On the other hand, Parking Lot Paving and Exterior Painting are highly correlated, although paving and painting have nothing to do with each other, other than the fact that both are weather-sensitive. If a bidder on this job were to assume that all the work packages are independent, he would seriously underestimate his risk of cost overrun on this project.

7.5 Practice Problems

Problem 7.5.1 Consider Practice Problem 3.5.1 from Chap. 3 (i.e. a small project comprised of six sequential activities). Calculate the covariance between the cost of each the six activities and the total project cost.

Problem 7.5.2 Weather is only one of many possible common causes that might affect the costs and durations of multiple activities on a project. A few other possible causes are listed below. Based on your experience, list another five or six possible common causes.

1. Weather
2. Power failure
3. Escalation in prices of steel or other basic commodities
4. Changes in government regulations or regulators
5. Shortages of construction craftsmen in the area

Problem 7.5.3 Suppose that w is another underlying common factor similar to z, with the following conditions:

$$E[w] = \bar{w}$$
$$E\left[(w - \bar{w})^2\right] = \sigma_w^2$$
$$E\left[(w - \bar{w})(u_k - \bar{u}_k)\right] = 0 \quad for \forall k$$
$$E\left[(w - \bar{w})(z - \bar{z})\right] = \rho_{wz}\sigma_w\sigma_z$$

That is, the underlying variable w is uncorrelated with the work package variables but may or may not be correlated with the other common factor z. Then, let:

$$x_j = m_j + u_j + b_j z + c_j w$$
$$x_k = m_k + u_k + b_k z + c_k w$$

By a derivation similar to that above, show the following, for all values of j and k:

$$\bar{x}_k = m_k + b_k \bar{z} + c_k \bar{w}$$

$$\sigma_{x_k}^2 = \sigma_{u_k}^2 + b_k^2 \sigma_z^2 + c_k^2 \sigma_w^2 + 2b_k c_k \rho_{wz} \sigma_w \sigma_z$$

$$\rho_{x_j,x_k} = \frac{\rho_{u_j u_k} \sigma_{u_j} \sigma_{u_k} + b_j b_k \sigma_z^2 + c_j c_k \sigma_w^2 + \left(b_j c_k + b_k c_j\right) \rho_{wz} \sigma_w \sigma_z}{\sigma_{x_j} \sigma_{x_k}}$$

Thus, with two common factors, the entire $N \times N$ correlation matrix for the work package costs can be generated from $2N + 3$ terms: N terms for the b_k coefficients, N terms for the c_k coefficients, two variances for the underlying factors, and one (possible) correlation between the two common factors.

References

Pearson ES, Tukey JW (1965) Approximate means and standard deviations based on distances between percentage points of frequency curves. Biometrika 52(3–4):533–546
Wilkinson L (2006) Revising the Pareto chart. Am Stat 60(4):332–334
Wright S (1921) Correlation and causation. J Agric Res 20(7):557–585

Chapter 8
Approximate Risk Function for Project Durations

Abstract In this chapter we discuss the application of the second moment method in scheduling networks. The issue we focus here is on finding the probability distribution for the project completion time when there are multiple network paths in the project and therefore the critical path is itself uncertain. We present an approximate method to this problem and discuss its validity in a larger managerial context.

Keywords Scheduling networks · Longest path · Approximate methods · Probability distribution

8.1 Introduction

The estimation of the probability distribution, and hence the risk of overrunning the budget, for project costs is relatively straightforward, given the usual cost model, in which the total project cost is the sum of the work package costs. In this case, as we have seen, the probability density function for the total project cost is approximately Normal, regardless of the probability distribution of the work package costs, whether or not the work package costs are correlated. Moreover, by the properties of the moments of sums of random variables, the mean of the distribution on total project costs is the sum of the work package means, and the variance of the total project cost distribution is the sum of the terms in the work package covariance matrix.

Networks, however, introduce another issue. The expected length of each path through the network is the sum of the expected durations of all the activities on that path, but the critical path (that is, the longest path) may change depending on the random values of the individual activity durations. The issue is to find the (approximate) probability distribution for the project completion time when there are multiple paths through the project network, and which path may be critical (that is, controls the project duration) is itself uncertain. This chapter explores this topic further, using an approximate approach.

© Springer Nature Switzerland AG 2020

I. Damnjanovic, K. Reinschmidt, *Data Analytics for Engineering and Construction Project Risk Management*, Risk, Systems and Decisions, https://doi.org/10.1007/978-3-030-14251-3_8

8.2   Project Paths Model

Suppose that the project network consists of a total of N activities. Let x_j be a random variate representing the length of activity j, and let the mean and standard deviation of this variate be given by (MacCrimmon and Ryavec 1964):

$$E\left[x_j\right] \equiv \bar{x}_j$$
$$E\left[\left(x_j - \bar{x}_j\right)^2\right] \equiv \sigma_j^{\,2}$$

Suppose that the correlation between the duration of activity j and the duration of activity k is ρ_{jk}, and so the $N \times N$ covariance matrix $\mathbf{V}$ for all project activities is:

$$\mathbf{V} = \begin{pmatrix} \sigma_1^{\,2} & \rho_{12}\sigma_1\sigma_2 & \cdots & \rho_{1k}\sigma_1\sigma_k & \cdots & \rho_{1N}\sigma_1\sigma_N \\ \rho_{12}\sigma_1\sigma_2 & \sigma_2^{\,2} & \cdots & \rho_{2k}\sigma_2\sigma_k & \cdots & \rho_{2N}\sigma_2\sigma_N \\ \rho_{1N}\sigma_1\sigma_N & \rho_{2N}\sigma_2\sigma_N & & \cdots & \rho_{Nk}\sigma_k\sigma_N & \cdots & \sigma_N^{\,2} \end{pmatrix}$$

Now assume that, *based on the mean durations* $\bar{x}_j$, the network algorithm computes the critical (longest) path, the next-to-longest, the third longest, etc., up to some reasonable number M of longest paths. Thus there are no remaining paths through the network longer than the shortest path in this set, based on the mean activity durations. Number the M longest paths such that path 1 is the longest, path 2 the second longest, path 3 the third longest, etc. Then, let A_1 be the set of all the activities in path 1, let A_2 be the set of all activities in path 2, A_k the set of activities in path k, etc. Let N_k be the number of activities in path k, etc. Of course, these sets are not mutually exclusive, because there may be many activities that appear in more than one path.

Then, define T_1 as a random variable representing the length of path 1. This is the sum of the (random) lengths of all N_1 activities in path 1, or,

$$T_1 = \sum_{i \in A_1} x_i$$

It is important that the network algorithm uses the mean activity durations, so that the mean length of path 1 as calculated by the algorithm is given by:

$$\bar{T}_1 = \sum_{i \in A_1} \bar{x}_i$$

Similarly, for paths 2, 3, …, k, …, and M.

The variance of the length of path 1 is obtained by the usual approach for finding the variance of a sum of random variables, as in the case of work package costs.

In this case a square $N_1 \times N_1$ path covariance matrix $\mathbf{V}_1$ is set up to contain only the N_1 activities on path 1. That is, we extract the variances and covariances for the N_1 activities on path 1 from the total network covariance matrix $\mathbf{V}$ defined above. Or, it may be easier when using spreadsheets to obtain $\mathbf{V}_1$ by deleting from the network covariance matrix $\mathbf{V}$ all the rows and columns corresponding to the activities that are *not* in path 1.

The variance of the length of path 1, var[T_1], is obtained by summing all the N_1^2 terms in the $\mathbf{V}_1$ covariance matrix. Of course, it is not actually necessary to construct a new matrix $\mathbf{V}_1$; one can compute the variance var[T_1] from the network variance matrix $\mathbf{V}$ simply by extracting the N_1^2 terms corresponding to the activities in path 1. Similar calculations for paths 2 though M produce the variances for all the separate paths through the network: var[T_2], var [T_3], ... , var [T_k], ... , var [T_M].

Of interest now is the correlation between any two paths in the set of M paths. Consider just paths 1 and 2, the longest and the second longest paths. The covariance of the lengths of path 1 and path 2 is given by:

$$\mathrm{cov}\left[T_1, T_2\right] = E\left[\left(T_1 - \bar{T}_1\right)\left(T_2 - \bar{T}_2\right)\right] = E\left[\left(\sum_{j \in A_1} x_j - \sum_{j \in A_1} \bar{x}_j\right)\left(\sum_{k \in A_2} x_k - \sum_{k \in A_2} \bar{x}_k\right)\right]$$

$$\mathrm{cov}\left[T_1, T_2\right] = E\left[\left\{\sum_{j \in A_1}\left(x_j - \bar{x}_j\right)\right\}\left\{\sum_{k \in A_2}\left(x_k - \bar{x}_k\right)\right\}\right]$$

The product of the terms in the summation signs will result in $N_1 \times N_2$ terms, $\left(x_j - \bar{x}_j\right)\left(x_k - \bar{x}_k\right)$ for $\forall j \in A_1$, $\forall k \in A_2$, corresponding to each of the N_1 activities in path 1 multiplied by each of the N_2 activities in path 2. There are two causes of correlation between path 1 and path 2:

If the same activity appears in both paths, then:

$$j = k, \text{ so that } E\left[\left(x_j - \bar{x}_j\right)\left(x_k - \bar{x}_k\right)\right] = E\left[\left(x_j - \bar{x}_j\right)^2\right] = \sigma_j^2$$

If there are two different activities on the two paths, then the lengths of these activities are correlated if there is a nonzero term in the correlation matrix [ρ_{jk}]:

$$j \neq k, \text{ so that } E\left[\left(x_j - \bar{x}_j\right)\left(x_k - \bar{x}_k\right)\right] = \rho_{jk}\sigma_j\sigma_k$$

Now, by definition,

$$\mathrm{cov}\left[T_1, T_2\right] = \rho_{T_1, T_2}\sqrt{\mathrm{var}\left[T_1\right]}\sqrt{\mathrm{var}\left[T_2\right]}$$

$$\rho_{T_1, T_2} = \frac{\mathrm{cov}\left[T_1, T_2\right]}{\sqrt{\mathrm{var}\left[T_1\right]}\sqrt{\mathrm{var}\left[T_2\right]}}$$

Correlations between path lengths will typically tend to be positive, because common activities on two paths will obviously contribute to positive correlation, unless the covariance between two particular activities is very large and negative. Negative correlation between activity durations means that there would be a tendency for a larger than expected duration for activity j on one path to be associated with a shorter than expected duration for activity $k \neq j$ on another path. This might happen, for instance, if the two activities occur at about the same time and have common resource requirements, and these resources are limited. Then, assigning more resources to activity k than expected would make that activity's duration shorter than expected, and would result in fewer resources than expected for activity j, which would make that activity take longer than expected, and the two activities could be negatively correlated.

As path 1 is, by the construction above, the critical path (based on the mean activity durations), the length of path 1 will at least heavily influence the total project completion time. Consequently, the correlations between T_1 and the remaining $M - 1$ paths are of most interest. Suppose the correlation between path 1 and path j is large; that is, ρ_{T_1, T_j} is high. High correlation means that there is a tendency for a greater than expected length on path j to be associated with a greater than expected length on path 1. But path 1 is the longest path by definition, and so $E[T_j] \leq E[T_1]$. Thus, if $E[T_j]$ is significantly less than $E[T_1]$, with high positive correlation between path j and path 1, it is unlikely that path j would ever be the critical path. For example, if path j were to take much longer than expected, then path 1 probably also takes much longer than expected, and therefore may still be the *critical* path. Hence, in these circumstances, path j can be ignored, as it is unlikely to contribute much uncertainty to the total project duration. The question remains, how large does the correlation have to be, to be considered to be large, and to justify dropping path j from further consideration? *As an approximation*, we might say that any correlation greater than some cutoff value could be considered highly correlated and therefore the correlated path could be neglected.

On the other hand, the correlation between path 1 and some path k could be small. This means that the durations on path 1 and on path k are independent, and path k could become the critical path in some circumstances, depending on the actual values of the random activity durations. Whether path k becomes the critical path also depends on the values for $E[T_1]$, $E[T_k]$, var $[T_1]$, and var$[T_k]$:

If $E[T_k] \ll E[T_1]$ and/or var$[T_k] \ll$ var $[T_1]$, then it is unlikely that path k will ever be critical. If this is true for all paths 2, 3, ..., M, then path 1 is a dominant path and will almost always be the critical path. Then the probability density function for the completion time of the project will be nearly the same as the probability density function for the length of path 1. As the length of path 1 is simply the sum of the durations of the activities on this path, the probability density function will be approximately Normal, using the same reasoning as we used when considering work package costs.

If $E[T_k]] \cong E[T_1]$ and var$[T_k] >$ var $[T_1]$, then there may be a reasonably high probability that path k is the critical path, even though, on the basis of the mean activity durations, path 1 is critical. If this is the case, then the probability density function for the project finish time can become highly skewed to the right.

The question remains, how small does the correlation have to be, to be considered to be approximately zero, and to justify treating the length of path k as independent of the length of path 1? *As an approximation*, we might say that any correlation less than some cutoff, say 0.3 or so, could be considered uncorrelated and therefore that the two paths in question are independent.

It is easy to see that if path 1 and path k are negatively correlated, then path k is also of concern, because then path k would tend to become longer when path 1 is shorter, increasing the likelihood that the critical path would shift to run through path k.

Suppose now we are interesting in assessing the risk that a project will overrun. Here we will look only at path 1 and path k, where these two paths are judged to be independent on the basis of the computation of ρ_{T_1, T_k}. Consider some finish time of interest, say t, computed from the project start at time 0. Then the probability of no overrun is the probability that neither path exceeds time t, which is just the probability that both paths are less than t. That is, we wish the probability:

$$\Pr\left[T_1 \leq t \cap T_k \leq t\right]$$

Then, the risk of an overrun (on either path 1 or path k) is just the probability that one or the other overruns, which is just 1.0 – the probability that neither overruns:

$$\Pr\left[overrun\right] = \Pr\left[T_1 > t \cup T_k > t\right] = 1 - \Pr\left[T_1 \leq t \cap T_k \leq t\right]$$

From elementary probability theory, the joint probability is given by the product of the conditional probability times the marginal probability:

$$\Pr\left[T_1 \leq t \cap T_k \leq t\right] = \Pr\left[T_1 \leq t | T_k \leq t\right]\Pr\left[T_k \leq t\right] = \Pr\left[T_k \leq t | T_1 \leq t\right]\Pr\left[T_1 \leq t\right]$$

In this expression, $\Pr[T_1 \leq t | T_k \leq t]$ is *the conditional probability* that the length of path 1 is less than t *given that* the length of path k is less than t. Similarly, $\Pr[T_k \leq t | T_1 \leq t]$ is the conditional probability that the length of path k is less than t given that the length of path 1 is less than t. Unfortunately, conditional probability distributions are difficult to determine. Therefore, we take recourse in the low correlation between path 1 and path k and make the *assumption* that low correlation implies that these two paths are approximately independent. With this assumption of independence, the above equation simplifies to:

$$\Pr\left[T_1 \leq t \cap T_k \leq t\right] \cong \Pr\left[T_1 \leq t\right]\Pr\left[T_k \leq t\right]$$

and the risk of an overrun is, approximately:

$$\Pr\left[overrun\right] = 1 - \Pr\left[T_1 \leq t \cap T_k \leq t\right] \cong 1 - \Pr\left[T_1 \leq t\right]\Pr\left[T_k \leq t\right]$$

Now, $\Pr[T_k \leq t]$ is just the cumulative probability distribution for T_k evaluated at the duration t, and similarly for $\Pr[T_1 \leq t]$. These values can easily be obtained if we assume some form of the cumulative probability distribution. As noted earlier, the path length T_1 is simply the sum of the N_1 random activity durations along path 1. We saw before that the total path length is also approximately Normal, whether the individual activities are independent or correlated. We have already computed the mean and the variance of path length T_1 above. The same arguments apply to T_k.

Let $\Phi(\tau)$ represent the value of the cumulative probability distribution for the unit Normal (the Normal with mean 0 and standard deviation 1) evaluated at point τ. Then we have, from the equation above,

$$\Pr \cong 1 - \Pr\left[T_1 \le t\right]\Pr\left[T_k \le t\right] = 1 - \Phi\left(\tau_1\right)\Phi\left(\tau_k\right)$$

$$\Pr \cong 1 - \Phi\left(\frac{\left[t - \bar{T}_1\right]}{\sqrt{\operatorname{var}\left[T_1\right]}}\right)\Phi\left(\frac{\left[t - \bar{T}_k\right]}{\sqrt{\operatorname{var}\left[T_k\right]}}\right)$$

in which the transformations

$$\tau_1 = \left(\frac{\left[t - \bar{T}_1\right]}{\sqrt{\operatorname{var}\left[T_1\right]}}\right); \tau_k = \left(\frac{\left[t - \bar{T}_1\right]}{\sqrt{\operatorname{var}\left[T_1\right]}}\right)$$

are made to convert to unit Normal variates.

It is then straightforward to compute these functions for many values of t and then to plot the cumulative distribution function of the probability of an overrun versus t. To make the tails more visible to the eye, it is often helpful to plot the probability density function, which is of course just the derivative of the cumulative distribution function. If this is done in the general case, it will be seen that there are certain combinations of $\bar{T}_1, \operatorname{var}\left[T_1\right], \bar{T}_k$ and $\operatorname{var}[T_k]$ that produce probability density functions that are nearly symmetrical and Normal, and there are other combinations of $\bar{T}_1, \operatorname{var}\left[T_1\right], \bar{T}_k$ and $\operatorname{var}[T_k]$ that produce probability density functions that are very highly skewed to the right. The latter condition occurs especially when $\sqrt{\operatorname{var}\left[T_k\right]}$ is large compared to $\bar{T}_1 - \bar{T}_k$.

This point can be seen more clearly by making the transformations

$$\tau_1 = \frac{\left[t - \bar{T}_1\right]}{\sqrt{\operatorname{var}\left[T_1\right]}}$$

$$t = \bar{T}_1 + \tau_1\sqrt{\operatorname{var}\left[T_1\right]}$$

$$\tau_k = \frac{\left[t - \bar{T}_1\right]}{\sqrt{\operatorname{var}\left[T_1\right]}} = \left[\frac{\bar{T}_1 + \tau_1\sqrt{\operatorname{var}\left[T_1\right]} - \bar{T}_k}{\sqrt{\operatorname{var}[T_k}}\right]$$

$$\tau_k = \frac{\left(\bar{T}_1 - \bar{T}_k\right)}{\sqrt{\operatorname{var}\left[T_k\right]}} + \tau_1\frac{\sqrt{\operatorname{var}\left[T_1\right]}}{\sqrt{\operatorname{var}\left[T_k\right]}}$$

Then, the cumulative probability of not overrunning is given by:

$$\Phi\{\tau\}\,\Phi\left\{\frac{\left(\overline{T}_1-\overline{T}_k\right)}{\sqrt{\mathrm{var}\left[T_k\right]}}+\tau\sqrt{\frac{\mathrm{var}\left[T_1\right]}{\mathrm{var}\left[T_k\right]}}\right\}$$

for all values $-\infty \leq \tau \leq \infty$.

This derivation can be extended to a greater number of paths, as long as they all can be considered to be independent of each other. That is, there might be three paths in the above expressions if all the possible correlations between path 1, path j, and path k were small. However, if path j and path k were not correlated with path 1, but were highly correlated with each other, there would be only two independent paths; either j or k would be dropped.

This chapter shows one way to get an approximate relation for the risk of a project schedule overrun without recourse to Monte Carlo simulation (McCabe 2003). The significant approximations are as follows:

All paths through the network that are highly correlated with the critical path length are neglected, on the basis that, if they were perfectly correlated, they would never become critical. However, some cutoff value of the correlation coefficient must be chosen, such that any computed correlation above this value will be considered to be equivalent to perfect dependence. This approximation may introduce error, as the neglected paths might have become critical under some rare combination of random variates.

All paths that have low correlations with the critical path are assumed to be independent of the critical path. However, some cutoff value of the correlation coefficient must be chosen, such that any computed correlation below this value will be considered to be equivalent to perfect independence. This approximation may introduce error, as the computation assumes that the joint probability is the product of the marginal probability distributions, which is true only if all the variables are independent.

If one defines a critical correlation ρ_{ind} such that any computed path correlation $\rho \leq \rho_{ind}$ is considered to be equivalent to $\rho = 0$, and a critical correlation ρ_{dep} such that any computed path correlation $\rho \geq \rho_{dep}$ is considered to be equivalent to $\rho = 1$, then there may be some computed values $\rho_{ind} < \rho < \rho_{dep}$ that fall into neither category. In the simple model given here, such situations cannot be handled. In cases in which this situation occurs, the value of ρ_{ind} should be adjusted upward and the critical value ρ_{dep} should be adjusted downward until the set of correlations lying between them is empty. For example, set $\rho_{ind} = \rho_{dep} = 0.5$.

Example 8.1

Consider a very simple project network shown in Table 8.1. The initial node is START and the completion is FINISH. The table below gives, for the activities 1 through 8, the precedences, the Most Likely duration (the mode), the Lower Bound duration (5-th percentile), and the Upper Bound duration (the 95-percentile). Durations are all in weeks.

Table 8.1 Project network for Example 8.1

Activity (j)	Predecessors	Successors	Lower bound (x_5)	Median (x_{50})	Upper bound (x_{95})	Mean (μ)	Standard deviation (σ)
1	START	2, 5	8	9	18	10	3
2	1	3	8	9	18	10	3
3	2	4	8	9	18	10	3
4	3, 5	FINISH	8	9	18	10	3
5	1	4	8	15	24	15	5
6	START	7	6	8	22	10	5
7	6	8	3	15	29	15	8
8	7	FINISH	6	8	22	10	5

The values for the means were estimated from the given estimates for the Lower Bound, Most Likely, and Upper Bound using the Pearson-Tukey formula:

$$\mu = 0.63x_{50} + 0.185\left(x_5 + x_{95}\right)$$

In this notation, x_k represents the estimated value of the random variable at the k-th percentile of the probability distribution. Therefore, x_{50} is the value of the random variable at the 50th percentile, hence the Median; x_5 is the estimated value of x at the 5th percentile, or Lower Bound; and x_{95} is the value of x at the 95th percentile, or Upper Bound. Keefer and Bodily (1983) found that the maximum percentage error using this formula in their experiments as less than 0.1%, and the average percentage error was about 0.02%.

The standard deviations in the table were estimated by the Extended Pearson-Tukey formula (1965):

$$\sigma = \frac{x_{95} - x_5}{3.25}$$

The correlation coefficients between the durations of the various activities are given by the following 8×8 correlation matrix:

$$
\begin{bmatrix}
 & 1 & 2 & 3 & 4 & 5 & 6 & 7 & 8 \\
1 & 1.00 & 0.20 & 0.20 & 0.20 & 0.20 & 0.00 & 0.00 & 0.00 \\
2 & 0.20 & 1.00 & 0.20 & 0.20 & 0.20 & 0.00 & 0.00 & 0.00 \\
3 & 0.20 & 0.20 & 1.00 & 0.20 & 0.20 & 0.00 & 0.00 & 0.00 \\
4 & 0.20 & 0.20 & 0.20 & 1.00 & 0.20 & 0.00 & 0.00 & 0.00 \\
5 & 0.20 & 0.20 & 0.20 & 0.20 & 1.00 & 0.00 & 0.00 & 0.00 \\
6 & 0.00 & 0.00 & 0.00 & 0.00 & 0.00 & 1.00 & 0.90 & 0.90 \\
7 & 0.00 & 0.00 & 0.00 & 0.00 & 0.00 & 0.90 & 1.00 & 0.90 \\
8 & 0.00 & 0.00 & 0.00 & 0.00 & 0.00 & 0.90 & 0.90 & 1.00
\end{bmatrix}
$$

The covariances for the durations of the various activities are computed from the correlation matrix and the standard deviations shown in the first table above. They are easily obtained from the correlation matrix by multiplying each correlation coefficient by the standard deviations of the activity in its row and the activity in its column: Doing so gives the following 8×8 variance-covariance matrix:

$$
\begin{bmatrix}
 & 1 & 2 & 3 & 4 & 5 & 6 & 7 & 8 \\
1 & 9.0 & 1.8 & 1.8 & 1.8 & 3.0 & 0 & 0 & 0 \\
2 & 1.8 & 9.0 & 1.8 & 1.8 & 3.0 & 0 & 0 & 0 \\
3 & 1.8 & 1.8 & 9.0 & 1.8 & 3.0 & 0 & 0 & 0 \\
4 & 1.8 & 1.8 & 1.8 & 9.0 & 3.0 & 0 & 0 & 0 \\
5 & 3.0 & 3.0 & 3.0 & 3.0 & 25.0 & 0 & 0 & 0 \\
6 & 0 & 0 & 0 & 0 & 0 & 25.0 & 36.0 & 22.5 \\
7 & 0 & 0 & 0 & 0 & 0 & 36.0 & 64.0 & 36.0 \\
8 & 0 & 0 & 0 & 0 & 0 & 22.5 & 36.0 & 25.0
\end{bmatrix}
$$

Critical path calculations are often, as in PERT, based on the most likely durations. The most likely durations are the modes, but here the medians are estimated, not modes. However, when the probability distributions are skewed to the right, as these are, the median values are greater than the modes. Thus, using the medians as single point estimates of the modes is conservative. Based on the *median* values for the activity durations from the table above, the three longest paths through the network (the only paths through this simple network) are:

Path 1:   START – 1–2–3–4 – FINISH;   duration = 36 weeks
Path 2:   START – 1–5–4 – FINISH;   duration = 33 weeks
Path 3:   START – 6–7–8 – FINISH;   duration = 31 weeks

However, based on the *mean* values for the activity durations, the three longest paths through the network (the only paths through this simple network) are:

Path 1:   START – 1–2–3–4 – FINISH;   duration = $\overline{T}_1$ = 40 weeks
Path 2:   START – 1–5–4 – FINISH;   duration = $\overline{T}_2$ = 35 weeks
Path 3:   START – 6–7–8 – FINISH;   duration = $\overline{T}_3$ = 35 weeks

The mean path lengths are longer than the sum of the median activity durations because the probability distributions for these activities are all skewed to the right; the means are greater than the medians. Note that the median path length is not equal in general to the sum of the medians of the activities along that path. In fact, it was argued above that the path length is approximately Normal, so the median path duration is equal to the mean path duration, hence the median path duration is greater than the sum of the activity medians.

For these three paths, the variance of each path is computed from the sum of the covariances of all the activities on that path. For illustration, the variance-covariance matrices are shown for each of these paths.

For Path 1, the variance-covariance matrix is obtained from the above matrix by deleting all the rows and columns for activities not in this path:

$$
\begin{array}{c|cccc}
 & 1 & 2 & 3 & 4 \\
\hline
1 & 9.0 & 1.8 & 1.8 & 1.8 \\
2 & 1.8 & 9.0 & 1.8 & 1.8 \\
3 & 1.8 & 1.8 & 9.0 & 1.8 \\
4 & 1.8 & 1.8 & 1.8 & 9.0
\end{array}
$$

Summing all the elements of this matrix gives the variance of the duration of Path 1 var$[T_1]$ as 57.6, so the standard deviation is 7.59 weeks.

For Path 2, the variance-covariance matrix is obtained from the total covariance matrix by deleting all the rows and columns for activities not in this path:

$$
\begin{array}{c|cccc}
 & 1 & 2 & 3 & 4 \\
\hline
1 & 9.0 & 1.8 & 1.8 & 1.8 \\
2 & 1.8 & 9.0 & 1.8 & 1.8 \\
3 & 1.8 & 1.8 & 9.0 & 1.8 \\
4 & 1.8 & 1.8 & 1.8 & 9.0
\end{array}
$$

Summing all the elements of this matrix gives the variance of the duration of Path 2 var$[T_2]$ as 58.6, so the standard deviation is 7.66 weeks.

For Path 3, the variance-covariance matrix is obtained from the total covariance matrix by deleting all the rows and columns for activities not in this path:

$$
\begin{array}{c|ccc}
 & 6 & 7 & 8 \\
\hline
6 & 25.0 & 36.0 & 22.5 \\
7 & 36.0 & 64.0 & 36.0 \\
8 & 22.5 & 36.0 & 25.0
\end{array}
$$

Summing all the elements of this matrix gives the variance of the duration of Path 3 var$[T_3]$ as 303, so the standard deviation is 17.41 weeks.

It is now possible to compute the correlation coefficients between paths 1 and 2. Using the equation derived above,

$$
\text{cov}[T_1, T_2] = E\left[\left\{\sum_{j \in A_1}\left(x_j - \bar{x}_j\right)\right\}\left\{\sum_{k \in A_2}\left(x_k - \bar{x}_k\right)\right\}\right]
$$

Here, Path 1 has four terms and Path 3 has three. The 12 product terms are then:

$$
\begin{bmatrix}
\text{Activity} & 1 & 4 & 5 \\
1 & E\left[(x_1-\bar{x}_1)^2\right] & E\left[(x_1-\bar{x}_1)(x_4-\bar{x}_4)\right] & E\left[(x_1-\bar{x}_1)(x_5-\bar{x}_5)\right] \\
2 & E\left[(x_1-\bar{x}_1)(x_2-\bar{x}_2)\right] & E\left[(x_2-\bar{x}_2)(x_4-\bar{x}_4)\right] & E\left[(x_2-\bar{x}_2)(x_5-\bar{x}_5)\right] \\
3 & E\left[(x_1-\bar{x}_1)(x_3-\bar{x}_3)\right] & E\left[(x_3-\bar{x}_3)(x_4-\bar{x}_4)\right] & E\left[(x_3-\bar{x}_3)(x_5-\bar{x}_5)\right] \\
4 & E\left[(x_1-\bar{x}_1)(x_4-\bar{x}_4)\right] & E\left[(x_4-\bar{x}_4)^2\right] & E\left[(x_4-\bar{x}_4)(x_5-\bar{x}_5)\right]
\end{bmatrix}
$$

Taking the expectation of each term gives the matrix:

$$
\begin{bmatrix}
\text{Activity} & 1 & 4 & 5 \\
1 & \sigma_1^{\,2} & \rho_{1,4}\sigma_1\sigma_4 & \rho_{1,5}\sigma_1\sigma_5 \\
2 & \rho_{1,2}\sigma_1\sigma_2 & \rho_{2,4}\sigma_2\sigma_4 & \rho_{2,5}\sigma_2\sigma_5 \\
3 & \rho_{1,3}\sigma_1\sigma_3 & \rho_{3,4}\sigma_3\sigma_4 & \rho_{3,5}\sigma_3\sigma_5 \\
4 & \rho_{1,4}\sigma_1\sigma_4 & \sigma_4^{\,2} & \rho_{4,5}\sigma_4\sigma_5
\end{bmatrix}
$$

Note that, even if all the correlation coefficients between activities were zero, the covariance of paths 1 and 2 would still contain the terms $\sigma_1^2 + \sigma_4^2$ because activities 1 and 4 appear in both paths. Replacing the terms in the above matrix by numbers:

$$
\begin{bmatrix}
\text{Activity} & 1 & 4 & 5 \\
1 & 3.0(3.0)=9.0 & 0.2(3.0)(3.0)=1.80 & 0.2(3.0)(5.0)=3.00 \\
2 & 0.2(3.0)(3.0)=1.80 & 0.2(3.0)(3.0)=1.80 & 0.2(3.0)(5.0)=3.00 \\
3 & 0.2(3.0)(3.0)=1.80 & 0.2(3.0)(3.0)=1.80 & 0.2(3.0)(5.0)=3.00 \\
4 & 0.2(3.0)(3.0)=1.80 & 3.0(3.0)=9.0 & 0.2(3.0)(5.0)=3.00
\end{bmatrix}
$$

Adding up all these terms gives $\mathrm{cov}[T_1, T_2] = 40.8$. From this covariance, the correlation coefficient between the two paths is:

$$
\mathrm{cov}[T_1,T_2] = \rho_{T_1,T_2}\sqrt{\mathrm{var}[T_1]}\sqrt{\mathrm{var}[T_2]}
$$

$$
\rho_{T_1,T_2} = \frac{\mathrm{cov}[T_1,T_2]}{\sqrt{\mathrm{var}[T_1]}\sqrt{\mathrm{var}[T_2]}} = \frac{40.8}{7.59(7.66)} = 0.70
$$

Similar reasoning shows that $\rho_{T_1,T_3} = \rho_{T_2,T_3} = 0$.

Suppose that the project manager wishes to set a schedule date for project completion such that the risk of overrunning this date is approximately 5%. What value should he commit to for the scheduled completion date? If he/she considers that the correlation 0.70 is large, then Path 2 is dependent on Path 1 and so Path 2 can be ignored (because its mean duration is less than that of Path 1, and has about the same standard deviation as Path 1. Path 3 is independent of path 1, so both paths 1 and 3 must be considered. Using the equation derived earlier,

$$\Pr \cong 1 - \Pr[T_1 \le t]\Pr[T_3 \le t]$$

The probability of overrunning any specified schedule date is shown in the following Fig. 8.1.

Here the project manager can read off the schedule date that has a probability of 5% (or any other value) of being overrun; this value is 64 weeks. Also shown in the figure are the plots for the probability of overrun for Path 1 and Path 3 taken separately. Path 3 has a lower expected value (35 weeks) than Path 1 (40 weeks), but the higher variance of Path 3 means that it has a higher probability of controlling the total project length at longer durations. That is, if the project manager considers only Path 1, schedule duration with a 5% chance of being overrun is 52 weeks, which is 12 weeks too early. The possibility that Path 3 might become the critical path requires the 64 week schedule; neglecting Path 3 because of its lower mean value is unconservative and can lead to serious schedule overruns.

What, however, if the project manager considers the 0.70 correlation between Path 1 and Path 2 to be small rather than large? With this view, all three paths would be independent. The equation above then becomes:

$$\Pr \cong 1 - \Pr[T_1 \le t]\Pr[T_2 \le t]\Pr[T_3 \le t]$$

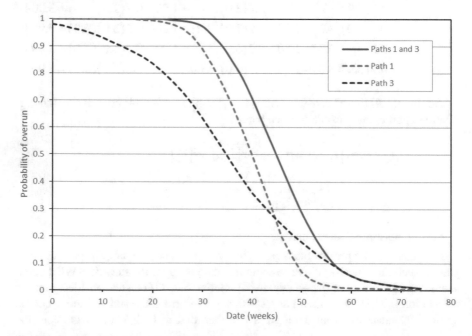

Fig. 8.1 Path 1 and Path 3 probability of overrunning

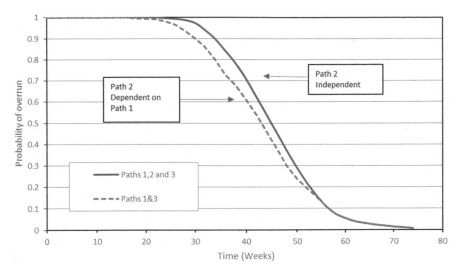

Fig. 8.2 Comparison with Path 2 (Dependent and Independent)

Figure 8.2 below compares the results when Path 2 is considered to be dependent on Path 1 (as in the previous figure) and the results when Path 2 is considered to be independent of the other paths.

The two solutions in the figure constitute bounds on the true solution. As the correlation coefficient between Path 1 and Path 2 lies in the interval $0 < 0.70 < 1$, so must the probability distribution lie between the results for zero correlation and perfect correlation. In the area of interest, that is, the upper tail, above about 55 weeks, there is no detectible difference between the two curves. Therefore, the result is not sensitive to the determination of whether a correlation of 0.70 is high (that is, can be taken as 1.0) or low (that is, can be taken as 0).

Figure 8.3 compares the results shown in the previous figure with the probability distribution for the longest path determined by a Monte Carlo simulation using correlated Normal random variables with the means, variances, and correlation matrix given earlier. The Monte Carlo simulation was run for 32,000 critical paths. As can be seen, the Monte Carlo results are close to those obtained by the method described in this chapter. The curves in the figure deviate somewhat, but are close in the tails, and the answer to the project manager's question, What is the schedule date with a 95% probability of being met, is again 64 weeks.

8.3   Practice Problems

Problem 8.3.1 The following table summarizes the precedence relationship among the activities on a project (see Table 8.2). Also, it provides 3-point estimates for the duration of each activity. Determine how many paths are in this network and which one is the critical i.e. the longest.

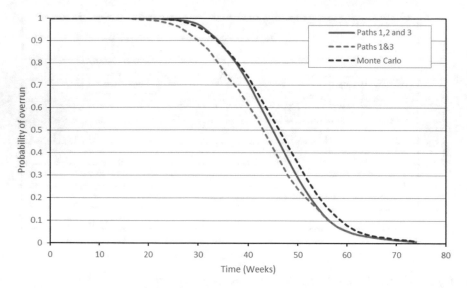

Fig. 8.3 Comparison with Monte Carlo simulation

Table 8.2 Problem data

Activity	Predecessor	Estimated duration		
		Optimistic	Most likely	Pessimistic
A	–	12	15	30
B	A	4	8	18
C	B, D	2	6	10
D	A	3	4	11
E	B, D	2	4	6
F	C, E	4	8	18

Find one-sided upper 95% confidence interval on project completion time. How would your answer change if the activities are correlated (see the correlation matrix below)?

$$\begin{bmatrix} \text{Activity} & \text{A} & \text{B} & \text{C} & \text{D} & \text{E} & \text{F} \\ \text{A} & 1.0 & & & & & \\ \text{B} & & 1.0 & 0.8 & 0.8 & 0.8 & 0.8 \\ \text{C} & & 0.8 & 1.0 & & & \\ \text{D} & & 0.8 & & 1.0 & & \\ \text{E} & & 0.8 & & & 1.0 & \\ \text{F} & & 0.8 & & & & 1.0 \end{bmatrix}$$

Table 8.3 Problem data

Path	Mean path length, correlation = 0.00	Probability of being the critical path	Mean path length, correlation = 0.90	Probability of being the critical path
1–3–6–8–9	55.92	0.36	47.52	0.33
1–4–5–7–9	57.35	0.34	57.70	0.44
1–2–5–7–9	57.33	0.30	61.11	0.23

Table 8.4 Problem data

| Activity | Predecessor | Normal | | Crashed | |
		Duration	Cost	Mean duration	Mean cost
A	–	6	10	2	38
B	–	4	12	4	12
C	–	4	18	2	36
D	A	6	20	2	40
E	B, D	3	30	2	33
F	C	10	10	6	50
G	F, E	6	20	2	100

Problem 8.3.2 A Monte Carlo simulation of a simple project network was executed for various degrees of correlation between the network activities. The results for three primary paths and two values of the correlation coefficients, 0.00 and 0.90, are tabulated below (see Table 8.3).

Assuming that this single example is representative of the general case, describe what appears to be happening with regard to the simulated critical paths as the activity-to-activity correlation is increased.

Problem 8.3.3 Susan was provided the following information about the project she is managing (see Table 8.4); it includes duration and cost of project activities for a normal schedule that utilizes the existing technologies and is predictable and accelerated schedule using innovative methods. Also, she was told that the activities can be accelerated/crashed in increments of 1 day with the cost that is proportional. For example, activity A can be crashed at most 4 days which will result in $28,000 of additional cost ($38K − $10K), and so forth.

Assuming that the cost of all crashed activities are random variables with COV = 0.3 what is the mean and standard deviation of the cost of the project if it's to be completed 2 day earlier?

References

Keefer DL, Bodily SL (1983) Three-point approximations for continuous random variables. Manag Sci 29(5):595–609

MacCrimmon KR, Ryavec CA (1964) Analytical study of the PERT assumptions. Oper Res 12(1):16–37

McCabe B (2003) Construction engineering and project management III: Monte Carlo simulation for schedule risks. In: Proceedings of the 35th conference on Winter simulation: driving innovation. Winter Simulation Conference, pp 1561–1565

Chapter 9
Contingency and Management Reserves

Abstract In this chapter we introduce the concept of contingency and management reserves. We provide the formal definition and a discussion on its use and misuse. We further provide the mathematical formulation of the contingency and the contingency factor, the consideration of contingency in program management, and the effect of correlation.

Keywords Contingency · Program management · Correlation · Probability distribution

9.1 Introduction

In discussions of risk, the term *contingency* is often used. Contingency levels should be set for each project based on acceptable risk, degree of uncertainty, and confidence levels for meeting baseline requirements. The process of evaluating potential project contingency funding requires the application of risk assessment and probabilistic estimating techniques. Contingency may be added by the person or organization making the original estimate, or by some person or organization higher in the organizational structure, by owners, clients, or sponsors. Contingency may be added at all management levels connected with a project. The estimate to which the contingency is added may itself contain contingency applied by lower levels in the project organization. Some people use the terms *allowance* or *management reserve*, in order to avoid the use of the term *contingency*, but it is not necessarily the case that all these are synonymous.

The dictionary definition of contingency is (Merriam-Webster 1983):

> Contingency … a (1): the condition that something may or may not occur: the condition of being subject to chance (2): the happening of anything by chance: fortuitousness … 2 a: something that is contingent: an event or condition occurring by chance and without intent, viewed as possible or eventually probable, or depending on uncertain occurrences or coincidences … b: a possible future event or condition or an unforeseen occurrence that may necessitate special measures <a reserve fund for *contingencies*> c: something liable to happen as a chance feature or accompaniment of something else.

© Springer Nature Switzerland AG 2020
I. Damnjanovic, K. Reinschmidt, *Data Analytics for Engineering and Construction Project Risk Management*, Risk, Systems and Decisions,
https://doi.org/10.1007/978-3-030-14251-3_9

Note the frequent appearance of the word "chance,' as well as its synonyms "possible," "probable," fortuitous," and "without intent." From the dictionary definition it might be supposed that a contingency would be for the purpose of covering "the happening of anything by chance ... uncertain occurrences or coincidences ... [or] an unforeseen occurrence" and therefore, the expenditure of the contingency would be a result of chance, and should happen only if the "unforeseen occurrence" actually occurred. In that case, the actual expenditure of the contingency would be itself "an event occurring ... by chance," and one would not expect the contingency to be exhausted in the normal course of activities. However, the term contingency is not understood in this way by many people; the term is often taken to refer to funds that will be completely expended in the course of the project, no matter what happens.

It is useful to distinguish between *systematic error* and *random error* in the estimation of costs, durations, and other factors (Taylor 1997). Systematic error refers to a bias or offset in our measurements or estimates. Systematic error is illustrated in Fig. 9.1a.

If we are hitting consistently below the aiming point, we need to adjust our sights. If we are surveying and measuring a baseline with a chain, and we fail to adjust for thermal expansion of the chain, then there may be a systematic error in all the readings, depending on the ambient temperature. The objective of engineering measurements is to identify and eliminate all sources of systematic errors. When all systematic errors have been corrected, then any errors remaining are random errors, as in Fig. 9.1b: One could also say that random errors are errors the causes of which have not (yet) been identified and therefore cannot (yet) be eliminated. The question is whether these residual errors are small enough that they do not lead to bad decisions and do not prevent the successful accomplishment of the project objectives.

Random errors may also result in compensation. If we feel that shooting low is much less desirable than shooting high, as in the case of the asymmetric target (see Fig. 9.1c), then we may want to adjust our aim in order to do better, even when the errors are random. This asymmetric situation occurs when we have different preferences for being high or low.

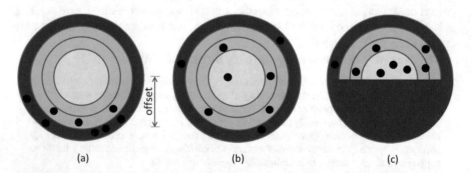

Fig. 9.1 Types of errors. (**a**) Systematic error, (**b**) random error, and (**c**) random error with asymmetric target

It may be that the errors, although random, are excessive, and should be reduced. This may require a change in the process. If a firm is subject to excessive random errors in cost estimates, it may need a new cost estimation process. This general subject is discussed at greater length in Chap. 12.

An example may help in illustrating the point. Suppose that someone tabulates the number of valves used on a substantial set of previous similar projects, and finds that, in every case, the number of valves actually installed was always 1.17 times the number of valves taken off of the engineering drawings by the quantity development function. Then a rule might be to plan on purchasing 17% more valves than the number of valves counted by quantity takeoff from the engineering drawings. This 17% allowance for undercounting is not a contingency in the sense of the dictionary definition, because there is nothing about it referring to events due to chance. In this example, it would be an empirical fact that valves are consistently undercounted, and the 17% factor is to offset this inherent bias or systematic error. Therefore, the additional 17% for the valves not counted would always be spent. Note that the rule does not say to add 17% to the number of valves in the preliminary cost estimate, because this cost estimate might already have some adjustment factor in it.

Suppose now that the above example is slightly different, that is, from the historical data, the investigator determines that the factor relating actual valves used to valves counted on drawings is a random number, with average or mean 1.17, and standard deviation 0.10. Then the systematic error or bias is 17% of the valve takeoff, but there is a random component as well, which depends on chance events, and hence is contingent. To account for both the systematic error and the random error, we have to add 17% allowance to the number of valves actually counted, and then add another contingency to account for chance (for example, design changes or damage to valves in storage). When talking about chance, or contingency by the above definition, we must make probabilistic statements. If we want to be sure that we order enough valves 95% of the time, then we need to add 33% to the number counted (17% for the mean or systematic bias and 1.645 times the standard deviation for the chance variation). Note that we would *expect* to have 16% of the counted valves left over, as these were ordered for protection against running out, and were *expected* to be unused. The term *expected* is used here in the probabilistic sense: *expected* means *on the average*. Therefore, the *expected number* of valves needed is 1170, but this is not at all inconsistent with actually using 1250, for example. Table 9.1 summarizes the procurement and use of valves on this project.

Table 9.1 Project valve use

Valves	Number
Valves actually counted on drawings	1000
Valves added to correct for undercount (systematic error)	170
Valves added to reduce the probability of running out of valves due to chance to 5% or less	164
Total number of valves to be ordered	1334
Expected number of valves to be installed	1170
Expected number of valves to be left over	164

In these notes, the proper allocation of contingency, whether denominated in dollars or in valves, depends on the risks to the project of running out of resources; contingency is a buffer against this eventuality. Suppose that Project A is a refinery in Houston. Then the penalty for running out of valves is relatively low, as the lead time in the valve supply chain is relatively short (unless these valves are some special type not in stock). However, suppose that Project B is an LNG facility on a deepwater platform far off the coast of equatorial Africa. Running out of valves might cause a work stoppage, and so replacement valves might have to be sent from Houston by chartered aircraft. Then the project manager for Project B should have a lower probability of running out of valves, and hence a larger contingency.

It is generally easier to deal with contingency in monetary terms, as money is fungible (can be used for various purposes), whereas we may not know what to do with 164 left-over valves (spare parts, perhaps?). The ambiguity arises because the term contingency is often used for both the amount needed to cover the systematic error and the amount needed to buffer against the risk attributable to chance. This usage can cause confusion, as some people, referring to contingency as systematic error or bias, expect that they will use it all up, whereas others, referring to random error or chance, expect that the contingency funds will be left over, and all funds not necessary to cover random or chance events ought to be returned to the project sponsor. Not surprisingly, those who claim the right to expend all contingency funds tend to be project managers, and those who expect to see at least some of the contingency allowance returned tend to be owners.

In practice, relatively few projects return leftover contingency funds to the sponsor unless the project sponsor's program manager holds the contingency from the start. In general, project managers may regard cost contingency funds as theirs to use; if the risks fail to materialize, the funds will be expended on something else, such as out-of-scope project improvements. Therefore, not only is there no consensus on how large a contingency should be, there is no agreement on the basic point of whether the contingency is an unassigned cost that is intended to be spent or insurance that is intended not to be spent.

Beyond these bare minima, there is little consensus about the specific meaning of the term *contingency*. The term management reserve is often also used, sometimes as a virtual synonym of contingency, sometimes in distinction to it. The EIA Standard (EIA 1998) does not define the term contingency, but does define budget and management reserve, as follows:

BUDGET AT COMPLETION – The total authorized budget for accomplishing the program scope of work. It is equal to the sum of all allocated budgets plus any undistributed budget. *(Management Reserve is not included.)* (Emphasis added.)

MANAGEMENT RESERVE – An amount of the total budget withheld for management control purposes rather than being designated for the accomplishment of a specific task or set of tasks.

However, virtually the same language is used by others to mean *contingency*.

The Guide to the Project Management Body of Knowledge (PMBOK® Guide) – Fourth Edition, (PMI 2008), defines both contingency reserves and management reserves:

Budget reserve analysis can establish both the contingency reserves and the management reserves for the project. Contingency reserves are allowances for unplanned but potentially required changes that can result from realized risks identified in the risk register. Management reserves are budgets reserved to unplanned changes to project scope and cost.... Reserves are not part of the project cost baseline, but may be included in the total budget for the project.

The Guide to the Project Management Body of Knowledge (PMBOK® Guide) – Fourth Edition, (PMI 2008), starts out well by stating the contingency exists "to account for cost uncertainty" (page 173) but then goes off-track by suggesting that contingency may be a percentage of the estimated cost or a fixed number:

Cost estimates may include contingency reserves (sometimes called contingency allowances) to account for cost uncertainty. The contingency reserve may be a percentage of the estimated cost, a fixed number, or may be developed by using quantitative analysis methods.

As more precise information about the project becomes available, the contingency reserve may be used, reduced or eliminated. Contingency should be clearly identified in schedule documentation. Contingency reserves are part of the funding requirements.

Later, *The Guide to the Project Management Body of Knowledge (PMBOK® Guide)* (PMI 2008), states (page 301):

Estimates are made of potential project schedule and cost outcomes listing the possible completion dates and costs with their associated confidence levels. This output, often expressed as a cumulative distribution, can be used with stakeholder risk tolerances to permit quantification of the cost and time contingency reserves. Such contingency reserves are needed to bring the risk of overrunning stated project objectives to a level acceptable to the organization.

For cumulative distribution, see the example in this chapter, below.

To provide more consistency, in this book the terms contingency and management reserve will be defined precisely. However, the reader is cautioned, as discussed above, that these definitions are not universally accepted.

We assume here a probabilistic view of costs and risks. That is, we assume that there is some underlying probability density function on the activity (or project) costs (or duration). This probability distribution may not be known, and is certainly not derived from data on relative frequencies, but is assumed to exist as a subjective probability distribution, although it may exist only in the minds of the project participants. Being subjective rather than objective, different people may have (radically) different ideas about what it is. Given this probabilistic viewpoint, contingency is, then, an amount of money (or time, in the case of project schedules) added to the mean (or expected value) of this probability distribution in order to bring the total of the expected value plus the contingency up to a certain acceptable risk of overrun (that is, the remaining probability that the actual cost or duration would be above the mean plus contingency). Contingency may be added (or even subtracted) at all levels connected with a project. The estimate to which the contingency is added may itself contain contingency, applied by lower levels in the project organization.

Note that by the definition used here, contingency is added to the mean. Note especially that the expected value is not necessarily the deterministic or single point value that may be given as the *estimate*. Although some people assume that the single point estimate may be taken as the mean of a probability distribution, there is no reason to believe that there is any direct connection between the point estimate and the expected value. The point estimate might be the mode of the distribution, or the median, or have no relationship whatsoever with a probability distribution. The estimate may already include some allowance for contingency.

9.2   Setting Contingencies

The only project risk that is affected by the level of project contingency is the risk that the project will run out of money or time before it is completed. In projects that have absolutely fixed spending limits; for example, a public agency that has to issue bonds to cover a project, the spending limit should contain enough contingency so that the likelihood of running out of funds is acceptably small, as it may not be possible to gain additional funding. In private sector projects, a project manager who runs over his budget may have to appeal to the CEO, to the board of directors, or to outside bankers for additional funds to complete the project. This necessity creates a decision point at which the owner or the funding or lending source may decide to terminate the project by refusing further funds. Therefore, the initial funding allocation may include contingency such that a request for additional funds would precipitate a management review of the feasibility of the project. But adding contingency raises the project budget, and hence in effect raises the project cost. Similarly, adding schedule contingency increases the schedule duration, which implies that the project must start earlier.

Setting the contingency is then a matter for some negotiation between the project manager, who naturally wants the contingency allowance set high, and the funding source, who may want the contingency set low to maintain greater control over the project. The smaller the contingency, the more often the higher level of management would be asked for additional funds. Although the proper set point along this axis is subject to dispute, there is some point at which the upper management level would be considered to be micromanaging. Basically, the higher level of management typically takes the view that there are uncertainties associated with estimating and with executing jobs; and that they have hired the lower level managers to manage these activities, and they should be allowed to go and do it, without constantly coming back for additional funding approval. This point will, of course, vary with the organization.

The defining equation for contingency as used here is:

$$B = \mu + a\sigma$$

B = Budget for a project or a work package
μ = Expected value of the project or WP cost
σ = Standard deviation of the project or WP cost
a = Contingency factor
$a\sigma$ = Contingency allowance

The contingency should be added to the expected value because the mean value is the only single number that gives an unbiased value uncontaminated by the variance (uncertainty). Any other number used as an estimate conflates the mean and some fraction of the standard deviation into a single figure. If all the cost estimates for all the work packages in a project were known to be the mean values of their individual distributions, the mean value of the total project would be the sum of these values. However, if the individual work package estimates are not the mean values of their distributions, then the sum of these estimates is not the mean value of the total cost (or duration). The total is the mean value plus some multiple of the standard deviation, but there is no information contained in it with which to estimate the standard deviation. Therefore, summing the estimates for all work packages without the discipline of using the mean value gives an estimate of the total cost that is not related to any identified level of uncertainty.

Typically, work package estimates, even at the lowest level, contain some contingency factor. There is built-in bias from the estimator, who may expect to be criticized more severely for an underestimate than for an overestimate. Also, there is a need to cover errors in estimating, which are usually errors of omission. That is, in taking off quantities from drawings, a quantity estimator may miss some, but is less likely to over count. Moreover, even if a computer does the quantity development, some instances may be missing from the drawings, and others may be added later, contributing to a bias toward undercounting, and adding contingency is a bias intended to offset this bias.

In the field, construction personnel will certainly take the position that the adverse impact of having material left over due to overestimation is far less than the adverse impact of running out before the job is finished. For these and other reasons, estimators may tend to add some contingency or safety factor. For similar reasons, the next level upward typically also adds contingency. Each person or organization that adds contingency does so to protect itself from the consequences of uncertainty, and these consequences are generally considered to be more dire if the number is underestimated rather than overestimated.

Therefore, by the hypothesis that each management level wishes to set the risk adjusted cost estimate at some preferred quantile of the underlying probability distribution, that management level must have some view of how much contingency has been already incorporated in the estimates they receive. If a manager wishes to set the risk adjusted cost estimate at approximately the 80th percentile, and the manager believes that the estimate is at the 50th percentile, then the difference in dollars (or time) between the 50th and the 80th percentiles must be added. If the manager believes the estimate has already been adjusted, up to the 70th percentile, for example, then less should be added. This process does not require that the manager know or state explicitly what the lower level contingency or risk is; only that the manager is familiar with the organization and how it works.

Conversely, if a manager believes that the estimates have already been adjusted up to, say, the 90th percentile, then the manager may cut the estimate. This may also occur if the manager's knowledge of the business is such that he has a different view of the underlying uncertainties than others have. Then, the manager may wish a risk adjusted cost estimate at the 80th percentile, and his subordinates or subcontractors may have already adjusted their estimates at the 80th percentiles, but their subjective probability distributions do not agree as to where the 80th percentile is. Or, a high-level contractor manager may feel that, regardless of the contingency or safety factor deemed desirable, competitive conditions won't permit it. That is, there is a risk in setting contingencies too low, but also a risk in setting them too high, if one has to bid for a contract fixed price or get financial authorization from the sponsor's management. In general, as estimates flow upward through different levels of organizations, we may expect that higher levels have better knowledge of strategic business or political conditions, and so may make different decisions about contingencies. This may be untrue in specific cases, however. At the owner's level, we may assume that knowledgeable owners can make better decisions about contingencies than inexperienced owners or owners that do not take the effort to be knowledgeable about project costs and durations.

9.3   Project Policy and Procedures

Project policy and procedures documents should include precise and consistent definitions for terminology. The term contingency is in particular need of a consistent definition, as it means different things to different people. Equally important are discussions and examples of approaches to setting budget contingency and setting schedule contingency. Contingency is not like value engineering, change control, or other cost control methods. Contingency is an allowance for error or a safety margin on budget overruns that is not based on reducing costs or risks, but on increasing the budget. By itself, contingency is not a cost control method, as its purpose is simply to have enough funds to pay for uncontrolled costs.

The definition of contingency as a percentage of the estimated cost to complete, instead of a percentage of the original estimate, is an improvement, but is a change from past practices in many cases. Project contingencies should be reported and reviewed in a consistent way, and this should be emphasized in policy documents in order to achieve consistency across all projects. Also, providing recommended or standard values or ranges for overall contingency allowances is a questionable practice, as it encourages project managers to use these values instead of performing project-specific risk assessments. A contingency percentage that might be adequate for some conventional infrastructure projects will be totally inappropriate for big science projects, waste remediation projects, and other one-of-a-kind or first-of-a-kind plants, for which the technology may be new and unproven or the volume and characterization of the wastes uncertain, and which may need to retain much larger contingencies even at the final design stage.

There are at least two distinct issues in setting contingency. One purpose of contingency is an allowance for unknowns in making estimates. Because these errors are predominantly errors of omission, some allowance must be added to cover them. But systematic undercounting reflects a bias. This is an uncertainty in the sense that the actual systematic error is unknown. This type of contingency is bottoms-up, and is estimated at the work package or activity level. Adding all these work package contingencies would be valid if they all represented systematic errors, but is not, however, justified by statistical analysis insofar as these contingencies represent random errors. Adding random errors can easily result in a very large number, which then becomes the project budget. However, there is another kind of contingency needed for risk mitigation. This is not due to omissions in making estimates or anything else at the activity level. It is needed to allow for unknowns at the overall project level. A construction project that is really a research and development project may use new technology, which requires more project-level contingency. A waste remediation project needs contingency to cover the possibility that the in-situ waste may differ from the original characterization, but this is not an activity-level contingency. This kind of distinction is not merely verbal; there is an important difference in how such risks are estimated and project policies and procedures ought to make this distinction clear.

Moreover, there is a difference in how these different contingency factors are managed. If the project's base budget includes those costs that are known and countable, with a separate allowance for the unknowns in estimating these costs, then one expects that over the life of the project all or most of this allowance will be transferred to budget, as these actual quantities and costs are identified. But project contingency or management reserve may cover risk factors that would have very high impact if they occurred but are also highly unlikely to occur. If the contingency allowance for a possible flood is not used because no flood occurs, then this contingency allowance should not be transferred to the base budget to cover overruns in other areas.

Who owns this contingency and what should happen to it if it is not necessary to expend it is a very important issue that should be addressed in the policies and procedures documents. It can be argued that management reserves for high consequence, low probability events should be held at the program level, not at the project level. Policies and procedures should address the different kinds of contingency, the need for contingency allowances, who controls them, and what should happen to them.

If a contract is for a fixed price, the contractor owns the contingency inside the bid price, and the contractor gets it if there is any left at the end. But this is not the case with cost-plus contracts. Even with a fixed-price contract, the owner needs to hold some contingency to cover potential change orders. Contingency policies need to distinguish between fixed-price, cost-plus, cost plus incentive fee, and other common types of contracts in the discussion of risk and contingency, and should state whether contingency is controlled at the project manager level or at the owner's program level.

Why does anyone add contingency? Herein, it is assumed that the organization adding the contingency wishes to insulate itself from some of the consequences of overruns. Suppose for the sake of specificity, that a management level sets the contingency at that amount that will set the budget (for some activity, work package, subcontract, or total project, depending on the level of management) at, say, the

80-th percentile. That is, the probability of overrunning the risk-adjusted budget is 20%. The probability of underrunning the risk-adjusted budget would seem to be 80%, but it is more like zero. That is, it is likely that the entire contingency will be spent on the activity or project. If the actual cost of the activity turns out to be more than the risk-adjusted budget, for whatever cause, then the level of the organization responsible for this activity will have to request additional funding (or time) from the next higher level (or the owner), or terminate the project for lack of funds. Conversely, if the actual cost of the activity turns out to be less than the risk-adjusted budget, for whatever cause, then the level of the organization responsible for this activity is unlikely to return the excess funds to the next higher level (or the owner). Therefore, the risk-adjusted cost estimate provides a floor on the cost (or time) but does not provide a ceiling.

Why then do people use contingency? If the activity budget was set at, say, the 50-th percentile of the (assumed) probability distribution on the total cost, then the management level responsible for that activity would have to go back to the next higher level for more funds (or time) about 50% of the time. This might seem excessive. Certainly the organization doing the work would feel that asking for more funding with this frequency would be excessive, and therefore would want to add contingency so that this necessity would arise less often. The next higher level of management, which presumably controls the funds for the lower levels, might not wish to be bothered so much either, and therefore also has some reason for permitting contingency.

Of course, there is no need to set the budget with contingency at the 50th percentile. One could use the 40th, 30th, 20th, or 10th percentile. The smaller the percentile, the more often the higher level of management would be asked for additional funds. Although the proper set point along this axis is subject to dispute, there is some point at which the upper management level would be considered to be micromanaging. Basically, the higher level of management typically takes the view that there are uncertainties associated with estimating and with executing jobs; and that it has hired the lower levels to manage some activities, and they should be allowed to go and do it, without constantly coming back for additional funding approval. This point will, of course, vary with the organization.

Why is the contingency here said to be added to the expected value? Because the mean value, as a single number, gives a precise and unbiased estimate of the expected value uncontaminated by the variance. Any other number used as an estimate conflates the mean and some fraction of the standard deviation into a single figure. If all the cost estimates for all the work packages in a project were known to be the mean values of their individual distributions, the mean value of the total project would be the sum of these values. However, if the individual work package estimates are not the mean values of their distributions, then the sum of these estimates is not the mean value of the total cost (or duration). The total is the mean value plus some multiple of the standard deviation, but there is no information on which to estimate the standard deviation. Therefore, summing the estimates for all work packages gives an erroneous, and usually high, estimate of the total cost.

In short, the view used here is that the term contingency applies to a single proj-ect, or to sub projects; the risk adjusted cost estimate (including contingency) may be exceeded but will never be underrun. Many people differ on whether the contin-gency, so -called, is supposed to be spent or not supposed to be spent. Here, the view is that it will be spent.

By distinction, management reserve is an amount of money (or time) controlled by a higher level of management that may be used at the discretion of that level, and may be moved between project and activities according to need. The fundamental attribute of management reserve is that it relies on having a number of projects that are statistically independent (or approximately so), so that the management reserve for the total program (of multiple projects) is not the sum of the management reserves for each project (or sub project).

Examples can help to illustrate this. Suppose that a construction project deter-mines that one significant risk is that of a 100-year flood occurring during the con-struction period. Suppose we say the probability of this is 0.01, and the estimated cost of the damage if it occurs is $50 million. Then the expected loss is 0.01 times 50,000,000 = $500,000. If this amount is added to the project budget as a contin-gency, then there are two possible outcomes:

The flood risk actually materializes, with probability 0.01, and the damage is $50 million, which is far above the $500,000 contingency allowance. The project will have to ask for $49,500,000 in additional funding to cover the loss.

The flood risk does not materialize, with probability 0.99, and the $500,000 con-tingency will be spent on something connected with the project.

That is, the net result of this approach is only to increase the project cost by $500,000.

However, suppose that instead of this contingency, the owner were to set aside a management reserve to cover the potential losses. One may expect that there are other projects with similar risks, and if they are statistically independent of this project (they are not all located on the same river, for example), then a loss on one will not be associated with losses on others; the owner will allocate the manage-ment reserves to the project requiring them. To do this, the owner needs deep pockets.

A private owner with inadequate funds to cover such management reserves might buy flood insurance. Based on the expected value, the premium for this insurance should be, from the owner's viewpoint, less than $500,000. This premium, whatever it is, becomes a part of the budget for the project, and is neither management reserve nor contingency, but simply an additional cost.

Or, the owner might take $500,000 and build a cofferdam around the project site to protect it from the 100-year flood. This is a scope increase, and is part of the budget, and is neither a contingency nor a management reserve, by the definitions used here – whether or not the flood occurs, the cofferdam is there, so the cost of the cofferdam is a project cost incurred to mitigate (avoid) the risk of a $50,000,000 loss due to flooding.

9.4   Contingencies and Management Reserves in Program Management

Often project contingencies are set too low, due to an absence of risk analysis, but just as important as the size of the contingency is, who should control it.

It is often stated that project budgets and contingencies should be based on risk assessments, that is, on probabilities. However, probabilistic statements are impossible to verify on the basis of a single observation. But if a *program* or contractor performs a large number of *projects*, statistical statements could in principle be verified over the population of all projects. As an illustration, if all project budgets are assigned contingencies such that the probability of overrunning is, say, 15%, then over a large number of projects one would expect that 15% would finish over budget but 85% would finish under budget. Does this actually happen? Quantitative comparisons are unfortunately limited, due to the absence of *post mortem* analysis of completed projects for lessons learned. However, even qualitatively, there may be serious questions about whether the assumptions made in assigning risk-based budgets are in fact validated by experience. The objective of this section is to examine some of these assumptions and the conclusions that follow from them, to evaluate whether they are even qualitatively justified. This section also tries to examine the issue of contingency at a higher level than the single project, from the viewpoint, say, of a program or contractor with a large number of projects. The following material is highly simplistic, but represents some elementary principles that might be kept in mind when seeking new approaches to risk and contingency management.

One inherent difficulty in the application of probability to projects is the fact that the probability distributions used in cost estimating (or scheduling) are not based on objective physical measurements or relative frequencies, which can in principle be measured and reproduced by independent observers. The probability distributions used for risk assessment generally are not based on adequate data, because adequate data often are not collected, but rather are based on judgments, experience, and other subjective factors, and cannot be objectively reproduced by different observers. That is, these probabilities are subject to bias and to manipulation. It is a natural human trait to try to protect oneself against the effects of uncertainties, and therefore individuals at all stages in the process of risk assessment and cost estimation may increase or decrease, consciously or unconsciously, safety margins of their own which are hard to identify.

To begin very simply, suppose there is one project, for which there is assumed to be a known, objective, probability distribution on the project cost. Let this probability distribution have the expected value (mean) μ and variance σ^2. Let the budget be given as the expected value plus some safety margin above the expected value. That is, let $a \geq 0$ be a factor such that the budget (B) for this project is specified as a standard deviations above the mean; that is, $B = \mu + a\sigma$. That is, a is set such that there is some acceptable probability that the actual cost will exceed the budget, and this probability is just the area under the probability density function above $B = \mu + a\sigma$.

To be more specific, assume that the project cost probability density is approximated by the Normal distribution (which is generally a good working assumption, if there are insufficient data to justify another specific distribution, such as the log Normal or the Erlang). If we set $a = 1$, for example, then the budget B is one standard deviation (σ) above the mean (μ). Then, by the Normal assumption, there is roughly a 16% chance that the actual cost will exceed the budget. If $a = 2$ the probability that B would be exceeded is about 2%, and if $a = 0$ the probability that $B = \mu$ would be exceeded is 50% (assuming the mean and the median are the same). As a numerical example, let $\mu = \$400$ and let $\sigma = \$1000$ (in thousands of dollars); this is a moderately risky project with coefficient of variation $\sigma/\mu = 0.25$. If we are satisfied with a budget with 84% probability of being adequate (that is, not being overrun), then set $a = 1$ and $B = \mu + a\sigma = \$4000 + \$1000 = \$5000$.

Now assume there is a second project with exactly the same statistical properties (to make the comparison easier). Consider a higher level of management, for example a program or a contractor performing fixed-price work, with cognizance over both projects. Simply adding together the budgets for the two individual projects would indicate that the higher management level needs a budget $B_T = 2B = \$10,000$.

However, the expected value of the total expenditure at the higher level is the sum of the expected values for the two projects, or $\mu_T = \mu + \mu = 2\mu = \$8000$. Also, *if the two projects are statistically independent*, the total variance at the higher level is just the sum of the variances, or $\sigma_T^2 = \sigma^2 + \sigma^2 = 2\sigma^2$. Thus, $\sigma_T = \sqrt{[2\sigma^2]} = \sigma\sqrt{2} = \$1414$. But, if we were to apply the same confidence factor to the total costs at the higher management level, namely a 84% probability of being sufficient, $a_T = 1$ (as used for each project) and so $B_T = \mu_T + a_T\sigma_T = \$8000 + 1(\$1414) = \$9414$, which is not the answer $B_T = 2B = \$10,000$ obtained above.

Therefore, adding together the individual budgets *for independent projects* (or Work Packages within a project) overestimates the required reserve margin, if it is desired to have the same probability of overrun at the higher management level as for the individual projects. Actually, the sum of the budgets for the two projects, $2B = \$10,000$, corresponds to a value for $a_T = \sqrt{2} = 1.414$ which in turn corresponds to a probability of 92% of underrunning, and 8% of overrunning (using the Normal distribution). The reason for this difference, obviously, is that at the higher management level there is an opportunity for an overrun in one project to be offset by an underrun in the other. This possibility does not occur at the single project level, so the higher management level needs less margin to provide the same level of confidence, *as long as the projects are statistically independent*.

Generalizing the above, now let there be N projects with exactly the same properties (to avoid having to make up new numbers and more complicated notation), such that all N projects report to the same higher level of management. The scope of this higher management level, say the company, is just the sum of all the projects under its cognizance. Adding the N individual project budgets given above would indicate that the higher level needs a budget of $NB = \$5000N$.

Again, the expected value of the total expenditure at the higher level is the sum of the expected values for the N projects, or $\mu_T = N\mu = \$4000N$. Also, *if all N projects are statistically independent*, the total variance at the higher level is the sum of the variances, or $\sigma_T^2 = N\sigma^2$. Thus, $\sigma_T = \sqrt{N\sigma^2} = \sigma\sqrt{N} = \$1000\sqrt{N}$. Then, if we apply the same confidence factor to the higher management level, namely a 84% probability of its budget being sufficient, $a_T = 1$ and

$$B_T = \mu_T + a_T\sigma_T = \$4000N + \$1000\sqrt{N}$$

which is much less than the conservative total budget $NB = \$5000N$.

Suppose that $N = 100$, say, which might be the order of magnitude of projects in a large industrial owner or a contractor. Then $\sqrt{N} = 10$. Then the budget at the higher management level corresponding to 84% probability of being sufficient, would be

$$B_T = \mu_T + a_T\sigma_T = \$4000N + \$1000\sqrt{N} = \$400,000 + \$10,000 = \$410,000.$$

Thus, the sum of the individual budgets for the N projects, $NB = \$500,000$, would greatly overestimate the required reserve margin *if the projects are all independent*, and would imply a value for $a_T = \sqrt{N} = 10$, so if the budget at the program or company level were set at NB it would be *ten standard deviations* above the expected value. This would correspond to an infinitesimal probability of overrunning and a virtual certainty of underrunning (standard tables for the Normal distribution do not even give values at the 10σ level).

If the company were satisfied with a safety margin of 84% (that is, the probability of exceeding its entire program budget for 100 jobs is 16%), then the budget at this level,

$$B_T = \mu_T + a_T\sigma_T = \$4000N + \$1000\sqrt{N} = \$410,000$$

if divided equally over the $N = 100$ projects would give a budget for each of $\dfrac{B_T}{N} = \dfrac{\$410,000}{100} = \$4100$ (recall that, for simplicity of illustration, all N projects have the same individual risk). This would correspond to a margin for each individual project of,

$$a\sigma = \frac{a_T\sigma_T}{N} = \frac{\left(a_T\sigma\sqrt{N}\right)}{N} = 0.1\sigma$$

which (by the Normal assumption) corresponds to a probability of 54%. That is, for the company to have a 16% risk of exceeding its total budget, each individual project would have a probability of 46% of exceeding its individual budget, if the projects *were all independent*. A probability of 16% of a company exceeding its budget may be too high (or it could explain why the failure rate for contractors is relatively high), but any reasonable figure greater than zero will still show that the sum of the individual project budgets is far too high, for N a large number *and the projects all independent*. Again, this is because, for large N, the assumption of randomness requires that surpluses on some projects will tend to offset overruns on others.

The discussion above is, obviously, just an illustration of the principle on which insurance is based: the larger the base over which risks are distributed, the smaller the margin needs to be. A possible project budgeting approach, based on this example, might be for the owner's program or the contractor (or other higher authority) to assign a budget to each project equal to its expected value and to hold all the reserve margin at the higher management level. Then projects that overrun (and about half would be expected to overrun in this scenario) would request additional funding from the contingency pool (management reserve) held at the higher management level. Projects that underrun (and about half would be expected to underrun) would be expected to return the excess to the higher management level to cover the overruns on the others. This method would substantially reduce the total amount of margin needed, compared to treating each project separately.

As continually reiterated above, the results here are crucially reliant on the assumption that *all projects (or work packages) are statistically independent*. If there is some dependency or correlation between the costs for different projects, then the beneficial effect of grouping risks is reduced. (As it is in insurance: an insurance company may make money selling fire insurance as long as fires are random and independent, but a major disaster such as an earthquake centered in one location could far exceed its reserves.) If project costs are positively correlated, then the variance at the higher management level is increased. If all correlation coefficients are +1.0, then the required budget at the higher level reverts to the sum of the individual project budgets. (Negative correlations would have the opposite effect, but it is easy to see that more projects would be positively correlated than negatively correlated, *unless the contractor takes specific steps to obtain jobs that are diversified, that is, negatively correlated*.) For most companies, whether project costs are statistically independent or dependent, or the value of the correlation coefficients between them, is presently unknown, although in principle at least this is subject to empirical determination. An extensive discussion of correlation and dependence is beyond the intent of this chapter; the main point to be stressed here about independence is that it cannot be simply assumed to be true, and assuming that it is true could be very unconservative if it is *not* true.

Examining the performance of large companies, however, leads to an apparent contradiction: risk analyses almost invariably assume that projects (or work packages) are independent, but if the program (or contractor) is the sum of its projects, then adding all the separate project budgets together is, in effect, acting as if the projects are statistically *dependent*. If projects are treated as completely independent and each budgeted for a 16% probability of overrunning, projects would underrun 84% of the time, and statistically 84% of all projects should return budget surpluses as profits to the (fixed-price) contractor or as savings to the owner. This doesn't happen.

An alternate hypothesis, based on some data from many different types of projects, is that the project costs are in fact consistently *underestimated*, such that the reserve margins are negative and actual project budgets are less than the expected values. Then each project budget might be *a* standard deviations *below* the mean, or $B = \mu - a\sigma$ (for $a \geq 0$). Using, as before, $\mu = \$4000$ and $\sigma = \$1000$ but $a = 0.25$, then a project budget one-quarter of a standard deviation below the expected value would be

$$B = \mu - a\sigma = \$4000 - 0.25(\$1000) = \$3750$$

which would have a probability of about 60% of overrunning and a probability of 40% of underrunning. The sum of the individual budgets for N projects would be $NB = \$3750N$.

But the expected value of the total expenditure at the higher corporate level would be $\mu_T = N\mu, \sigma_T^2 = N\sigma^2$, and $\sigma_T = \sigma\sqrt{N}$. Taking $N = 100$, if the higher-level budget is the sum of the individual project budgets,

$$B_T = NB = \$375,000 = \mu_T - a_T\sigma_T = \$4000N - a_T\left(1000\sqrt{N}\right) = \$400,000 - \$10,000a_T$$

This implies $a_T = 2.5$, so the higher-level budget is 2.5 standard deviations *below* its expected value. This corresponds to a probability greater than 99% that the higher-level organization would exceed its total budget and less than 1% that it could meet its budget. Under these assumptions, if each project is budgeted below its expected cost, then the sum of the individual budgets for the N projects, $B_T = NB = \$375,000$, greatly *underestimates* the required reserve margin, and the higher management level (almost) always overruns. Id enough projects overrun, this can lead to bankruptcy if the contractor or owner is a corporation. This simple example illustrates one reason why the failure rate for construction contractors is higher than that in other businesses: if, by chance, a number of projects overrun, the contractor fails.

Under what circumstances might project budgets be biased on the low side (less than the expected values)? Some possibilities might be:

• Projects are pushed by proponents, who recognize that their probability of getting funded decreases with increasing project cost estimates. They also recognize that, even if a project is underfunded, some funding is better than none, and they expect to go back to the owner or funding agency to rebaseline the budget once the project is under way and past the point of no return. So, there may be major incentives for project proponents to lowball the cost estimates, and disincentives for accurate estimating. (Gresham's Law says that bad money drives out good, so we may say that bad estimates drive out good ones. From an evolutionary perspective, unbiased estimators don't get their projects funded and hence they die out.) The above analysis shows that even a small lowball bias at the project level would lead to virtual certainty that budgets would be overrun at a higher management level.

• Another possibility is that project estimates are unbiased at the project level, but are arbitrarily reduced at a higher corporate level, on the basis that estimates are too large to be competitive or contain too much contingency (aka fat). Or, trying to do more projects than the funds available would support, program managers may simply divide their fixed resources among their projects regardless of project estimates. This may be characteristic of some public agencies, in which projects are politically mandated but program funds are inadequate. This behavior is also a self-reinforcing, positive feedback loop.

Another, not mutually exclusive, hypothesis is an asymmetry in how funds are handled. In the above analysis, it was assumed that cost overruns in some projects are (statistically) offset by underruns in others; hence reserve margins can be (proportionately) lower when taken over many projects. To take a different view, suppose that every project that overruns its budget appeals for relief for a higher program level, and gets it, but that projects that underrun hold on to all or part of the budget surplus instead of passing it back. This may be more characteristic of cost-plus contracts.

The discussion above has used the term "projects" as the lowest level entities, summing up to "programs" at the corporate level. An identical analysis could be made with "Work Packages" replacing "projects" as the basic elements. The statistical results for large numbers would be reinforced, because a typical major project could have 100–1000 Work Packages, so $N = 1000$ in the above examples. Work Packages would then sum up, in two or more steps, to projects, which would sum up to programs, etc. This would give a hierarchy of five or more levels from Work Packages up to owner or corporate management. However: although it may be plausible (although not probable) that all projects in one company's program are statistically independent, it is not possible that all Work Packages in a single project could be statistically independent (although they are often assumed to be, without any analysis). Therefore, account would have to be taken of correlations or dependencies between Work Packages, which is an interesting subject that has been insufficiently investigated.

The discussion above applied to costs. Similar observations and conclusions can be reached by substituting "duration" for "cost" and "schedule" for "budget." Actually, there are some new and interesting issues and results when considering durations and schedules rather than cost estimates and budgets, but these are again beyond the intent of this chapter.

9.5   Correlation Effect

As previously noted, the variance of the sum of a number of correlated work packages is the sum of all the covariances in the covariance matrix. Sometimes in these notes, for expository reasons, the covariance matrix is simplified by assuming that all work packages have the same standard deviation σ. A simplified correlation matrix may be the following, in which all pairs of work packages have the same correlation, ρ, as in the 10 by 10 correlation matrix below:

$$\begin{bmatrix}
 & 1 & 2 & 3 & 4 & 5 & 6 & 7 & 8 & 9 & 10 \\
1 & 1.0 & \rho & \rho & \rho & \rho & \rho & \rho & \rho & \rho & \rho \\
2 & \rho & 1.0 & \rho & \rho & \rho & \rho & \rho & \rho & \rho & \rho \\
3 & \rho & \rho & 1.0 & \rho & \rho & \rho & \rho & \rho & \rho & \rho \\
4 & \rho & \rho & \rho & 1.0 & \rho & \rho & \rho & \rho & \rho & \rho \\
5 & \rho & \rho & \rho & \rho & 1.0 & \rho & \rho & \rho & \rho & \rho \\
6 & \rho & \rho & \rho & \rho & \rho & 1.0 & \rho & \rho & \rho & \rho \\
7 & \rho & \rho & \rho & \rho & \rho & \rho & 1.0 & \rho & \rho & \rho \\
8 & \rho & \rho & \rho & \rho & \rho & \rho & \rho & 1.0 & \rho & \rho \\
9 & \rho & \rho & \rho & \rho & \rho & \rho & \rho & \rho & 1.0 & \rho \\
10 & \rho & \rho & \rho & \rho & \rho & \rho & \rho & \rho & \rho & 1.0
\end{bmatrix}$$

If the number of work packages is N, then the variance and the standard deviation of the total cost for this pattern of correlations may be written:

$$Var(T) = \sigma_T^2 = \sigma^2 N\left[1 + \rho(N-1)\right]$$
$$\sigma_T = \sigma\sqrt{N\left[1 + \rho(N-1)\right]}$$

Figure 9.2 shows the variance of the total project cost $\dfrac{\sigma_T^2}{\sigma^2}$ versus σ^2 for several values of ρ. (Multiply the ordinates by σ^2 to obtain the true values for the variance of the total project cost.)

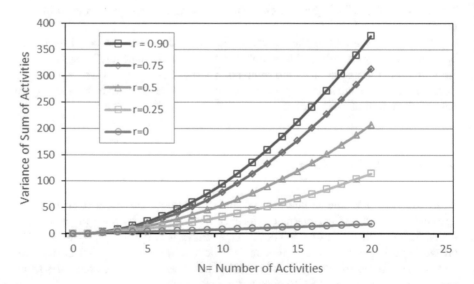

Fig. 9.2 Variance of total project costs

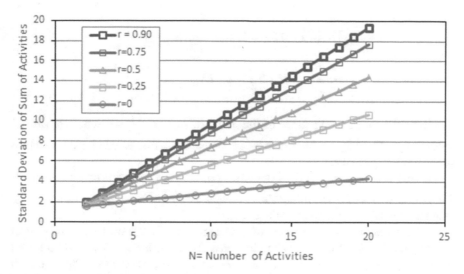

Fig. 9.3 Standard deviation of total project costs

Figure 9.3 shows the standard deviations of the total project cost $\dfrac{\sigma_T}{\sigma}$ versus N for several values of ρ. (Multiply the ordinates by σ to obtain the true value for the standard deviation of the total project cost.)

Another simple correlation pattern that uses exponentially-decaying correlations is:

$$
\begin{bmatrix}
 & 1 & 2 & 3 & 4 & 5 & 6 & 7 & 8 & 9 & 10 \\
1 & 1.0 & \rho & \rho^2 & \rho^3 & \rho^4 & \rho^5 & \rho^6 & \rho^7 & \rho^8 & \rho^9 \\
2 & \rho & 1.0 & \rho & \rho^2 & \rho^3 & \rho^4 & \rho^5 & \rho^6 & \rho^7 & \rho^8 \\
3 & \rho^2 & \rho & 1.0 & \rho & \rho^2 & \rho^3 & \rho^4 & \rho^5 & \rho^6 & \rho^7 \\
4 & \rho^3 & \rho^2 & \rho & 1.0 & \rho & \rho^2 & \rho^3 & \rho^4 & \rho^5 & \rho^6 \\
5 & \rho^4 & \rho^3 & \rho^2 & \rho & 1.0 & \rho & \rho^2 & \rho^3 & \rho^4 & \rho^5 \\
6 & \rho^5 & \rho^4 & \rho^3 & \rho^2 & \rho & 1.0 & \rho & \rho^2 & \rho^3 & \rho^4 \\
7 & \rho^6 & \rho^5 & \rho^4 & \rho^3 & \rho^2 & \rho & 1.0 & \rho & \rho^2 & \rho^3 \\
8 & \rho^7 & \rho^6 & \rho^5 & \rho^4 & \rho^3 & \rho^2 & \rho & 1.0 & \rho & \rho^2 \\
9 & \rho^8 & \rho^7 & \rho^6 & \rho^5 & \rho^4 & \rho^3 & \rho^2 & \rho & 1.0 & \rho \\
10 & \rho^9 & \rho^8 & \rho^7 & \rho^6 & \rho^5 & \rho^4 & \rho^3 & \rho^2 & \rho & 1.0 \\
\end{bmatrix}
$$

If the number of work packages is N, then the variance and the standard deviation of the total cost for this pattern of correlations may be written:

$$Var(T) = \sigma_T^2 = \sigma^2 \left[N + 2\sum_{k=1}^{N-1}(N-k)\rho^k \right]$$

$$\sigma_T = \sigma \sqrt{\left[N + 2\sum_{k=1}^{N}(N-k)\rho^k \right]}$$

Figure 9.4 shows the variance of the total project cost $\dfrac{\sigma_T^2}{\sigma^2}$ versus N for several values of ρ. (Multiply the ordinates by σ^2 to obtain the true values for the variance of the total project cost.)

Figure 9.5 shows the standard deviations of the total project cost $\dfrac{\sigma_T}{\sigma}$ versus N for several values of ρ. (Multiply the ordinates by σ to obtain the true value for the standard deviation of the total project cost.)

An example illustrates the effect of correlations on the assessment of the project or program risk. Consider the example stated above, with $N = 10$; $\mu = 4000$ for all the work packages; $\sigma = 1000$ for all the work packages; $\rho = 0.00$ for the case with independent work packages; and $\rho = 0.75$ using the exponentially decaying correlation pattern above.

Figure 9.6 compares the risk function or probability density function for the case with correlated work packages and the case with all independent work packages. Figure 9.6 shows that the mean values for the two cases are identical, but the standard deviations are greatly different, the standard deviation of the total project cost being much greater for the correlated case than for the independent case.

Figure 9.7 below plots the same information in the form of the cumulative probability distribution functions for the correlated and uncorrelated cases. In this figure it is easily seen that, if the project owner wishes to set a project budget at the

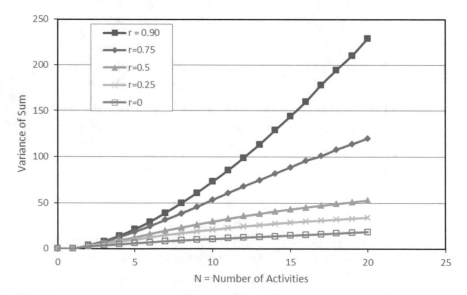

Fig. 9.4 Variances vs number of activities; exponential decay

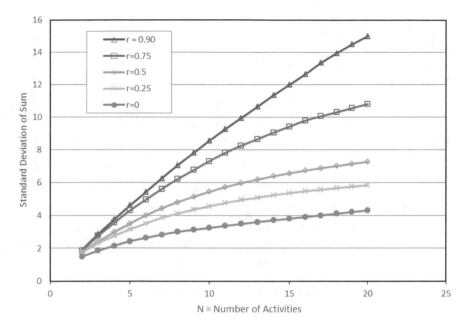

Fig. 9.5 Standard deviations vs. number of activities; exponential decay

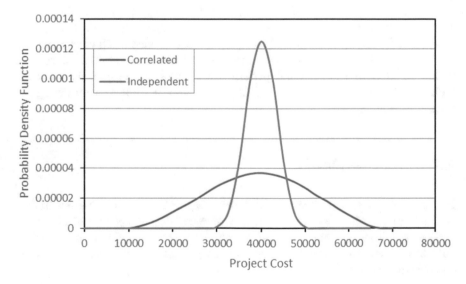

Fig. 9.6 Project total cost probability density functions

95% confidence level, the budget including contingency would be about $45,000 if all the N work packages were independent. However, if the N work packages are correlated, with $\rho = 0.75$ as shown earlier, then the required budget would be about $57,000. Conversely, if the project manager assumed that all the work packages were independent, and set a budget with contingency at the 95% level of $45,000,

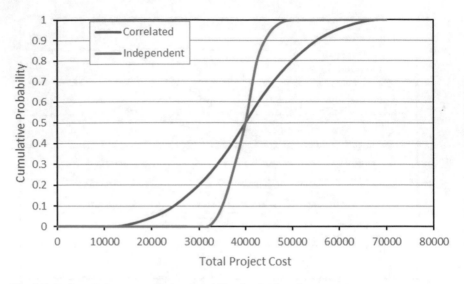

Fig. 9.7 Project total cost cumulative probability density functions

the risk that this budget would be overrun is, from the graph below, about 32%. That is, in this example, if the project manager assumes that all work packages are independent, then he believes that there is only one chance in 20 of overrunning his budget of $45,000. However, if the work packages are actually correlated (using the decaying exponential pattern given), then the project manager has under-estimated the risk and there is really one chance in three of overrunning this budget amount.

9.6 Practice Problems

Problem 9.6.1 Consider Problem 3.5.1 from Chap. 3. Assuming that uncertainty comes from the schedule only (i.e. the cost of materials and labor are deterministic and set at their mean value), calculate the required contingency funds to meet 95% confidence level. How would your result change if all the activities are positively and negatively correlated?

Problem 9.6.2 A state highway engineer has to determine contingency funds for asphalt paving program for next planning period. There 12 jobs to be executed and the engineering estimate of the total tons of asphalt to be placed are available and shown in Table 9.2. Assuming that the estimates are mean values and that the typical coefficient of variation for asphalt placing jobs is 0.2, calculate contingency

Table 9.2 Problem data

Job index	Job title	Job number	Tons of asphalt to be placed Estimated (mean)
1	Guilford Road Resurfacing	1035	11,936
2	Mebane Oaks Road and Highway 119	1036	9,900
3	State Highway 49 at Trollingwood Road	1037	31,900
4	State Highway 49 at Orange Street	1038	12,753
5	Alamance at Guilford Road	1040	15,389
6	Davidson County Resurfacing	20,452	62,039
7	City of Reidsville	20,454	3,143
8	US 52 Northbound Lanes Shoulder	20,461	11,900
9	US 52 Southbound Lanes Shoulder	20,462	12,450
10	US 220 at Guilford Road	20,466	7,941
11	City of Thomasville	20,469	2,842
12	Business 40 and US 431 Ramps	20,474	4,112
Sum			186,305

level (in tons of asphalt) at the program level that meet confidence interval of 95%. How would your results change if the asphalt quantities are positively or negatively correlated? What can cause quantities to be correlated? If the agency's budget allows only for 40,000 tons extra, what is the probability that the program will run out of money?

References

EIA Standard (1998) Earned value management systems, EIA-748. Electronic Industries Alliance, Arlington

Merriam-Webster, Inc (1983) Webster's ninth new collegiate dictionary. Merriam-Webster

Project Management Institute PMI (2008) A guide to the project management body of knowledge, 4th edn. Project Management Institute, Newtown Square

Taylor J (1997) Introduction to error analysis, the study of uncertainties in physical measurements. University Science Books, Mill Valley

Part III
Risk Monitoring and Reassessment in Project Execution

Chapter 10
Bayesian Revision of Probability Estimates

Abstract In this chapter we introduce Bayesian methods for updating model parameters based on new project-specific data. We provide formal introduction of Bayes theorem to update parameters of probability distributions used in project risk analysis. We discuss examples of updating the probability of arrival of machine breakdown and new change orders using Bayes theorem.

Keywords Bayes law · Estimation of parameter distribution · Poisson process

10.1 Introduction

Generating models is an essential element of risk management. After all, it is difficult to manage risks if you don't what the risks are. Lacking clairvoyance, our premise is that the past is some kind of guide to the future – although by no means a perfectly reliable guide. Therefore our first aim in modeling is to create a representation of the process that is applicable to the past. Developing a model is to devise the form of the model or models that we think are potential candidates for this representation. There is no unique form of model for any situation, because a model must also meet the test of parsimony – it should be as complex as necessary to represent the process, but not more so. Clearly, a model has to run (on a computer or in the brain) faster than the real world – a model that is so complicated that its execution lags the real process is of no use to us. (Although such a model may be of use to scientists – early computer models of the weather ran more slowly than actual atmospheric processes).

After formulating a possible form for a model, the next step is to fit the parameters of the model to the available historical data. Therefore, a model has to contain at least one free parameter that can be adjusted to fit the data. The step of fitting the parameters to the model requires that we have some definition of what we mean by fitting, and what constitutes a good fit versus a bad fit. In a large number of cases, best fit means choosing the parameters to minimize the sum of the squares of the deviations of the model predictions from the actual data. Here, predictions means

© Springer Nature Switzerland AG 2020

I. Damnjanovic, K. Reinschmidt, *Data Analytics for Engineering and Construction Project Risk Management*, Risk, Systems and Decisions,
https://doi.org/10.1007/978-3-030-14251-3_10

backcasting, not forecasting: that is, using the model to predict the past, not the future. Of course, no one can make a living by predicting the past, so this step is only to establish the validity of the proposed model when compared to the measured results. Hence, this step is called model validation or model verification.

To provide forecasting potential our models should also be able to make the use of the data that project itself generates; in other words, we should be able to revise our "generic" value of the estimated parameters with the parameters that fit the newly observed – project data. One may argue this is still backcasting, but it is a backcasting based on the actual project circumstances, not on the "average" circumstances of all projects done in past. To provide this model update based on new project-specific data we rely on Bayes' theorem.

In probability theory and statistics, Bayes' theorem (also known as Bayes' law or Bayes' rule) describes the probability of an event, based on conditions that might be related to the event. Bayes law for events X and Y can be derived as follows (DeGroot and Schervish 2012):

$$P\{X \cap Y\} = P\{X|Y\} P\{Y\} = P\{Y|X\} P\{X\}$$

$$P\{X|Y\} \quad = \frac{P\{X \cap Y\}}{P\{Y\}}$$

$$P\{Y|X\} \quad = \frac{P\{X \cap Y\}}{P\{X\}} = \frac{P\{X|Y\} P\{Y\}}{P\{X\}}$$

One way of looking at this formulation is as an expression of the conditional probability distribution of cause given the observed evidence using the converse conditional probability of observing evidence given the cause:

$$P\left(Cause|Evidence\right) = P\left(Evidence|Cause\right) \frac{P\left(Cause\right)}{P\left(Evidence\right)}$$

Bayes theorem can be easily extended to update the parameters of a model given observed outcomes

$$P\{D \cap \Theta\} = P\{D|\Theta\} P\{\Theta\} = P\{\Theta|D\} P\{D\}$$

$$P\{D|\Theta\} = \frac{P\{D \cap \Theta\}}{P\{\Theta\}}$$

$$P\{\Theta|D\} = \frac{P\{D \cap \Theta\}}{P\{D\}}$$

where, $P\{D|\Theta\}$= Conditional probability that outcome D would be observed given parameters Θ.

Table 10.1 Prior probability distribution of θ_j

Index j	Parameter value θ_j	Prior probability $P[\theta_j]$
1	0.48	0.20
2	0.49	0.20
3	0.50	0.20
4	0.51	0.20
5	0.52	0.20
Sum		1.00

$P\{\Theta|D\}=$ Posterior conditional probability that Θ is the value of the parameters given observed outcomes D.

If we have two sets of observations $D_1 =$ first set of observations (data) and $D_2 =$ second set of observations (data) then

$$P\{\Theta|D_2\} = \frac{P\{D_2 \cap \Theta\}}{P\{D_2\}} = \frac{P\{D_2|\Theta\}P\{\Theta\}}{P\{D_2\}} = \frac{P\{D_2|\Theta\}P\{\Theta|D_1\}}{P\{D_2\}}$$

Where, $P\{\Theta|D_2\}=$ posterior conditional probability Θ is the value of the parameters given observed outcomes D_1 and D_2.

For example, let's consider a situation when one would like to determine if the coin unbiased based on consecutive observations of the coin flip outcomes. Let $\theta =$ probability of heads be our initial uncertainty about the value of the parameter. Here $\theta = 0.5$ implies that the coin is unbiased. Table below summarizes our prior estimates. Since we are unsure what estimates are more likely than the others we will assume that they are all equally likely; this is often referred as noninformative prior (see Table 10.1).

Figure 10.1 illustrates probability mass function for a uniform noninformative prior (tabulated in Table 10.1).

After one flip of the coin; $D =$ Tails, given that

$$P\{\theta_j|D\} = \frac{P\{D|\theta_j\}P\{\theta_j\}}{\sum\limits_{j=1}^{5} P\{D|\theta_j\}P\{\theta_j\}}$$

Now we have the following results (See Table 10.2).

After five flips of the coin, $D =$ Tails-Tails-Tails-Tails-Tails, our estimates change. Table 10.3 shows the posterior mass function.

Figures 10.2 and 10.3 show estimated probability given the Run of eight consecutive Tails and alternating Tail-Head sequence.

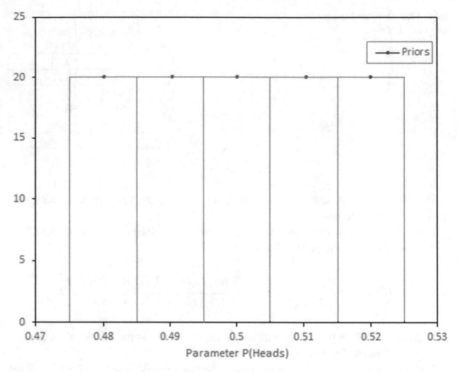

Fig. 10.1 Probability mass function for a uniform noninformative prior

Table 10.2 Posterior probability distribution of θ_j after D = Tails

Index j	Parameter value θ_j $\theta_j = P\{Heads\}$	Prior probability $P[\theta_j]$	$P\{D = Tails|\theta_j\}$	$P\{Tails|\theta_j\}P\{\theta_j\}$	Posterior probability$P\{\theta_j|D = Tails\}$
1	0.48	0.20	0.52	0.104	0.208
2	0.49	0.20	0.51	0.102	0.204
3	0.50	0.20	0.50	0.100	0.200
4	0.51	0.20	0.49	0.098	0.196
5	0.52	0.20	0.48	0.096	0.192
Sum		1.00		0.500	1.000

Table 10.3 Posterior probability distribution of θ_j after D = Tail-Tails-Tails-Tails-Tails

Index j	Parameter value θ_j $\theta_j = P\{Heads\}$	Prior probability $P[\theta_j]$	$P\{D = Tails|\theta_j\}$	$P\{Tails|\theta_j\}P\{\theta_j\}$	Posterior probability$P\{\theta_j|D = Tails\}$
1	0.48	0.20	$(0.52)^5$	0.00760408	0.2414
2	0.49	0.20	$(0.51)^5$	0.00690051	0.2191
3	0.50	0.20	$(0.50)^5$	0.00625000	0.1984
4	0.51	0.20	$(0.49)^5$	0.00564950	0.1793
5	0.52	0.20	$(0.48)^5$	0.00509608	0.1618
SUM		1.00		0.03150017	1.000

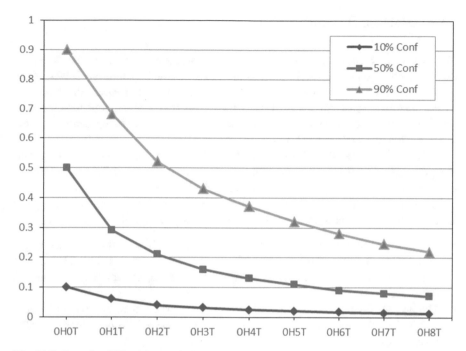

Fig. 10.2 Bayesian P(Heads) run of tails

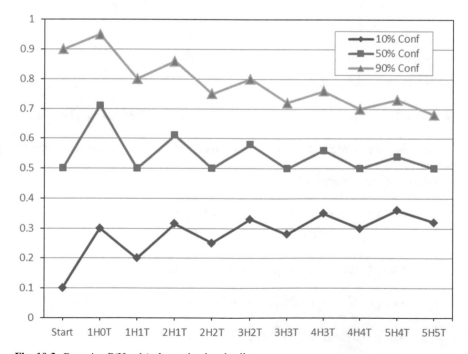

Fig. 10.3 Bayesian P(Heads) alternating head-tails

10.2   Poisson Process

The model most commonly used to represent systems of random events, independent of each other, occurring in time, is the Poisson model (Kingman 1992). Many real processes in engineering project management have been successfully modeled as Poisson processes. The Poisson model assumes that events occur at some rate, and the process has no memory (that is, the time of an event is completely unrelated to the timing of any previous events.

A derivation of the Poisson process can be found in virtually every undergraduate probability textbook. An abbreviated version is presented here primarily to expose some of the assumptions that underlie the Poisson model.

Let T be a time interval, with the starting point of the interval at any random point in time. The derivation is predicated on the following assumptions or stipulations:

- The probability of observing exactly n events in the time interval T is the same, no matter where the starting point of the time interval is located. That is, the Poisson process is a memoryless equilibrium process, the same everywhere in time.
- The probability that exactly one event will be observed in an interval dT is λdT.
- The probability that two or more events in the interval dT is proportional to $(dT)^2$, which is of higher order than dT.

Let $P\{n, T\}$ be the probability that exactly n events occur in time period T. Then exactly n events can occur in time period $T + dT$ in the following ways:

- n events in time T and no events in time dT;
- $n - 1$ events in time T and one event in time dT;
- $n - 2$ events in time T and two events in time dT; etc.

Then the probability of this is:

$$P\{n,T+dT\} = P\{n,T\}P\{0,dT\} + P\{n-1,T\}P\{1,dT\} + P\{n-2,T\}P\{2,dT\} + \ldots$$

Note that the conditions with n events and with $n - 1$ events are mutually exclusive.

The following probability identity holds:

$$1 = P\{0,dT\} + P\{1,dT\} + P\{2,dT\} + \ldots$$
$$\therefore P\{0,dT\} = 1 - P\{1,dT\} - P\{2,dT\} + \ldots$$

Substituting $P\{0, dT\}$ into the equation above gives:

$$P\{n,T+dT\} = P\{n,T\}\big[1 - P\{1,dT\} - P\{2,dT\} + \ldots\big] + P\{n-1,T\}P\{1,dT\} + \ldots$$

Rearranging and dividing by dT gives:

$$\frac{P\{n,T+dT\}-P\{n,T\}}{dT}=\frac{P\{1,dT\}}{dT}\Big[P\{n-1,T\}-P\{n,T\}\Big]+P\{1,dT\}+\ldots$$

Under the assumption that the probability distribution is differentiable, the term on the left of the above equation is the derivative of $P\{n,T\}$ with respect to T. The higher-order terms on the right hand side go to zero in the limit, and using the identity.

$$P\{1,dT\}=\lambda dT$$

gives the equation

$$\frac{dP\{n,T\}}{dT}+\lambda P\{n,T\}=\lambda P\{n-1,T\}$$

Solution of this differential equation recursively for values of n starting with $n=0$ gives the Poisson probability mass function:

$$P\{n,T\}=\frac{e^{-\lambda T}\left(\lambda T\right)^{n}}{n!}$$

The cumulative probability distribution must be found by summing up the probabilities for $n=1,2,3$, etc.

10.3   Failure Rates Using the Poisson Distribution and Bayes' Law

Consider the case in which events, such as equipment breakdowns, occur independently in time, with $N(t)$ the number of events in time t. As noted earlier that, if the hazard rate is constant, if the process has no memory, and if at most one event can occur in infinitesimal time Δt, then the probability of exactly $N(t)=m$ events in any finite time interval t is given by the Poisson distribution:

$$P_{m}\left(t\right)=\frac{e^{-\lambda t}\left(\lambda t\right)^{m}}{m!}\text{ for }m=0,1,2,3,4,\ldots$$

Of course, to use this equation one must have information on the value of the failure rate λ (equipment breakdowns per day) or the reciprocal, $1/\lambda$, the mean time between breakdowns (MTBF). Often, the only data available on the breakdown rate are obtained from observing the equipment itself. To escape this dilemma, we combine subjective information possessed by the project manager and construction engineer, based on experience on past projects with similar equipment with data collected on the present job for the specific equipment of interest.

To do this, we can apply Bayes' Law to update our estimate of the failure rate as additional failures are observed. Here, the parameter λ is treated as a random variable and is described by a probability distribution, not a deterministic value. By the discrete form of Bayes' Theorem, if X and Y are events,

$$P\{X \cap Y\} = P\{X\} P\{Y|X\} = P\{Y\} P\{X|Y\}$$

The variables X and Y may be considered logical values; that is, they have either the value True or the value False. The meaning of the terms above is as follows:

- $P\{X\}$, the marginal probability of X is the probability that X has value True, regardless of the value taken on by Y.
- $P\{Y\}$, the marginal probability of Y is the probability that Y has value True, regardless of the value taken on by X.
- $P\{X \cap Y\}$, the joint probability of event X and event Y, is the probability that events X and Y are both True at the same time.

There are four logical possibilities, as given in Table 10.4.

From a relative frequency viewpoint, one would say that, of all the occurrences of event X and event Y, as given in the above truth table,

- $P\{X \cap Y\}$ is the relative number of times both X and Y have the value True, or the frequency of being in the cell on the northwest corner in the table.
- $P\{X|Y\}$, the conditional probability of X given Y, is the probability one would assign to X being True if it were known for a fact that Y is True.
- $P\{Y|X\}$, the conditional probability of Y given X, is the probability one would assign to Y being True if it were known for a fact that X is True.

If knowledge of Y has no effect on the assignment of probability that X is True, then:

$$P\{X|Y\} = P\{X\}$$
$$P\{X \cap Y\} = P\{X|Y\} P\{Y\} = P\{X\} P\{Y\}$$

The last equation is the condition of independence; X and Y are said to be statistically independent if $P\{X|Y\} = P\{X\}$ and $P\{Y|X\} = P\{Y\}$.

Suppose now that Y is actually a set of mutually exclusive and collectively exhaustive events,

$$Y_1, Y_2, \ldots, Y_k, \ldots, Y_n$$

Table 10.4 Logical outcomes of X and Y

	Y = True	Y = False
X = True	X is true and Y is true	X is true and Y is false
X = False	X is false and Y is true	X is false and Y is false

Assume that the occurrence of one or another of these events is a necessary condition for the occurrence of event X. That is, one of these events must be True in order for X to be True. Then the Bayes' equation may be written:

$$P\{X \cap Y_k\} = P\{X\}P\{Y_k \mid X\} = P\{Y_k\}P\{X \mid Y_k\}$$

We are interested in the conditional probability that some Y_k is True, given that we observe that X is True, so the above equation is rewritten as:

$$P\{Y_k \mid X\} = \frac{P\{Y_k\}P\{X \mid Y_k\}}{P\{X\}}$$

$$P\{Y_k \mid X\} = \frac{P\{Y_k\}P\{X \mid Y_k\}}{P\{X \mid Y_1\}P\{Y_1\} + P\{X \mid Y_2\}P\{Y_2\} + \ldots + P\{X \mid Y_k\}P\{Y_k\} + \ldots + P\{X \mid Y_n\}P\{Y_n\}}$$

This is the probability that one of the events Y_k is True, given that we have observed X. We assume that the machine or process in question has an underlying failure rate, λ, which we cannot observe directly. We can directly observe, however, the failures or breakdown events. So, we now use the Bayes equation to infer a value for the failure rate, given that we observe a series of breakdowns (time between failures).

A formulation using probability density functions for λ as a continuous random variate is possible, but we use the discrete form here because the derivation is straightforward, and its application is simple to implement. Suppose then that λ is considered a discrete random variate that can take on any one of n discrete values, λ_k, $1 \leq k \leq n$; then, let Y_k represent the specific event $\lambda = \lambda_k$.

Also let X represent the event that exactly $N(t) = m$ breakdowns are observed in the interval from time 0 (when we start counting) to time t. Then we wish to find the probability that the underlying breakdown rate has the value λ_k, conditional on the fact that we have observed m breakdowns in time t. Note that we are assuming a constant underlying failure rate.

Bayes' law (above) becomes, substituting in the above general case:

- Y_k is the event $\lambda = \lambda_k$; that is, Y_k is true if λ_k is the underlying failure rate
- X is the event that m failures occur in time t; this is the event that is directly observed
- $X|Y_k$ is the event that we observe m failures in time t, conditional on the underlying failure rate being λ_k, and so $P\{N(t) = m \mid \lambda = \lambda_k\}$ is the probability that we would observe exactly m failures in time t if the underlying failure rate were λ_k
- $P\{X\} = P\{N(t) = m\}$ is the probability that we would observe exactly m failures in time t, taken over all values of λ_k
- $P_0\{\lambda_k\}$ is the prior probability (that is, prior to observing any breakdowns) we assign to the event that the underlying failure rate is λ_k

With these substitutions, Bayes' equation becomes:

$$P\{\lambda = \lambda_k \cap N(t) = m\} = P\{\lambda = \lambda_k\} P\{N(t) = m \mid \lambda = \lambda_k\}$$
$$= P\{N(t) = m\} P\{\lambda = \lambda_k \mid N(t) = m\}$$

This can be rewritten as:

$$P_1\{\lambda = \lambda_k \mid N(t) = m\} = \frac{P\{N(t) = m \mid \lambda = \lambda_k\} P_0\{\lambda = \lambda_k\}}{P\{N(t) = m\}}$$

That is, the posterior (after the fact) probability that λ is some particular value λ_k, given that we have observed exactly $N(t) = m$ events in time t, is $P\{\lambda = \lambda_k | N(t) = m\}$. This probability can be determined for all values of k from the above equation, in which:

- $P\{N(t) = m | \lambda = \lambda_k\}$ is the probability of getting exactly m breakdowns in time t, given that the underlying breakdown rate is λ_k. This probability is obtained from the Poisson equation:

$$P\{m \mid \lambda_k\} = P_m(t) = \frac{e^{-\lambda_k t} (\lambda_k t)^m}{m!}$$

- $P\{N(t) = m\} = \sum_{k=1}^{n} P\{N(t) = m \mid \lambda = \lambda_k\} P_0\{\lambda = \lambda_k\}$ is the total probability of observing exactly m failures, summed over all values of λ.
- $P_0\{\lambda = \lambda_k\}$ is the prior (before the fact) probability assigned to λ_k.

Then the Bayesian procedure (for discrete values) is as follows.

- Set some number n of possible discrete values of λ
- Set discrete values for all λ_k
- Estimate some prior distribution $P_0\{\lambda = \lambda_k\}$ on the probability of each λ_k, with the restriction that $\sum_{j=1}^{n} P_0\{\lambda = \lambda_j\} = 1$

(Note that, in the Bayes equation, if the prior probability is zero for some value of λ_k, such that $P_0\{\lambda_k\} = 0$, then, $P_1\{\lambda_k\} = 0$; that is, the posterior probability will always be zero for that λ_k. Note also that the mean value of the prior distribution is

$$\bar{\lambda} = \sum_{j=1}^{n} \lambda_j P_0\{\lambda_j\})$$

- At the first breakdown, $m = 1$, note the time t_1. Then, for all values of j, compute:

$$P\{m=1|\lambda_j\} = \frac{e^{-\lambda_j t_1}\left(\lambda_j t_1\right)^1}{1!} = \lambda_j t_1 e^{-\lambda_j t_1}$$

- The total probability of exactly one event in time t_1 is the sum of the terms $P\{m = 1|\lambda_j\}P_0\{\lambda_j\}$ for all values of j:

$$P\{m=1\} = \sum_{j=1}^{n}\frac{e^{-\lambda_j t_1}\left(\lambda_j t_1\right)^1 P_0\left\{\lambda_j\right\}}{1!} = t_1\sum_{j=1}^{n}\lambda_j e^{-\lambda_j t_1} P_0\left\{\lambda_j\right\}$$

- Application of the general Bayes' equation gives:

$$P\{\lambda=\lambda_k|N(t)=1\} = \frac{P\{N(t)=1|\lambda=\lambda_k\}P\{\lambda=\lambda_k\}}{P\{N(t)=1\}}$$

$$P_1\{\lambda_k|m=1\} = \frac{\lambda_k t_1 e^{-\lambda_k t_1} P_0\left\{\lambda_k\right\}}{t_1\sum_{j=1}^{n}\lambda_j e^{-\lambda_j t_1} P_0\left\{\lambda_j\right\}}$$

This last equation, applied to all k, gives the posterior probability of each λ_k, based on the prior distribution $P_0\{\lambda_k\}$ and the observed time at the first breakdown. Note that now the mean value of the breakdown rate, based on the revised (posterior) set of probabilities, is: $\bar{\lambda} = \sum_{j=1}^{n}\lambda_j P_1\left\{\lambda_j|m=1\right\}$.

When the next event, $m = 2$ at $t = t_2$, occurs, we update the probability mass function for λ by using the previous posterior distribution $P_1\{\lambda_k|m = 1\}$ as the prior, computing the new posterior distribution $P_2\{\lambda_k|m = 2\}$. There are two equivalent methods for doing this.

Method A. Use the original prior distribution $P_0\{\lambda_k\}$ as the prior, updated by the Poisson equation for $m = 2$. (That is, two events occur in time interval t_2). Then:

$$P_2\{\lambda=\lambda_k|N(t)=2\} = \frac{P\{N(t)=2|\lambda=\lambda_k\}P_0\{\lambda=\lambda_k\}}{P\{N(t)=2\}}$$

$$P_2\{\lambda_k|m=2\} = \frac{e^{-\lambda_k t_2}\left(\lambda_k t_2\right)^2 P_0\left\{\lambda_k\right\}}{\sum_{j=1}^{n}e^{-\lambda_j t_2}\left(\lambda_j t_2\right)^2 P_0\left\{\lambda_j\right\}}$$

This process continues for $m = 3,4,5$, etc., using the Poisson equation for m events but the original prior $P_0\{\lambda_k\}$ at each update.

Method B. Set $\Delta t_2 = t_2 - t_1$ and use the computed posterior distribution $P_1\{\lambda_k|m = 1\}$ as the prior, based on one observed event in time t_1, updated by the

Poisson equation for $m = 1$, the probability that one additional event occurs in time Δt_2). Then:

$$P_2\{\lambda = \lambda_k | N(t_2) = 2\} = \frac{P\{N(\Delta t_2) = 1 | \lambda = \lambda_k\} P_1\{\lambda = \lambda_k\}}{P\{N(\Delta t_2) = 1\}}$$

$$P_2\{\lambda_k | N(t_2) = 2\} = \frac{e^{-\lambda_k \Delta t_2} (\lambda_k \Delta t_2) P_1\{\lambda_k\}}{\displaystyle\sum_{j=1}^{n} e^{-\lambda_j \Delta t_2} (\lambda_j \Delta t_2) P_1\{\lambda_j\}}$$

Example 10.1

A certain process was started, and the prior probability distribution was taken as the uninformative prior, that is, one in which all discrete values of λ_k are equal. Here, the user is in effect saying that he cannot distinguish between different values of the failure rate from $\lambda_1 = 0.05$ to $\lambda_{20} = 0.20$. As there are 20 possible values for λ, each has prior probability 0.05.

The first observed period from start to the first failure is 9.8 days, resulting in the following plot of the posterior probability for each of the λ_k (see Fig. 10.4; the uniform prior is the dotted line, the posterior is the heavy solid line).

The second failure occurs after an interval of 11.7 days, so the new posterior distribution is shown below (as the heavy solid line; the posterior after one failure is the light line): (see Fig. 10.5).

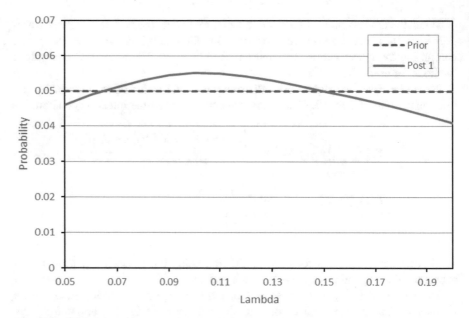

Fig. 10.4 Prior and posterior distributions after one observation

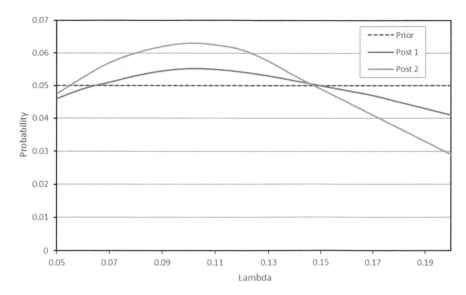

Fig. 10.5 Prior and posterior distributions after two observations

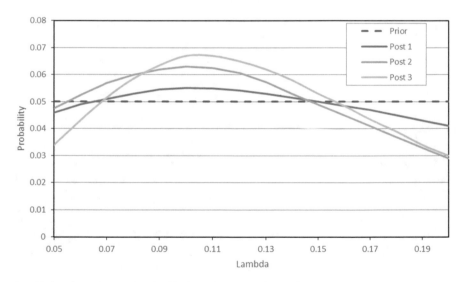

Fig. 10.6 Prior and posterior distributions after three observations

The third failure occurs after an interval of 7.0 days; the revised posterior is shown in Fig. 10.6.

After 12 failure periods, the times between failures are given in Table 10.5:

The resulting posterior distributions after each of the 12 periods are shown in Fig. 10.7.

Table 10.5 Time between failure data

Period	Time between failures, days
1	9.8
2	11.7
3	7.0
4	5.7
5	14.4
6	7.7
7	6.5
8	6.3
9	4.4
10	9.6
11	11.3
12	7.7

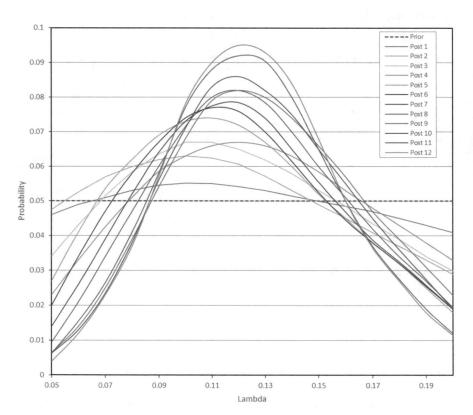

Fig. 10.7 Prior and posterior distributions after 12 observations

10.3.1   Continuous Probability Density Function

As noted before, random events, such as earthquakes, machine breakdowns, and other phenomena occur independently in time. Call $N(t)$ the number of events observed in time t. Then $N(t)$ is called a *counting process*. It was noted earlier that, if the hazard rate is constant, if the process has no memory, and if at most one event can occur in infinitesimal time Δt, then the probability of exactly $N(t) = m$ events in any finite time interval t can be approximated by the Poisson distribution:

$$P_m(t) = \frac{e^{-\lambda t}(\lambda t)^m}{m!} \text{ for } m = 0,1,2,3,4,\ldots$$

Of course, to use this equation one must have information on the value of the failure rate λ (earthquakes per year, equipment breakdowns per day) or the reciprocal, $\frac{1}{\lambda}$, the mean time between events (MTBF). Often, the only data available on the breakdown rate for some construction equipment are obtained from observing the equipment itself. However, subjective information possessed by the construction engineer, based on experience on past projects with similar equipment, can be combined with data collected on the present job for the specific equipment of interest.

As previously, we assume that the machine or process in question has an underlying failure rate, λ, which we cannot observe directly. We can directly observe, however, the failures or breakdowns. So, we now use the Bayes equation to infer a probability distribution on the value for the failure rate, given that we observe a series of breakdowns (time between failures). A formulation using probability density functions for λ as a continuous random variate, rather than discrete-valued, assumes that λ can take on any positive value, $0 < \lambda < \infty$.

Suppose that exactly $N(t) = m$ breakdowns are observed in the interval from time 0 (when we start counting) to time t. Then we wish to find the probability that the underlying breakdown rate lies in some interval $[\lambda, \lambda + \Delta\lambda]$, conditional on the fact that we have observed m breakdowns in time t. Note that we are assuming a constant underlying failure rate. $P\{N(t) = m\}$ is the Poisson probability that we would observe exactly m failures in time t, conditional on λ. This is a counting process, so we assume that m is a nonnegative integer.

The prior *probability density function* on the failure rate (that is, prior to observing any breakdowns) is assumed to be the Erlang (or Gamma) probability. (The reason for this choice will be seen below). The Erlang formula (for α an integer) is:

$$p(\lambda) = \frac{\lambda^{\alpha-1}e^{-(\lambda/\beta)}}{(\alpha-1)!\beta^\alpha}; \lambda > 0$$

Here the random variable is λ, and α and β are the parameters of the Erlang distribution.

The mean value and the variance of this probability density function are

$$E[\lambda] = \alpha\beta$$
$$\text{var}[\lambda] = \alpha\beta^2$$

Note that, if $\alpha = 1$, then the Erlang distribution reduces to the exponential distribution:

$$p(\lambda) = \frac{\lambda^{\alpha-1} e^{-(\lambda/\beta)}}{(\alpha-1)! \beta^\alpha} = \frac{1}{\beta} e^{-(\lambda/\beta)}$$
$$E[\lambda] = \beta$$
$$\text{var}[\lambda] = \beta^2$$

After some integration of the Bayes function, it can be shown that the posterior probability density function for λ using the Erlang prior distribution (conditional on having observed m events in time t) is:

$$p(\lambda|m,t) = \frac{\lambda^{m+\alpha-1} \exp\left[-\lambda\left(t + \frac{1}{\beta}\right)\right] \left(t + \frac{1}{\beta}\right)^{m+\alpha}}{(m+\alpha-1)!}; \lambda > 0$$

This function is also an Erlang probability distribution, with updated parameters, such that the parameter α in the prior is replaced by $m + \alpha$ in the posterior, and $\frac{1}{\beta}$ in the prior is replaced by $t + \frac{1}{\beta}$ in the posterior. The Erlang distribution is said to be *conjugate* because an Erlang prior distribution generates an Erlang posterior distribution, and vice versa (Forbes et al. 2011).

The mean and variance of this probability density function are

$$E[\lambda] = \frac{(m+\alpha)}{\left(t + \frac{1}{\beta}\right)}$$

$$\text{var}[\lambda] = \frac{(m+\alpha)}{\left(t + \frac{1}{\beta}\right)^2}$$

Then the Bayesian inference procedure is summarized as follows:

- Estimate the mean and variance $E[\lambda]$, var$[\lambda]$ of the prior (Erlang) distribution;
- Determine the prior values of the Erlang parameters α, β from

$$\alpha = \frac{E[\lambda]^2}{\text{var}[\lambda]}$$

$$\beta = \frac{\text{var}[\lambda]}{E[\lambda]}$$

- Adjust the mean and variance estimates such that α is an integer. If α is not an integer, then the term $(\alpha - 1)!$ in the Erlang density functions must be replaced by the gamma function $\Gamma(\alpha)$.
- The prior probability distribution on λ is then

$$p(\lambda) = \frac{\lambda^{\alpha-1} e^{-(\lambda/\beta)}}{(\alpha-1)! \beta^{\alpha}}; \lambda > 0$$

- After any number of events, m, note the time t. Then the posterior probability distribution on λ, after m observed events in time t, is given by

$$p(\lambda|m,t) = \frac{\lambda^{m+\alpha-1} \exp\left[-\lambda\left(t+\dfrac{1}{\beta}\right)\right]\left(t+\dfrac{1}{\beta}\right)^{m+\alpha}}{(m+\alpha-1)!}; \lambda > 0$$

This can be applied any number of times, with the posterior being the prior in the next cycle. Note that, as m is the number of observed events since time 0, the interpretation of our expert judgment, as expressed in the prior probability distribution, is equivalent to having observed α breakdowns in our prior experience before time 0. This interpretation depends on α being an integer.

10.3.2   Derivation of the Posterior Probability Density Function

We assumed, in the above argument, that the prior probability density function on the rate of occurrence of events λ is given by the Erlang function (for α an integer):

$$p(\lambda) = \frac{\lambda^{\alpha-1} e^{-(\lambda/\beta)}}{(\alpha-1)! \beta^{\alpha}}; \lambda > 0$$

We also assumed that the arrival of events is Poisson, so:

$$f\left(N(t)=m|\lambda\right)=\frac{e^{-\lambda t}\left(\lambda t\right)^{m}}{m!}\text{ for }m=0,1,2,3,4,\dots$$

As usual when using Bayes' Law, we write the joint probability distribution (in this case the continuous probability density function) as the product of the two distributions above:

$$f\left[N(t)=m\cap\lambda\right]=\left[\frac{e^{-\lambda t}\left(\lambda t\right)^{m}}{m!}\right]\left[\frac{\lambda^{\alpha-1}e^{-(\lambda/\beta)}}{(\alpha-1)!\beta^{\alpha}}\right]$$

Then the marginal distribution on m is obtained by integrating λ out (integration replaces summation because here λ is a continuous variable rather than discrete, as before). Then:

$$f\left[N(t)=m\right]=\frac{t^{m}}{m!(\alpha-1)!\beta^{\alpha}}\int_{0}^{\infty}\exp\left[-\lambda\left(t+\frac{1}{\beta}\right)\right]\lambda^{m+\alpha-1}d\lambda$$

Integrating then gives the marginal density function:

$$f\left[N(t)=m\right]=\frac{t^{m}\left(m+\alpha-1\right)!}{m!(\alpha-1)!\beta^{\alpha}\left(t+\dfrac{1}{\beta}\right)^{m+\alpha}}$$

Then the posterior density function for λ, conditional on the observation $N(t) = m$, is the quotient of the joint distribution divided by the marginal:

$$p\left[\lambda|N(t)=m\right]=\frac{f\left[N(t)=m\cap\lambda\right]}{f\left[N(t)=m\right]}$$

$$p\left[\lambda|N(t)=m\right]=\frac{\exp\left[-\lambda\left(t+\dfrac{1}{\beta}\right)\right]\lambda^{m+\alpha-1}\left(t+\dfrac{1}{\beta}\right)^{m+\alpha}}{(m+\alpha-1)!}$$

Example 10.2

Consider the prior distribution as the Gamma Distribution with the failure data the same as in the previous example ($E[\lambda]$ = 0.2; var[λ] = 0.0064; $\sigma[\lambda]$ = 0.08; COV[λ] = 0.4; α = 6.25; β = 0.032; $1/\beta$ = 31.25). Figure 10.8 shows, for each event, the revised posterior probability distribution on the rate.

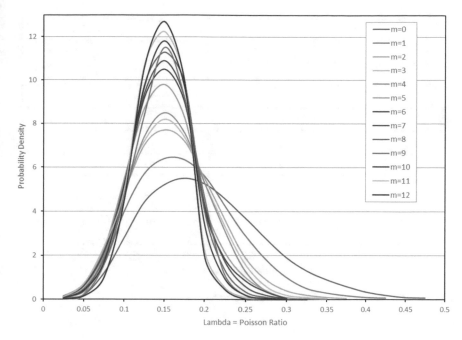

Fig. 10.8 Bayesian revision with gamma distribution

10.4   Project Change Orders Using the Poisson Model and Bayes' Law

Suppose that, on some project, a contractor is concerned about change orders. Change orders most often increase the cost or the duration of the project, or both. Consequently, the contractor is worried that the number of change orders he might get on this project would exceed the resources available. He decides that he needs some contingency factor to cover the possibility of an unexpected number of change orders.

This contingency could be expressed in dollars, representing a management reserve on the project costs to cover change orders, or expressed in time, representing a contingency in the project duration to allow for delays due to change orders. Touran (2003) considers both types of contingency, and applies probability distributions to the costs and delays of each change order to quantify these contingencies. Here we deal only with contingency expressed in change orders. For example, the project manager may wish to increase the number of engineers, construction personnel, or others assigned to the project in order to process these change orders.

Suppose the schedule for the project mentioned above is 50 weeks, and the project manager estimates, based on judgment, that the change order rate on the project would be one change every 4 weeks on the average. Hence, $\lambda = 0.25$ change orders

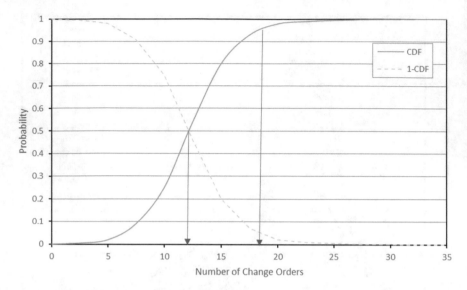

Fig. 10.9 Cumulative poisson distribution using mean rate

per week, $T = 50$ weeks, and $\lambda T = 12.5$ change orders over the life of the project. Figure 10.9 shows a plot of the cumulative Poisson distribution for these parameters, and the complementary cumulative, $1 - CDF$, which gives the probability that there will be more than X change orders during the project. From the figure it is seen that, if the project manager wants to be 95% confident about his ability to handle these change orders, he must have a contingency plan adequate for 18 major change orders over the life of the project. This implies an average rate of 0.36 change orders per week, about 50% more than the project manager's expected value. The contingency is the number the project manager chooses to prepare for, above the expected value. That is, the contingency at the 95% confidence level is $18-12 = 6$ change orders over the life of the project. Note that the impact of the change orders on the project duration is not considered in this example, so any project delays due to change orders would increase the length of time that the project is exposed to additional change orders.

The contingency determine here is dependent on the project manager's attitude to risk. At the 95% confidence level, there is only a 5% chance that the project manager will be unpleasantly surprised by receiving more than 18 change orders. If the project manager will accept a greater risk of surprises, then the contingency may be reduced. At the 50% confidence level, for example, the value for the number of change orders is equal to the mean, and contingency is zero if one defines contingency to be the difference between the value used for planning and risk management and the expected value. If the project manager chooses a 50% confidence level, then he is risk neutral; risk averse project managers will choose some confidence level above 50%.

The Poisson model assumes that the change order rate is constant in time, although it is possible to use an average rate if the rate is not believed to be constant. For example, the project manager might expect that the rate of receipt of change orders will not be constant but will vary like an inverted U, with a peak some time during the project duration. Note that it would be extremely difficult, based only on the data generated on one job, to discriminate between a constant rate and a variable rate, due to the random nature of the change orders. The Poisson model has the limitation that it has only one parameter, λ, and by estimating the mean time between change orders $\left(\dfrac{1}{\lambda}\right)$ we automatically set the variance as well $\dfrac{1}{\lambda^2}$. On the other hand, this means that we only have to estimate one parameter.

Given the above model, the project manager wishes to estimate the underlying rate at which change orders are issued based on observations of the times at which new change orders are received. That is, the assumption is that the change orders follow the Poisson process, and that there is some rate λ that governs this process, which we don't know but can try to estimate from some available information, in addition to experience. Consequently, there are two errors the project manager can make:

- The process is assumed to be Poisson but really is not, because the arrival of a change order is not independent of those previously issued. This is *model misspecification error*, the error due to using the wrong model. For example, in the Poisson model, the mean time between events is $\dfrac{1}{\lambda}$ and the variance of the interevent time is $\dfrac{1}{\lambda^2}$. Hence, the coefficient of variation for a Poisson process is always 1. Consequently, if we estimate the mean and standard deviation of the times between events and the coefficient of variation is not 1, then the process is not Poisson.
- The process is in fact approximately Poisson in nature, but the project manager estimates the rate incorrectly. This is *model-fit error*.

To revise our initial estimate of the rate, based on information gained form the on-going project, we will use Bayesian analysis, as discussed previously, to update our beliefs about the rate of change orders every time one is received. Hence, this is a *learning process*. For conceptual simplicity, we will here use the discrete form of Bayes' Rule. Those readers who have taken more advanced courses in Bayesian statistics will wonder why we do not use a more elegant closed-form solution. There is no doubt that there are better solutions to this problem, but the point here is to demonstrate that the Bayesian analysis can be applied in a relatively simple and straightforward way, and those who are aware of more elegant approaches can use them and see how close the answers given here are.

As mentioned above, for the simple Poisson model we will assume that there is a finite set of discrete values for the rate. Call these values λ_j, $1 \leq j \leq J$. Let n represent the number of change orders observed in time t, where t is measured from the start of the project. Then the Poisson probability distribution for n change orders in time t, given that λ is the change order rate, is:

$$P[n|\lambda] = \frac{e^{-\lambda t}(\lambda t)^n}{n!}$$

If λ_j is some assumed particular value of λ, then, as we discussed before, the Poisson distribution, conditional on $\lambda = \lambda j$, is:

$$P[n|\lambda = \lambda_j] = \frac{e^{-\lambda_j t}(\lambda_j t)^n}{n!}$$

Bayes' Law relates the joint distribution of n and λ_j to the conditional distributions at every time t, when n is the number of change orders in time t, by:

$$P[n \cap \lambda_j] = P[n|\lambda_j]P[\lambda_j] = P[\lambda_j|n]P[n]$$

Rearranging the above equation:

$$P[n|\lambda_j]P[\lambda_j] = P[\lambda_j|n]P[n]$$

$$P[\lambda = \lambda_j|n] = \frac{P[n|\lambda_j]P[\lambda_j]}{P[n]}$$

The denominator in the equation above is the likelihood of receiving n change orders, which is just the numerator summed over all possible values of j:

$$P[n] = \sum_{j=1}^{J} P[n|\lambda_j]P[\lambda_j]$$

Then one possible Bayesian approach for estimating the value of the change order rate from the observations is as follows:

- Assign values for the set of discrete breakdown rates $[\lambda_j]$.
- Estimate a priori probabilities $P[\lambda_j]$ based on experience, judgment, etc.
- When a change order is received, at time t, update the total number of change orders $n = n + 1$.
- Using the updated values of n and t, compute $P[n|\lambda = \lambda_j]$ from the Poisson equation for all values of j.

- Using $P[n|\lambda_j]$ from the last computation, and the prior estimates of $P[\lambda_j]$, compute
$$P[n] = \sum_{j=1}^{J} P[n|\lambda_j]P[\lambda_j].$$

- Compute $P[\lambda = \lambda_j|n]$ for all values of j. These are the updated posterior λ estimates for the probability of each discrete change order rate.
- Go to step 3.

- The best single point value for λ is the mean value obtained from

$$E[\lambda] = \sum_{j=1}^{J} \lambda_j P[\lambda_j]$$

Assume that the job has just begun, and there are as yet no change orders. We have no experience with this client, so how do we set the prior probabilities?

- We can use our experience on other jobs with other clients.
- We can ask other people who are familiar with this client.
- We can ask other people who are familiar with similar types of projects.

It may be easier for people to estimate the Poisson parameter by thinking of the time between change orders $\left(\dfrac{1}{\lambda}\right)$ rather than the change order production rate, λ.

Suppose that we do ask other people, and based on their experience and ours, we decide that:

- The least value for the mean time between change orders that is credible is 2 weeks, or $\lambda = 0.50$ change orders per week, approximately ($\lambda T \cong 25$ change orders).
- The greatest value that one can assume for the mean time between change orders is 40 weeks, or $\lambda = 0.025$ change orders per week, approximately ($\lambda T \cong 1.25$)
- This range is divided into 20 discrete values, as given by the following relationship: $\lambda_j = 0.025j$, $1 \le j \le 20$; $0.025 \le \lambda_j \le 0.50$.
- If we plead ignorance or any prior information about the probability distribution on the λ_j, we assume a priori that each of these values is equally probable, or $P[\lambda_j] = 1/20$, $1 \le j \le 20$. This is the *noninformative prior*, and implies that our estimates of λ will be based on the actual observations of change orders on this job, not experience on any other jobs.

Now we use the actual data as reported in the Bayesian revision algorithm. The first three change order arrives at times 3, 4, and 10 after the start of the project, as in the Table 10.6.

Figure 10.10 shows the posterior probability distributions on the rate λ after each of the three change orders. Also shown is the uniform (uninformative) prior distribution. After the first change order, the mode (most likely value) of λ is about 0.325 but he distribution is very flat after only one change order. After the second change

Table 10.6 First three change order arrivals	Change order number n	Time interval	Time since start
	1	3	3
	2	1	4
	3	6	10

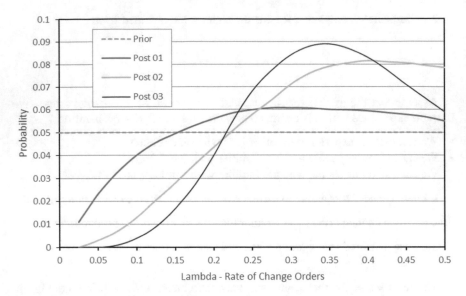

Fig. 10.10 Bayes' results based on observations of three change orders

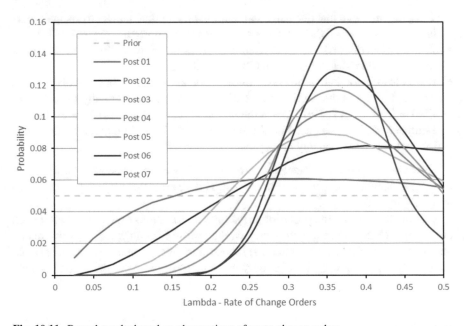

Fig. 10.11 Bayes' results based on observations of seven change orders

order, the mode moves up to about 0.425, and after the third change order the mode is down somewhat to 0.350.

Figure 10.11 shows a plot of the results after seven change orders. Time has progressed to 24 weeks out of 50 scheduled (see Table 10.7).

Table 10.7 First seven
change order arrivals

Change order number n	Time interval	Time since start
1	3	3
2	1	4
3	6	10
4	1	11
5	2	13
6	1	14
7	10	24

The probability density curves are getting noticeably tighter, as the additional information reduces the variance in (uncertainty of) the value of the parameter λ. The most likely value (mode) of λ is still, however, about 0.35.

10.5 Practice Problems

Problem 10.5.1 Consider "Truck Arrival Time" on the project site in Problem 4.7.3 from Chap. 4. Assume that prior to collecting data project engineer assumed that truck arrivals can be modeled using a homogeneous Poisson process with an arrival rate defined with a uniform prior with the minimum of 0.2 and the maximum of 1. How did the project engineer's estimates of the arrival rate changed with each new observation?

Problem 10.5.2 The new office building project has been using a pile construction subcontractor, AABB, with a single pile drilling machine, as site restrictions prevent more than one machine from operating at a time. After a period of time on the job, the project manager for the general contractor has raised some questions about the productivity of this subcontractor, and in particular about the reliability of its equipment, and whether it breaks down more often than would be expected, to meet the schedule date. Your assignment as field construction engineer is to review the performance of the AABB subcontractor so far and recommend any decisions or actions that should be taken for the remainder of the job.

The project records show the number of meters of concrete piles in each day that the subcontractor has been on the job. The table below shows the number of meters of piles driven on each day after mobilization by the subcontractor. The days with zero production represent days in which the pile driver is believed to have been broken down and under repair. To simplify the problem conceptually, assume that:

- Every morning after a day of production, an attempt is made to start up the equipment and it either works or it doesn't. If it works, it does a day's work and if it doesn't it is sent to repair for the entire day, producing nothing.

- Every morning after a day under repair, an attempt is made to start up the equipment and it either works or it doesn't. If it works, it does a day's work and if it doesn't it remains under repair and out of service for the entire day.

Assume there are two separate processes going on, which affect the productivity, and build two separate probability models, concerning:

- Reliability of the equipment, with respect to time between breakdowns and time to repair; and
- Productivity of the equipment and crew given that the pile driver is working.

Assuming that (a) the critical (i.e. maximum) failure rate that would make the project still complete on time is 0.2; and (b) the data from previous projects indicate that the mean and standard deviation of the arrival rate is 0.15 and 0.05 respectively, calculate the probability that the equipment failure rate will not result in project delays after each new observation. Assume Erlang distribution can be used to represent distribution of the arrival rate parameter.

Similarly, assuming that (a) the critical (i.e. minimum) productivity rate that would make the project still complete on time is 50 m per day per; and (b) the data from previous projects indicate the mean and standard decision of the productivity is 60 and 15 respectively, calculate the probability that the productivity rate will not result in project delays after each new observation. Assume Normal distribution can be used to represent distribution of the productivity rate; and true standard deviation is $\sigma = 20$. Posterior distributions for unknown mean (assuming Normal distribution) are shown in Table 10.8.

$$\sigma_1 = \left(\frac{1}{\sigma_0^2} + \frac{1}{\sigma^2 / n} \right)^{-1}$$

$$\mu_1 = \sigma_1 \left(\frac{\mu_0}{\sigma_0^2} + \frac{\overline{x}}{\sigma^2 / n} \right)$$

Table 10.8 Problem data

Days since start	Meters of piles reported	Number of piles	Days since start	Meters of piles reported	Number of piles
1	22.10	1	45	60.2	3
2	40.00	2	46	42.3	2
3	20.00	1	47	95.4	4
4	43.30	2	48	23	1
5	23.85	1	49	70.4	3
6	44.50	2	50	23	1
7	24.90	1	51	24	1
8	0	0	52	70.8	3
9	0	0	53	47.5	2
10	0	0	54	70.7	3

(continued)

Table 10.8 (continued)

Days since start	Meters of piles reported	Number of piles	Days since start	Meters of piles reported	Number of piles
11	0	0	55	72	3
12	0	0	56	71	3
13	0	0	57	48	2
14	0	0	58	0	0
15	0	0	59	0	0
16	47	2	60	24.3	1
17	0	0	61	72.4	3
18	22	1	62	47.85	2
19	0	0	63	69	3
20	20.1	1	64	46	2
21	46	2	65	46	2
22	23.5	1	66	0	0
23	69.7	3	67	0	0
24	45.6	2	68	0	0
25	46	2	69	0	0
26	46	2	70	46	2
27	0	0	71	0	0
28	46.5	2	72	23	1
29	0	0	73	0	0
30	22.8	1	74	23	1
31	40.6	2	75	0	0
32	69.5	3	76	0	0
33	70.4	3	77	46	2
34	24.1	1	78	0	0
35	48	2	79	0	0
36	48	2	80	67	3
37	47.8	2	81	22	1
38	48.4	2	82	66	3
39	69.8	3	83	66	3
40	67.8	3	84	44	2
41	24	1	85	40	2
42	71	3	86	66	3
43	23	1	87	65	3
44	23.3	1	88	69	3

References

DeGroot MH, Schervish MJ (2012) Probability and statistics. Pearson Education, Boston
Forbes C, Evans M, Hastings N, Peacock B (2011) Statistical distributions. Wiley, Hoboken
Kingman JFC (1992) Poisson processes, vol 3. Clarendon Press, Oxford
Touran A (2003) Probabilistic model for cost contingency. J Constr Eng Manag 129(3):280–284

Chapter 11
Managing Contingency and Budget to Complete

Abstract In this chapter we discuss methods for updating and managing project contingency as the outcomes work packages become known. More specifically we provide bivariate and multivariate formulation with a number of examples to illustrate different situations in which the presented methods can be implemented.

Keywords Budget to complete · Contingency · Correlation · Bayesian update

11.1 Introduction

Projects are learning experiences, and project organizations should learn as much as possible about risk and performance from the execution of the project under way. This chapter addresses the reassessment of project risks and the revision of budget and schedule contingencies as a project progresses, based on elementary probability theory. Projects are characterized as networks of activities or work packages, and as noted elsewhere in this book, the costs and durations of these activities may be (and most likely are) correlated. If the costs of work packages are correlated, then information about the actual cost of one completed activity or work package conveys information about the probability distributions of the others. The process that does this revision of the estimates is known as Bayes' law.

Using Bayes' law (discussed in more detail in Chap. 10), and the assumption that the activity costs and durations are Normally distributed, it is then possible to re-compute the probability distribution on the total project cost and duration every time a work package is completed. The revised risk functions are then available for use by project or program management to assess the best predictions of the final project cost and duration. This process starts with estimating the risk function on cost and duration during project planning, and continues with the revisions to these risk functions as the project progresses, until it completes.

If the revised predictions of total cost and duration exceed the acceptable values, the project manager or sponsor may take some steps, up to and including termination of the project in mid-course. This chapter does not deal with possible risk mitigation actions if that should occur. Here, the focus is on project budgets and contingencies. The principle is that the project manager sets a budget, which includes an amount for contingency, appropriate to the project when the project begins, and that this contingency is reset, up or down, as the project evolves. That is, if actual costs on the early work packages are less than the original expected values, then the project manager may believe that the risk of cost overrun is less than originally anticipated and some of the contingency (or management reserve) can be released to other projects where it is needed. Conversely, if the initial actual costs are higher than originally estimated, then more contingency may be needed, or, if this is not possible, then the risk that the project will overrun the existing budget (including contingency) will increase above the allowable risk.

Figure 11.1 shows a generic illustration of the method. At the project go-ahead, the estimated expected project cost is 70, plus a contingency or management reserve of 30, for a total budget of 100. Here, contingency has been set at the 95% confidence level; that is, there is only a 5% likelihood that the total budget will be exceeded. Actual costs for work packages, as they are obtained, are higher than the expected values, so the mean estimated cost at completion rises, as shown in the figure. However, the overall project budget, the sum of the expected cost at completion plus the contingency, remains at 100, while the management reserve falls. This is because, in this case, the increased actual costs are offset by a reduction in risk

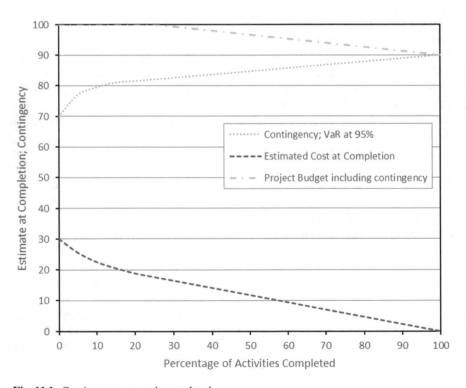

Fig. 11.1 Contingency vs. work completed

due to the progress in completing the project. At about 30% complete, the project's risks have been reduced to such an extent that not only can the contingency be reduced, but the total project budget can be reduced. This can free up resources for other projects.

Cost and schedule contingencies should be established at the outset of projects, in order to set project budgets and schedules with allowances for uncertainties, and this chapter addresses how to revise these initial contingencies as new information is being obtained during the project execution. People with project experience acknowledge that the performance on the earlier activities can be diagnostic of the performance to be expected for the rest of the project. What is needed is a usable method for doing this diagnosis.

Of course, even as a project progresses, the remaining work continues to have risks and therefore the project continues to need contingency, until all the work packages are complete, and there is no further need for contingency. This chapter describes an approach for setting and managing project budgets and schedules, including the revisions of contingencies, throughout the project duration, using a consistent definition of contingency. The method given here adopts a dynamic approach for revising the estimates of remaining project risks throughout the project duration. This method uses past results on the project to determine the uncertainties and contingencies on the future cost to complete. The method is founded on the most elementary project models: additive models for total project cost (the sum of all the individual work package costs) and for total project duration (the sum of the activity durations on the critical path). Although other models are possible, additive models are the most familiar. As will be seen, the method treats each work package identically, and is therefore suited for automatic operation. At any time during the project, the total budget is the sum of the sunk costs (the actual costs of the work packages completed), plus the expected value of the estimated cost to complete, plus the remaining contingency required to meet an acceptable level of risk.

Chapter 9 has discussed methods for setting contingencies in budget and schedule in order to reflect the existence of uncertainty or variability in knowledge about future costs, resources, and durations. At this point, one might reasonably raise the question: Should these initial contingencies be revised as the project progresses, given that information is being obtained in the process about how well the early stages of the project are doing compared to the estimates? When counting the votes on election night, analysis of the early returns is believed to give a good idea about the final results, and similarly many project managers and others with project experience acknowledge that the performance on the early work packages and activities can be diagnostic of the performance to be expected for the rest of the project. The problem is, none of these authorities gives a method for doing this diagnosis.

In the method presented here, it is assumed that work package or task costs (and activity durations) are not necessarily independent, but may in fact be correlated, and these dependencies permit information from past work package costs to be used to modify the estimates of future work package costs. Consider two work packages, say WP1 and WP2, numbered from the last to the first, such that WP2 is completed before WP1 finishes. Consider that you make an estimate of the cost of WP1 and WP2 before either is started. Then, after WP2 is complete, but before WP1 finishes, you are given the actual cost of WP2 and asked to make a re-estimate of the cost of WP1. Would you take the known cost of WP2 into consideration, or would you stay with your original estimate for WP1, regardless of the actual cost of WP2? If information about the actual cost of WP2, no matter what it is, would not cause you to change your estimate of the cost of WP1, then you believe that these two work packages are independent. Conversely, if information on the actual cost of WP2 would cause you to revise your estimate of the cost of WP1, either up or down, then you believe that the two work packages are correlated (either positively or negatively). Note that correlation does not necessarily imply causality – that something about WP2 *causes* the cost of WP1 to be higher or lower. It may be that WP2 and WP1 are related only through some third factor, perhaps even a hidden factor.

In the approach discussed here, as each work package is completed, the correlation between work package costs is used to re-estimate the future work package costs and cost to complete, and to recompute the contingency required to cover the risks for the remainder of the project. That is, the total project budget, the cost to complete, and the remaining contingency are recomputed on the basis of project cost experience, and continually adjusted as the project progresses. Because actual cost performance can provide information that reduces the uncertainty about future costs, it may be true that the required contingency, and hence the project budget including contingency, declines over the project lifetime, providing unused contingency funds that may be reallocated to other projects. Of course, it may also be true that the required contingency goes up, because the additional information about actual performance shows that the original contingency was inadequate. In this circumstance, additional contingency may be needed, and if it is not available, then the risk of over-running the budget may be higher than the original risk assessment or the risk desired.

The method described here is based on the assumption that the project manager holds or controls all the contingency. Each work package (or network activity) may draw upon the contingency as it needs to, but no predetermined amount of contingency is assigned to each work package. Therefore, the cost of the work does not expand to fill the contingency – at the work package level or at the project level. If the initial project contingency is not needed to cover the risk of overrun, then the project budget may be reduced and the available amount used for other projects in the owner's program. The objective here is to define one method for rationally assigning and managing contingencies over the entire project or program. To do so, the project manager must be able to reallocate contingencies among work packages as necessary.

Note that the method discussed here does not apply to certain events, such as natural causes (sometimes call Acts of God), the occurrences of which are described by the Poisson distribution (for example, Markov processes). Such Poisson events are by definition independent; the time between arrivals is Exponentially distributed, and knowledge of the time since the last event says nothing about the time of arrival of the next event.

11.2 Bivariate Case

To understand the method, it is assumed here for simplicity of exposition that the project under consideration consists of only two work packages. The general case with N work packages will be presented later. Let x_1 and x_2 be random variates representing the costs of work packages 1 and 2, respectively. (They could also be activity durations.) Let μ_1, μ_2 be the mean values for the work package costs and let σ_1, σ_2 be the standard deviations of the work package costs. That is, μ_1 and σ_1 are the parameters in the marginal distribution for x_1, which is assumed here to be approximated by the Univariate Normal distribution:

$$f_1(x_1) = \frac{1}{\sqrt{2\pi}\sigma_1} \exp\left\{-\frac{1}{2}\left(\frac{x_1 - \mu_1}{\sigma_1}\right)^2\right\}$$

A similar expression holds for the marginal distribution of x_2. Now, let ρ_{12} represent the correlation coefficient between the two work package costs. Assume that the joint probability density function for the work package costs is Bivariate Normal, the expression for which is:

$$f_{1,2}(x_1,x_2) = \left(\frac{1}{2\pi\sigma_1\sigma_2\sqrt{1-\rho_{12}^2}}\right)\exp\left\{-\frac{1}{2(1-\rho_{12}^2)}\left[\left(\frac{x_1-\mu_1}{\sigma_1}\right)^2 \right.\right.$$
$$\left.\left. -\frac{2\rho(x_1-\mu_1)(x_2-\mu_2)}{\sigma_1\sigma_2} + \left(\frac{x_2-\mu_2}{\sigma_2}\right)^2\right]\right\}$$

The marginal distribution of the random variate x_1 may be obtained by integrating the expression for the joint distribution, given above, over all values of x_2 in the interval $[-\infty, +\infty]$. The result of this integration is the Univariate Normal density function, as given before:

$$f_1(x_1) = \frac{1}{\sqrt{2\pi}\sigma_1} \exp\left\{-\frac{1}{2}\left(\frac{x_1 - \mu_1}{\sigma_1}\right)^2\right\}$$

That is, the marginal distribution of the cost for work package 1 is Normal with mean μ_1 and standard deviation σ_1. Similarly, the marginal distribution of the random variate x_2 may be obtained by integrating the expression above for the joint distribution over all values of x_1 in the interval $[-\infty, +\infty]$. The result of this integration is:

$$f_2(x_2) = \frac{1}{\sqrt{2\pi}\sigma_2} \exp\left\{-\frac{1}{2}\left(\frac{x_2 - \mu_2}{\sigma_2}\right)^2\right\}$$

That is, the marginal distribution of the cost for work package 2 is Normal with mean μ_2 and standard deviation σ_2, and the two marginal or univariate distributions are related through the bivariate distribution through the parameter ρ_{12}. These same relationships will hold for the more general case of the Multivariate Normal distribution, discussed later. In the general case, if one has a N-dimensional multivariate distribution, one can find the marginal distribution for any single variate by integrating out all the other $N - 1$ variates. Or, one may obtain the bivariate distribution for any pair of variates by integrating out all the remaining $N - 2$ variates, and so on.

11.2.1   Prior Distribution of Total Project Costs

The total project cost is a random variate T, because it is determined by the sum of the individual work package costs, which are random variates:

$$T = x_1 + x_2$$

[If the variables are sequential activities in a project network, then the sum T is the length of the path composed of these two activities. This may be generalized to the sum of the sequential activities on each path from the project start to completion. This method can be used to find the uncertainty on the length of any path through the network, but it does not necessarily define the critical path length].

The *a priori* expected value of the total project cost is:

$$E[T] = \mu_1 + \mu_2$$

The covariance matrix is given as usual by

$$V = \begin{pmatrix} \sigma_1^{\,2} & \rho_{12}\sigma_1\sigma_2 \\ \rho_{12}\sigma_1\sigma_2 & \sigma_2^{\,2} \end{pmatrix}$$

The *a priori* variance of the total cost is the sum of all the terms in the covariance matrix, or

$$\mathrm{var}[T] \equiv \sigma_T^2 = \sigma_1^2 + 2\rho_{12}\sigma_1\sigma_2 + \sigma_2^2$$

Having determined the first two moments, $E[T]$ and var[T], of the total cost, the next step is to set the budget based on these parameters such that the risk of overrunning this budget is acceptable. The total project budget is presumed to be set such that the risk, or probability, that this budget will be exceeded is some predetermined value α, which may depend on the type of project, the owner's aversion to risk, etc. Equivalently, if B_0 is the *a priori* budget including contingency, before any work packages have been executed, the probabilities of under-running and over-running this value are, respectively:

$$\Pr\{T \le B_0\} = 1 - \alpha$$
$$\Pr\{T > B_0\} = \alpha$$

To relate B_0 and α, it is necessary to assume some probability density function for T. If all the work package costs were independent, then T would be Normally distributed, by the well-known Central Limit Theorem. But the independent case is not very interesting, for, as will be seen below, a known value for one variate says nothing about the values of all the others, and is not very realistic, either. The Central Limit Theorem is not applicable if the variates are correlated, but this fact does not imply that T is not Normally distributed; it only implies that the Normality of T cannot be proven theoretically. There are many true statements that cannot be proven mathematically, and this is one of them.

Extensive Monte Carlo computer simulations have shown that the empirical probability distributions for the sum of a number of correlated variates are indistinguishable from Normal distributions, for all values of the correlations. That is, empirically if not theoretically, it is valid to assume the Normal distribution for T. Then the *a priori* budget B_0 is set as the expected value of the sum T plus some multiple k of the standard deviation of the total cost:

$$B_0 = E[T] + k\sqrt{\text{var}[T]} = \mu_1 + \mu_2 + k\sigma_T$$
$$B_0 = \mu_1 + \mu_2 + k\sqrt{\sigma_1^2 + 2\rho_{12}\sigma_1\sigma_2 + \sigma_2^2}$$

The appropriate value of k is determined from the tables of the Normal distribution, corresponding to the probability of exceedance α determined by the client. For example, if $\alpha \cong 0.15$, then $k \cong 1$ standard deviation; or, if $\alpha \cong 0.05$, then $k \cong 1.645$.

Another way to look at this is as follows. The integral of the probability density function for the total cost T from the value $T = B_0$ to the value $T = +\infty$ is, of course, α, the probability that there is a cost overrun:

$$\int_{B_0}^{+\infty} f(t)\,dt = \alpha$$

In a similar way, one could compute the expected value of the cost overrun, which is the integral from $T = B_0$ to $T = +\infty$ of the product of the cost times the probability density (Jorion 2001):

$$E[\text{cost overrun}] = \int_{B_0}^{+\infty} tf(t)\,dt = Value - at - Risk \equiv VaR$$

Or, one may compute the conditional expected value of the overrun, given that there is an overrun, from the ratio:

$$E[\text{cost overrun|overrun}] = \frac{\int_{B_0}^{+\infty} t f(t)\, dt}{\int_{B_0}^{+\infty} f(t)\, dt} = \frac{1}{\alpha} \int_{B_0}^{+\infty} t f(t)\, dt$$

This integral, the Value-at-Risk (VaR), is the conditional expected value of the cost overruns on the project, whereas α is the probability of a cost overrun, no matter how large. If one chooses to use this approach, then B_0 is determined such that the VaR is an acceptable value.

Once the budget B_0 is determined from the acceptable risk level, the *contingency* is defined as the increment to be added to the mean value to reach the desired risk level,

$$B_0 = E[T] + \text{contingency, or}$$

$$\text{contingency} = B_0 - E[T] = k\sqrt{\sigma_1^2 + 2\rho_{12}\sigma_1\sigma_2 + \sigma_2^2}$$

The contingency, as defined here, is added to the expected cost of the project; it is not added to the project estimate, the most likely cost, the median cost, or any other value than the mean, because project work package cost estimates may contain hidden contingencies of unknown magnitude.

11.2.2   Posterior Distribution of Project Costs

Now, suppose that the work packages are performed in sequence, and that work package 2 has been completed, and the cost is known for certain to be X_2. (In this exposition, it is always assumed that the values of the costs, once the work is completed, are known exactly, with no errors. Permitting the case in which the costs are reported inaccurately would add excessive complexity to this introductory argument.) The notation here uses lower case letters for random variates, and upper case letters for specific values (or realizations) of these variates. Then the conditional probability distribution on cost x_1, given that x_2 is some specific value X_2, is $f_{1|2}(x_1|x_2)$, which is determined from the joint probability density function and the marginal distribution by a fundamental law of probability, also often known as Bayes's Law or Bayes's Theorem (Gelman et al. 2013):

$$f_{1,2}(x_1, x_2) \equiv f_{1|2}(x_1|x_2) f_2(x_2) \equiv f_{2|1}(x_2|x_1) f_1(x_1)$$

That is, the joint probability density function on two variables is the product of one conditional distribution and one marginal distribution. The joint probability distribution was given above as the Bivariate Normal, and it was also shown above that the marginal distributions can be derived by integrating the joint distribution. With this information, it is possible to solve for the conditional probability density function as:

$$f_{1|2}\left(x_1\,|x_2\right)=f_{1,2}\left(x_1,x_2\right)/\,f_2\left(x_2\right)$$

It can be readily shown by this division that the conditional probability density function $f_{1|2}(x_1|x_2)$ is Normal with parameters

$$E\left[x_1|x_2=X_2\right]=\mu_1+\rho_{12}\left(\frac{\sigma_1}{\sigma_2}\right)\left(X_2-\mu_2\right)$$

$$\mathrm{var}\left[x_1|x_2=X_2\right]=\sigma_1^2\left(1-\rho_{12}^2\right)$$

With $x_2 = X_2$ known, the cost to complete the project is x_1, which is still a random variable. If x_1 and x_2 are independent, then $\rho_{12} = 0$ and knowing the value $x_2 = X_2$ provides no information about the remaining cost. This can be easily seen by substituting $\rho_{12} = 0$ in the two equations above; the result is $E[x_1|x_2 = X_2] = \mu_1$ and $\mathrm{var}\left[x_1|x_2=X_2\right]=\sigma_1^2$, the *a priori* values. Knowing the value of $x_2 = X_2$ does reduce the project risk, of course, as now there is no risk associated with the cost of work package 2, and this reduced risk can be reflected in the contingency needed after $x_2 = X_2$ has been determined.

On the other hand, if x_1, x_2 are dependent, that is, $\rho_{12} \neq 0$, then knowing that $x_2 = X_2$ provides new information about the probability distribution of the remaining work package, that is, the remaining cost to complete. For example, if x_1, x_2 are positively correlated, that is, $\rho_{12} > 0$ and if the actual cost of work package 2 is less than its expected value, that is, $X_2 < \mu_2$, then $E[x_1|x_2 = X_2] < \mu_1$. That is, the *a posteriori* expected cost of the remaining work package is less than the *a priori* expected value.

Of course, different results may be obtained, as can be seen by substituting $\rho_{12} < 0$ or $X_2 > \mu_2$ in the above expressions. If the correlation coefficient is negative, then the values of x_1 and x_2 tend to move in opposite directions, and a value $X_2 < \mu_2$ leads to $E[x_1|x_2 = X_2] > \mu_1$. However, any dependence, positive or negative, that is, $\rho_{12} \neq 0$, reduces the variance of the remaining work package cost, compared to the *a priori* value, regardless of the value of $x_2 = X_2$, because $\mathrm{var}[x_1|x_2 = X_2]$ depends on ρ_{12}^2. That is, *any* information about x_2 reduces the uncertainty in x_1, if they are correlated. Hence, *if there is dependence*, information about the actual cost of work completed can be used to generate better estimates of the cost to complete, with lower variance. If there is no dependence, then the actual cost of the work completed provides no information about the cost of the work remaining.

To explore this fact, assume that the project manager has a consistent view of contingency, in that the budget for the project after the completion of work package 2 should be set such that the probability of overrunning this amount is the same probability, α, that was used in the *a priori* situation. That is, whatever risk factor α was appropriate at the beginning of the project remains appropriate throughout the project life cycle. This means, of course, that the value of k also remains constant, even though the standard deviation changes. Therefore, after the completion of work package 2, a new budget value, B_2, is set, such that:

B_2 = cost of work completed + expected cost to complete + contingency.

Hence, using the above expressions:

$$B_2 = X_2 + E\left[x_1 \big| x_2 = X_2\right] + k\sqrt{\mathrm{var}\left[x_1 \big| x_2 = X_2\right]}$$

$$B_2 = X_2 + \mu_1 + \rho_{12}\left(\frac{\sigma_1}{\sigma_2}\right)(X_2 - \mu_2) + k\sigma_1\sqrt{1 - \rho_{12}^2}$$

Thus, knowledge of the actual cost of work package 2 may result in a change to the total project budget (shown here as a *budget reduction*) of $B_0 - B_2$:

$$B_0 - B_2 = \left[\mu_1 + \mu_2 + k\sqrt{\sigma_1^2 + 2\rho_{12}\sigma_1\sigma_2 + \sigma_2^2}\right] -$$
$$\left[X_2 + \mu_1 + \rho_{12}\left(\frac{\sigma_1}{\sigma_2}\right)(X_2 - \mu_2) + k\sigma_1\sqrt{1 - \rho_{12}^2}\right]$$

$$B_0 - B_2 = \mu_2 - X_2 + \rho_{12}\left(\frac{\sigma_1}{\sigma_2}\right)(\mu_2 - X_2) + k\left[\sqrt{\sigma_1^2 + 2\rho_{12}\sigma_1\sigma_2 + \sigma_2^2} - \sigma_1\sqrt{1 - \rho_{12}^2}\right]$$

This budget differential (reduction) can be seen to consist of three parts:

- $\mu_2 - X_2$, which represents the direct saving due to an underrun on work package 2 (assuming the case that $\mu_2 > X_2$);

- $\rho_{12}\left(\frac{\sigma_1}{\sigma_2}\right)(\mu_2 - X_2)$, which represents the expected saving on work package 1,

  based on the reported under-run on work package 2 and the correlation between the two work packages. It also depends on the ratio of the two standard deviations: the greater the standard deviation of work package 1 relative to work package 2, the greater the saving; and

- $k\left[\sqrt{\sigma_1^2 + 2\rho_{12}\sigma_1\sigma_2 + \sigma_2^2} - \sigma_1\sqrt{1 - \rho_{12}^2}\right]$, which represents a reduction in the

  contingency necessary to cover the risk, corresponding to a reduced uncertainty caused by the information provided by the fact that one work package is complete. Note that this term is independent of the actual reported cost, X_2.

This situation might have one of two possible outcomes:

- The budget could be reduced, and the difference could be returned to the client for use on other projects.
- The budget could be held constant. This would correspond to a greater contingency and a reduction in the risk factor α, as long as scope changes are not permitted to consume the available budget.

Of course, if the experience on the work completed was unfavorable, that is, if μ_2 < X_2, or if the correlation were negative, then the differential might be negative, and if increasing the budget is not acceptable to the client, then the available contingency would go down. This would increase the risk of a cost overrun above the original maximum desired value α.

Note that it is not necessary to assume that the only possible policy is to maintain a constant value for the risk of cost overrun α throughout the entire project. The project manager might, for example, wish to have the risk of a cost overrun decrease over the life of the project, if possible. That is, he might be willing to start a project with a risk of cost overrun of, say, $\alpha = 0.25$ *a priori*, with the goal that this should be reduced to, say, $\alpha = 0.05$ when the project is three-quarters complete. Other strategies may also be imagined.

Example 11.1

To illustrate the points above, assume the following *a priori* data:

$$\mu_1 = \mu_2 = \$100k$$
$$\sigma_1 = \sigma_2 = \$40k$$
$$\alpha = 0.05$$

Then, the coefficient of variation for both work packages is 40%, and the value of k corresponding to the risk 0.05 is $k = 1.645$. Consider two conditions:

(a) Independence, $\rho_{12} = 0.0$
(b) High correlation, $\rho_{12} = 0.9$

In both conditions, the *a priori* expected value of the total project cost, $E[T] = \$200K$. However, the covariance matrices differ for the two dependence conditions:

$$V_a = \begin{pmatrix} 1600 & 0 \\ 0 & 1600 \end{pmatrix} \text{ and } V_b = \begin{pmatrix} 1600 & 1440 \\ 1440 & 1600 \end{pmatrix}$$

By summing the terms in the covariance matrices, the *a priori* variances of the total project costs for the two dependency conditions are:

- Condition A: var[T] = 3200, so $\sigma_T = \$57K$ and coefficient of variation = 28%.
- Condition B: var[T] = 6080, so $\sigma_T = \$78K$ and coefficient of variation = 39%.

The contingency $k\sigma_T$ at the 5% level for the independent condition is $93K, and for the correlated condition is $128K. This gives budgets including contingency of $293K for condition a, and $328K for condition B. Clearly, in this case, the dependency between the work package costs raises the required budget with contingency at the 5% level by a significant amount, $35K.

Suppose that the actual cost of work package 2 is reported to be $x_2 = X_2 = \$100K = \mu_2$. As the actual value is exactly the mean, it implies that the best estimate for the cost to complete is the *a priori* mean of x_1. For the independent condition A, $\text{var}\left[x_1 | x_2 = X_2\right] = \sigma_1^2$ for any value of X_2, so the covariance matrix for the cost to complete is just $V = 1600$. Hence, for the independent condition, the total *a posteriori* project budget to maintain a constant risk factor is:

$$B_2 = X_2 + \mu_2 + k\sigma_T = 100 + 100 + 1.645\sqrt{1600} = \$266K$$

The required budget with contingency has gone down from \$293K to \$266K, which implies that \$27K of contingency could be reallocated to other projects.

In the correlated condition, b, the best estimate for the cost to complete is also the *a priori* mean of x_1. For this condition, the conditional variance depends on the values of σ_1 and ρ_{12}; that is, $\text{var}\left[x_1 | x_2 = X_2\right] = \sigma_1^2\left(1 - \rho_{12}^2\right) = 1600(1.00 - 0.81) = 304$ for any value of X_2, so the covariance matrix for the cost to complete is just $V = 304$. Hence, for the dependent condition, the total *a posteriori* project budget is:

$$B_2 = X_2 + \mu_1 + k\sigma_T = 100 + 100 + 1.645\sqrt{304} = \$229K$$

The required budget with contingency has gone down from \$328K to \$229K, which implies that \$99K could be returned to the sponsor. Hence, the initial budget was higher for the dependent condition, but the budget reduction after work package 2 is also much higher, due to the reduction in the uncertainty caused by the dependency. As a result, condition B. now has a lower budget than the independent condition.

Suppose instead that the actual cost of work package 2 is reported to be lower that the expectation, say $x_2 = X_2 = \$80K < \mu_2 = 100$. For the independent condition A, the actual value of X_2 has no influence on the cost to complete or the remaining uncertainty. Hence, for the independent condition, the total *a posteriori* project budget is:

$$B_2 = X_2 + \mu_1 + k\sigma_T = 80 + 100 + 1.645\sqrt{1600} = \$246K$$

In this case, the required budget with contingency has gone down from \$293K to \$246K, which implies that \$47K could be put to more productive uses, without affecting the risk on this project, which is still 5%.

In the positively correlated condition B, the reported value of x_2 below its mean implies that the best estimate for the cost to complete is now less than the *a priori* mean of x_1. For this condition, $\mu_{1|2} = \mu_1 + \left(\dfrac{\rho_{12}\sigma_1}{\sigma_2}\right)(X_2 - \mu_2) = 100 + 0.9(80 - 100) = \$82K$. The covariance matrix for the cost to complete is still $V = 304$. Hence, for the dependent condition, the total *a posteriori* project budget is:

$$B_2 = X_2 + \mu_{\sharp 2} + k\sigma_T = 80 + 82 + 1.645\sqrt{304} = \$191K$$

Due to the favorable result on work package 2, and the reduction in uncertainty due to dependence, the required budget with contingency has gone down from \$328K to \$191K, which implies that \$137K could be otherwise allocated. As a result, condition B now has a much lower budget than the independent condition A.

Suppose now that the actual cost of work package 2 is reported to be $x_2 = X_2 = \$127K > \mu_2 = 100$. For the independent condition A, the total *a posteriori* project budget is:

$$B_2 = X_2 + \mu_1 + k\sigma_T = 127 + 100 + 1.645\sqrt{1600} = \$293K$$

The over-run on work package 2 used up some of the project contingency, but the required *a posteriori* budget with contingency now happens to be exactly the same as the original *a priori* budget. The risk is still 5%.

In the correlated condition B, the reported value of x_2 above its mean implies that the best estimate for the cost to complete is now more than the *a priori* mean of x_1. For this condition, $\mu_{\sharp 2} = \mu_1 + \left(\dfrac{\rho\sigma_1}{\sigma_2}\right)(X_2 - \mu_2) = 100 + 0.9(127 - 100) = \$124K$.

The variance of the cost to complete is unchanged. Hence, for the dependent condition, the total *a posteriori* project budget is now:

$$B_2 = X_2 + \mu_{\sharp 2} + k\sigma_T = 127 + 124 + 1.645\sqrt{304} = \$280K$$

Even with the unfavorable result on work package 2, the reduction in uncertainty due to the dependence means that the required budget with contingency has gone down from \$328K to \$280K, which implies that \$48K could be put to other uses.

As a final example, suppose that the actual cost of work package 2 is reported to be $x_2 = X_2 = \$152K > \mu_2 = 100$. For the independent condition, the total *a posteriori* project budget is:

$$B_2 = X_2 + \mu_2 + k\sigma_T = 152 + 100 + 1.645\sqrt{1600} = \$318K$$

The required budget with contingency is now more than the original *a priori* budget. This means that the owner must add \$25K to the original budget to maintain the risk at the original 5%. More likely, however, the original budget, \$293K, will remain the same and the \$25K will be taken out of the contingency. That is, the result on work package 2 used up $25K of contingency, which reduces the total contingency ($293K − $252K) to $41K. This value corresponds to $k = 1.025$, and a risk of project overrun of 15%.

In   the   correlated   condition $\mu_{1|2} = \mu_1 + \left(\dfrac{\rho_{12}\sigma_1}{\sigma_2} \right)(X_2 - \mu_2) = 100 + 0.9(152 - 100) = \$147K$.   Hence,   for   the dependent condition, the total *a posteriori* project budget is now:

$$B_2 = X_2 + \mu_{1|2} + k\sigma_T = 152 + 147 + 1.645\sqrt{304} = \$328K$$

In this case, the reduction in uncertainty due to the dependence has offset the very unfavorable result on work package 2, and so the required budget with contingency is the same as the original budget; all the overrun on work package 2 has come out of contingency and the risk remains at 5%.

One might pose the question: Does the risk of a project necessarily decline over the life of the project? Here, risk is taken to mean the probability of a cost over-run, α. Consider the case that this risk remains constant. For α to remain constant, then k must remain constant. Then, from the equations above, one can determine the values that X_2 can take on such that the *a priori* and the *a posteriori* budgets are the same, and the factor k remains the same. That is,

$$B_2 = B_0$$

$$\therefore \left[\mu_1 + \mu_2 + k\sqrt{\sigma_1^2 + 2\rho_{12}\sigma_1\sigma_2 + \sigma_2^2} \right] = \left[X_2 + \mu_1 + \rho_{12}\left(\frac{\sigma_1}{\sigma_2} \right)(X_2 - \mu_2) + k\sigma_1\sqrt{1 - \rho^2} \right]$$

$$\left[(X_2 - \mu_2) + \rho_{12}\left(\frac{\sigma_1}{\sigma_2} \right)(X_2 - \mu_2) \right] = k\left[\sqrt{\sigma_1^2 + 2\rho_{12}\sigma_1\sigma_2 + \sigma_2^2} - \sigma_1\sqrt{1 - \rho_{12}^2} \right]$$

$$(X_2 - \mu_2)\left[1 + \rho_{12}\left(\frac{\sigma_1}{\sigma_2} \right) \right] = k\left[\sqrt{\sigma_1^2 + 2\rho_{12}\sigma_1\sigma_2 + \sigma_2^2} - \sigma_1\sqrt{1 - \rho_{12}^2} \right]$$

$$X_2 = \mu_2 + \frac{k\left[\sqrt{\sigma_1^2 + 2\rho_{12}\sigma_1\sigma_2 + \sigma_2^2} - \sigma_1\sqrt{1 - \rho_{12}^2} \right]}{\left[1 + \rho_{12}\left(\frac{\sigma_1}{\sigma_2} \right) \right]}$$

The value of X_2 from the equation above marks the boundary between the regime in which the risk is increasing and the regime in which the risk is decreasing. Even in the simple bivariate case, the value of this expression depends on almost all the parameters: μ_2, σ_1, σ_2, ρ_{12}, k.

Substituting into the equation for the parameters of condition A, with $\rho_{12} = 0$, gives the value $X_2 = \$127K$. That is, if the actual reported cost of work package 2 is exactly $127K$, then the risk of over-running with a constant budget remains constant; if the actual value is less than \$127K, then the risk decreases; if the actual value is more than \$127K then the risk increases.

Substituting into the equation for the parameters of condition B, with $\rho_{12} = 0.9$, gives the value $X_2 = \$152K$. That is, if the actual reported cost of work package 2 is exactly \$152K, then the risk of over-running with a constant budget remains constant; if the actual value is less than \$152K, then the risk decreases; if the actual value is more than \$152K then the risk increases.

The conclusion from this is that whether the risk decreases or increases is a complex function of virtually all the parameters of the problem, but can be determined easily by a computer calculation.

11.3   General Multivariate Case

Of course, projects have many more than two work packages. The general expression gives the joint multivariate Normal probability density function in N random variates $x_1, x_2, x_3, \ldots , x_N$ (Hald 1952). Let

$\mathbf{x} = N \times 1$ column vector of work package costs (random variables) $\{x_1, x_2, x_3, \ldots , x_N\}^T$
$\boldsymbol{\mu} = N \times 1$ column vector of mean values of work package costs $\{\mu_1, \mu_2, \mu_3, \ldots , \mu_N\}^T$
$\mathbf{V} = N \times N$ covariance matrix
$|\mathbf{V}| = $ determinant of the covariance matrix.
$\mathbf{V}^{-1} = $ inverse of the covariance matrix.

then,

$$f_X(x) = \frac{1}{(2\pi)^{\frac{N}{2}} \sqrt{|V|}} \exp\left\{-\frac{1}{2}(x-\mu)^T V^{-1}(x-\mu)\right\}$$

Note that the equation just above puts a restriction on the covariance matrix $\mathbf{V}$: it must be invertible. And in order that its inverse exist, $\mathbf{V}$ must be positive definite. That is, any square symmetric matrix is not necessarily a valid covariance matrix; to be a valid covariance matrix, a matrix must be square, symmetric, and positive definite.

To use this equation, we first consider the situation before any actual cost reports are available; this is the *a priori* condition, and we compute the required contingency to meet the defined level of risk as before. That is, we compute the mean value of the total project cost as:

$$E[T] = \sum_{i=1}^{N} \mu_i$$

and the variance of the total cost is the sum of all the terms in the covariance matrix:

$$\mathrm{var}[T] = \sum_{i=1}^{N} \sum_{j=1}^{N} V_{ij}$$

From the mean and the standard deviation of T we compute the risk of overrunning any chosen value of the cost from the tables of the Normal distribution.

Suppose that the work packages are numbered (or renumbered) such that work package N is completed first. Then the process is to use the known value $x_N = X_N$ to revise the means and variances of the other $N - 1$ work packages as appropriate, using the approach described earlier. Then this is repeated for the next work package to be completed, say $x_{N-1} = X_{N-1}$, and so on until the only remaining work package is x_1.

In the case with N random variables, we first partition the $\mathbf{x}$ and $\boldsymbol{\mu}$ vectors. Partition x into a vector of $N - 1$ values, called x_1, plus a scalar, called x_N; and similarly for μ_1, a $(N - 1)$ vector, and μ_N, a scalar:

$$\mathbf{x} = \left\{ \begin{array}{c} \mathbf{x}_{1((N-1)x1)} \\ \mathbf{x}_{N(1x1)} \end{array} \right\}_{(Nx1)} ; \boldsymbol{\mu} = \left\{ \begin{array}{c} \boldsymbol{\mu}_{1((N-1)x1)} \\ \boldsymbol{\mu}_{N(1x1)} \end{array} \right\}_{(Nx1)}$$

We then partition the $N \times N$ covariance matrix $\mathbf{V}$ in a conformal way:

$$\mathbf{V}_{(N \times N)} = \left[\begin{array}{cc} \mathbf{V}_{11((N-1)x(N-1))} & \mathbf{V}_{1N((N-1)x1)} \\ \mathbf{V}_{N1(1x(N-1))} & \mathbf{V}_{NN(1x1)} \end{array} \right]_{(NxN)}$$

By this partitioning, x_N is a univariate Normal random variable, with mean μ_N and variance V_{NN}. And, x_1 is distributed as $(N - 1)$ multivariate Normal with mean $(N - 1)$ vector $\boldsymbol{\mu}_1$ and covariance $((N - 1) \times (N - 1))$ matrix $\mathbf{V}_{11}$.

As soon as the actual value $x_N = X_N$ is observed, the conditional distribution of x_1 given $x_N = X_N$, or $x_1 | x_N = X_N$, is multivariate $((N - 1) \times (N - 1))$ Normal with mean $(N - 1)$ vector $\boldsymbol{\mu}_{1|N}$ and $((N - 1) \times (N - 1))$ covariance matrix $\mathbf{V}_{11 | N}$, given by the equations:

$$\boldsymbol{\mu}_{1|N} = \boldsymbol{\mu}_1 + \mathbf{V}_{1N} V_{NN}^{-1} (X_N - \mu_N)$$
$$\mathbf{V}_{11|N} = \mathbf{V}_{11} - \mathbf{V}_{1N} V_{NN}^{-1} \mathbf{V}_{N1}$$

In the a posteriori condition, we know the reported value. From this, we compute the revised conditional mean vector and covariance matrix from the equations above. Using these, we compute the required contingency to meet the defined level of risk by computing the *a posteriori* mean of the total remaining cost, by:

$$E[T | x_N = X_N] = \sum_{i=1}^{N-1} \mu_{1|Ni}$$

and the variance of the total remaining cost as the sum of all the terms in the conditional covariance matrix:

$$\text{var}\left[T\,\middle|\,x_N = X_N\right] = \sum_{i=1}^{N-1}\sum_{j=1}^{N-1} V_{11|Ni,j}$$

From the known value X_N and the mean and the standard deviation of $Tlx_N = X_N$ we compute the risk of overrunning any chosen value of the cost from the tables of the Normal distribution.

For the next cycle, we reset $N \leftarrow N - 1$ and repeat the process, step by step, until $N = 2$, at which point we have reached the situation discussed in the earlier part of this text.

The recursive operations stated in the general case are actually quite straightforward, but the difficulties of notation make the equations appear more complex than they really are. Perhaps the most difficult computation is the determination of whether the given original covariance matrix is invertible; that is, positive definite. The process above does not actually require the inversion of the whole covariance matrix, but it does require that the matrix be invertible in order that the multivariate joint Normal distribution should exist. Obviously, one can determine whether the covariance matrix is invertible by trying to invert it; the process either succeeds or fails. Unfortunately, if the inversion process fails, it provides little guidance on what to do about the problem. The computation of the eigenvalues of the covariance matrix, to determine if it is positive definite, and to adjust the covariances if it isn't, is discussed elsewhere in this book (see Chap. 4).

The process in the general multivariate case is illustrated below by the trivariate problem (that is, for $N = 3$). (Space limitations make it difficult to show larger matrices.) Here, we will start with three work packages, and reduce the trivariate case to the bivariate case, as illustrated in Part a of this chapter.

Therefore, we consider three work package costs: x_1, x_2, and x_3. To help distinguish the parameters in the revision process, here we use the following notation:

a priori means for each work package: m_1, m_2, and m_3
a priori standard deviations for each work package: s_1, s_2, and s_3
a priori correlation coefficients for each pair of WPs: r_{12}, r_{13} and r_{23}

The third work package will be finished first, so we wish to use the actual value reported for this work package to revise the estimates for the remaining two. The *a priori* covariance matrix, using the above notation, is:

$$\begin{bmatrix} \begin{bmatrix} v_{11} & v_{12} \\ v_{21} & v_{22} \end{bmatrix} & \begin{bmatrix} v_{13} \\ v_{23} \end{bmatrix} \\ \begin{bmatrix} v_{31} & v_{32} \end{bmatrix} & \begin{bmatrix} v_{33} \end{bmatrix} \end{bmatrix} = \begin{bmatrix} V_{11} & V_{12} \\ V_{21} & V_{22} \end{bmatrix} = \begin{bmatrix} \begin{bmatrix} s_1^2 & s_1 s_2 r_{12} \\ s_1 s_2 r_{12} & s_2^2 \end{bmatrix} & \begin{bmatrix} s_1 s_3 r_{13} \\ s_2 s_3 r_{23} \end{bmatrix} \\ \begin{bmatrix} s_1 s_3 r_{13} & s_2 s_3 r_{23} \end{bmatrix} & \begin{bmatrix} s_3^2 \end{bmatrix} \end{bmatrix}$$

Note the partitioning of the covariance matrix between rows 2 and 3 and columns 2 and 3; this will be used later. To perform the a priori analysis, we compute the expected value and the variance of the total cost, T:

$$E[T] = m_1 + m_2 + m_3$$

$$\mathrm{var}[T] = s_1^2 + s_2^2 + s_3^2 + 2s_1 s_2 r_{12} + 2s_1 s_3 r_{13} + 2s_2 s_3 r_{23}$$

With this mean and variance, we determine the *a priori* budget B_0 at the risk level α:

$$B_0 = E[T] + k\sqrt{\mathrm{var}[T]}$$

$$B_0 = m_1 + m_2 + m_3 + k\sqrt{s_1^2 + s_2^2 + s_3^2 + 2s_1 s_2 r_{12} + 2s_1 s_3 r_{13} + 2s_2 s_3 r_{23}}$$

In this notation, the work packages are numbered backward, so that work package N is the first to complete and work package 1 is the last. Thus, when work package 3 reports, we observe the actual value $x_3 = X_3$. We now use the equation above,

$$\mathbf{\mu_{1N}} = \mathbf{\mu_1} + \mathbf{V_{1N}} V_{NN}^{-1} (X_N - \mu_N)$$

with $N = 3$ to compute the revised means for the two remaining work packages. Let $\mu_{1|3}$ be the revised expected value for work package 1, conditional on obtaining the actual cost X_3 for work package 3, and let $\mu_{2|3}$ be the revised mean for work package 2, given $x_3 = X_3$. Then the $N - 1 = 2$ equations for these revised expected values are:

$$\mu_{1|3} = m_1 + v_{13} v_{33}^{-1} (X_3 - m_3)$$
$$\mu_{2|3} = m_2 + v_{23} v_{33}^{-1} (X_3 - m_3)$$

Note that, in this process, the only term actually inverted is the scalar term V_{NN}. Substituting into the above equations, the revised means are:

$$\mu_{1|3} = m_1 + \frac{s_1}{s_3} r_{13} (X_3 - m_3)$$

$$\mu_{2|3} = m_2 + \frac{s_2}{s_3} r_{23} (X_3 - m_3)$$

Now use the general equation given above for the revised covariance matrix,

$$\mathbf{V_{11|N}} = \mathbf{V_{11}} - \mathbf{V_{1N}} V_{NN}^{-1} \mathbf{V_{N1}}$$

With $N = 3$, this becomes

$$\mathbf{V}_{11|3} = \mathbf{V}_{11} - \mathbf{V}_{13}\mathbf{V}_{33}^{-1}\mathbf{V}_{31}$$

$$\begin{bmatrix} v_{11|3} & v_{12|3} \\ v_{21|3} & v_{22|3} \end{bmatrix} = \begin{bmatrix} v_{11} & v_{12} \\ v_{21} & v_{22} \end{bmatrix} - \begin{bmatrix} v_{13} \\ v_{23} \end{bmatrix} v_{33}^{-1} \begin{bmatrix} v_{13} & v_{23} \end{bmatrix}$$

$$\begin{bmatrix} v_{11|3} & v_{12|3} \\ v_{21|3} & v_{22|3} \end{bmatrix} = \begin{bmatrix} v_{11} - v_{13}^2 / v_{33} & v_{12} - v_{13}v_{23} / v_{33} \\ v_{12} - v_{13}v_{23} / v_{33} & v_{22} - v_{23}^2 / v_{33} \end{bmatrix}$$

Substituting in the values for the a priori variances gives:

$$\begin{bmatrix} v_{11|3} & v_{12|3} \\ v_{21|3} & v_{22|3} \end{bmatrix} = \begin{bmatrix} s_1^2\left(1-r_{13}^2\right) & s_1 s_2\left(r_{12} - r_{13}r_{23}\right) \\ s_1 s_2\left(r_{12} - r_{13}r_{23}\right) & s_2^2\left(1-r_{23}^2\right) \end{bmatrix}$$

However, we know from the bivariate case discussed earlier that the covariance matrix, after work package 2 has reported its costs, must be given by the following, in which σ_1 and σ_2 represent the revised standard deviations for work packages 1 and 2, respectively, and ρ_{12} represents the correlation coefficient between the costs of work package 1 and work package 2:

$$\mathbf{V}_{22|3} = \begin{bmatrix} \sigma_1^2 & \sigma_1\sigma_2\rho_{12} \\ \sigma_1\sigma_2\rho_{12} & \sigma_2^2 \end{bmatrix} = \begin{bmatrix} v_{11|3} & v_{12|3} \\ v_{21|3} & v_{22|3} \end{bmatrix}$$

$$\therefore \quad \begin{bmatrix} \sigma_1^2 & \sigma_1\sigma_2\rho_{12} \\ \sigma_1\sigma_2\rho_{12} & \sigma_2^2 \end{bmatrix} = \begin{bmatrix} s_1^2\left(1-r_{13}^2\right) & s_1 s_2\left(r_{12} - r_{13}r_{23}\right) \\ s_1 s_2\left(r_{12} - r_{13}r_{23}\right) & s_2^2\left(1-r_{23}^2\right) \end{bmatrix}$$

From the main diagonal terms in this expression, we can immediately determine the revised variances,

$$\sigma_1^2 = s_1^2\left(1-r_{13}^2\right)$$

$$\sigma_2^2 = s_2^2\left(1-r_{23}^2\right)$$

Substituting these values into the off-diagonal terms (which are of course equal, by symmetry), gives

$$\sigma_1\sigma_2\rho_{12} = s_1 s_2 \sqrt{\left(1-r_{13}^2\right)\left(1-r_{23}^2\right)}\rho_{12} = s_1 s_2\left(r_{12} - r_{13}r_{23}\right)$$

$$\therefore \rho_{12} = \frac{\left(r_{12} - r_{13}r_{23}\right)}{\sqrt{\left(1-r_{13}^2\right)\left(1-r_{23}^2\right)}}$$

We now have revised values for the expected values, variances, and correlation coefficient for the case with two work packages, which is just the situation covered in the bivariate case. We compute the revised mean and variance of the total cost as before:

$$E[T] = \mu_1 + \mu_2 + X_3$$
$$\text{var}[T] = \sigma_1^2 + \sigma_2^2 + 2\rho_{12}\sigma_1\sigma_2$$

In terms of the original parameters for the *a priori* case of three work packages, these are,

$$E[T] = m_1 + m_2 + \frac{s_1}{s_3}r_{13}(X_3 - m_3) + \frac{s_2}{s_3}r_{23}(X_3 - m_3) + X_3$$

$$\text{var}[T] = s_1^2(1 - r_{13}^2) + s_2^2(1 - r_{23}^2) + 2s_1s_2(r_{12} - r_{13}r_{23})$$

Example 11.2
To illustrate the trivariate example, assume the following *a priori* data:

$$m_1 = m_2 = m_3 = \$100K$$
$$s_1 = s_2 = s_3 = \$40K$$
$$\rho_{12} = \rho_{13} = \rho_{23} = 0.8$$
$$\alpha = 0.15; \; k = 1.0$$

Inserting these numerical values into the *a priori* equations given above,

$$E[T] = m_1 + m_2 + m_3 = \$300K$$

$$\text{var}[T] = s_1^2 + s_2^2 + s_3^2 + 2s_1s_2r_{12} + 2s_1s_3r_{13} + 2s_2s_3r_{23} = 12480$$

With this mean and variance, we determine the *a priori* budget B_0 at the risk level $\alpha = 0.15$:

$$B_0 = E[T] + k\sqrt{\text{var}[T]} = 300 + 1.0(111.7) = \$411.7K$$

This implies a contingency of $111.7K or 37% of the expected value of the total cost, corresponding to a 15% probability of a cost overrun.

Now suppose that work package 3 is completed at a cost of $95K. The revised expected values for the two work packages remaining are, from the above equations,

$$\mu_{1|3} = m_1 + \frac{s_1}{s_3} r_{13} \left(X_3 - m_3 \right) = 100 - (0.8)5 = \$96K$$

$$\mu_{2|3} = m_2 + \frac{s_2}{s_3} r_{23} \left(X_3 - m_3 \right) = 100 - (0.8)5 = \$96K$$

The revised variances are, similarly:

$$\sigma_1^2 = s_1^2 \left(1 - r_{13}^2 \right) = 576$$
$$\sigma_2^2 = s_2^2 \left(1 - r_{23}^2 \right) = 576$$

And, the revised correlation coefficient is:

$$\rho_{12} = \frac{\left(r_{12} - r_{13} r_{23} \right)}{\sqrt{\left(1 - r_{13}^2 \right)\left(1 - r_{23}^2 \right)}} = 0.44$$

The revised expected value and variance of the total cost, with one work package known, are:

$$E[T] = \mu_1 + \mu_2 + X_3 = 96 + 96 + 95 = \$287K$$
$$\text{var}[T] = \sigma_1^2 + \sigma_2^2 + 2\rho_{12}\sigma_1\sigma_2 = 1664$$

This give an overall coefficient of variation for the total cost of 0.14, compared to 0.37 in the *a priori* case. The revised budget is

$$B_3 = \$287K + 1.0(40.8) = \$327.8K$$

This represents a substantial reduction from the *a priori* budget, including contingency, of $411.7K.

Note that the coefficient of variation of the cost for each work package in the original case is $40/100 = 0.4$, which falls to $24/96 = 0.25$ for the two work packages after work package 3 reports. Also, the original correlation coefficients, 0.8, fall to 0.44 after one work package is known. Although the specific results of course vary according to the initial parameters and the number of work packages, this behavior is typical. As more information becomes available on actual costs, the variability and the correlation between the remaining variables decrease.

11.3.1   *Discussion*

Unfortunately, there are few data on how project managers actually manage project contingencies. The analysis above shows that a consistent contingency strategy can be developed based on maintaining a constant risk of cost overruns throughout the project life. In the case that tasks or work packages are correlated, the project manager can learn, from experience on early work packages, to modify the predictions of costs on later work packages and therefore to reduce the uncertainty in project total cost. In this approach, a reduction in uncertainty reduces the need for contingency. Then, given moderately favorable cost experience, the project manager may be able to reduce the contingency amount and to release money from the contingency pool to other uses.

In this approach, contingency is never parceled out to individual tasks or work packages; it is always retained by the project manager and thus is available for reallocation to other work packages that overrun, or for return to the owner, or for other uses. In this second moment model, contingency is related to uncertainty, or ignorance about the true costs in the future. Contingency is not an appropriate way to deal with risks of extraordinary events, with very high impact but very low likelihood, which may or may not occur.

It is often recommended that, if there is some high magnitude risk, with very low probability p and very high cost C, then one should apply a contingency equal to the expected loss, pC. This view is not taken here, on the basis that a contingency so derived is never of any use. If the event never occurs, which almost always will be true, with probability $\Pr = (1 - p) \cong 1$, then one has simply increased the cost of the project by pC. On the other hand, if the event occurs, the contingency in reserve, $pC \ll C$, is always inadequate to cover the need. That is, there are other and better ways to handle rare events than through contingency reserves.

The first priority is to assure that there are enough funds in reserve to get the project done. The risk of overrunning cannot be made zero, but it can be analyzed as shown above and reduced to some acceptable value, given as α. In some projects, poor cost performance coupled with excessive risk values will mean that all the original contingency is used up, and then some. However, there will be projects with adequate original contingency and favorable cost experience, and these projects will be able to free up contingency reserves for other uses.

The best use of released contingency reserves would be to return them to the owner, or to whomever is funding the project, who can then allocate these funds to other projects or other uses. That is, the contingency funds belong to the owner or sponsor, not to the project, to use in whatever manner it wishes, and if the contingency is not being used, it should be returned to its rightful owner.

Example 11.3
The second moment method provides a simple, convenient way to adjust the risks, and hence the required contingencies to cover the risks, as a project proceeds and evidence is obtained on how well (or badly) it is going, compared to the initial estimates. The objective of this approach is to react as soon as possible to information

on recent project performance that confirms or disconfirms the current estimates. The key parameter is the expected cost at completion (or, the expected time at completion). If the best estimate of the cost at completion, updated with the most recent progress information, is higher than the original estimate, then, assuming no scope changes, more contingency may be required or some program management corrective action may be needed to bring the project back on target. Conversely, if the updated best estimate of the cost at completion is the same as or lower than the original estimate, then the contingency required can be decreased and this contingency released to the program manager, as needed elsewhere. In the approach discussed here, the estimates of all future work packages are updated as the actual costs for each completed work package become available.

This point is illustrated by an example, very much simplified for exposition. To keep the example small enough to present here, we consider a project of only six work packages; real projects might have hundreds of work packages. Also, to keep it simple, we assume that the expected cost (the mean, the median, and the mode) for each work package is $100,000, and the uncertainty in each work package cost is given by the coefficient of variation, which is assumed in this example to be 40% for every work package. As the coefficient of variation is the standard deviation divided by the expected value (times 100), the estimated values for the standard deviations for all work packages are $40,000 each. Then the best estimate of the total cost at completion is the sum of the expected values for the work packages, or $600,000.

The correlation matrix used in this example is as follows:

$$
\begin{bmatrix}
WP & 1 & 2 & 3 & 4 & 5 & 6 \\
1 & 1.000 & 0.900 & 0.810 & 0.729 & 0.656 & 0.590 \\
2 & 0.900 & 1.000 & 0.900 & 0.810 & 0.729 & 0.656 \\
3 & 0.810 & 0.900 & 1.000 & 0.900 & 0.810 & 0.729 \\
4 & 0.729 & 0.810 & 0.900 & 1.000 & 0.900 & 0.810 \\
5 & 0.656 & 0.729 & 0.810 & 0.900 & 1.000 & 0.900 \\
6 & 0.590 & 0.656 & 0.729 & 0.811 & 0.900 & 1.000
\end{bmatrix}
$$

These correlations were computed from the formula:

$$\rho_{j,k} = \rho_o^{|k-j|}; \rho_o = 0.90$$

Then the second moment method described above gives a computed value for the standard deviation of the total cost of $217,830 (the detailed equations and calculations are not reproduced here). This means that the coefficient of variation of the total project cost is 36.3%, which is less than the coefficient of variation for each of the six work packages taken separately. Assume that we wish to set the contingency at the 90% confidence limit; which is to say that the budget is to be set such that

Table 11.1 Revised best estimate of cost

Work package	1	2	3	4	5	6	Total cost
Best estimate of cost	$140,000 Actual	$135,960	$132,360	$129,120	$126,210	$123,590	$787,240

there is a 10% chance or less that the budget will be overrun, and a 90% chance that the budget will not be overrun. The normal factor for one-sided 90% confidence is 1.283, so the required budget at the 90% confidence value is $600,000 plus 1.283($217,830) = $879,480.

Now suppose that the first work package is completed for an actual cost of $140,000, or 40% higher than the expected value of $100,000. Given this information, the revised best estimates for the remaining work packages, based on the method descried earlier, are as given in Table 11.1.

The fact that the first work package was completed for substantially more than the best prior estimate of $100,000, and the correlation between the work package costs, means that there is some evidence that the cost estimates on this project may be low, and accordingly we should revise our estimates of the costs of the remaining work packages upward. The table above shows these revised cost estimates. The best estimates for the remaining work packages, after obtaining the information about the actual cost of work package one, vary from $135,960 to $123,590. Consequently, the best estimate of the total cost at completion is now $787,240, up $187,240 from the initial estimate of $600,000. Assuming that the sponsor does not increase the budget, so that the budget with contingency remains constant at the original value of $879,480, the remaining contingency is only $879,480 − $787,240 = $92,240. This is positive, but less than the required contingency to cover the remaining costs at the 90% confidence level. That is, step one has actually used up some of the project contingency, even though no specific contingency was assigned to step one, and now the probability of overrunning the budget is more than 10%. In fact, the probability of overrunning the established budget with contingency is now 20%, and the budget corresponds to the 80% confidence level. The revised values for cost at completion and budget with contingency after step one are plotted in Fig. 11.2 below. (At this point, of course, only the step one results are known.)

The original best estimate of the cost at completion, the required contingency, and the 90% confidence value for the budget including contingency are shown on the axis for zero steps complete. When the first activity (step) is completed, for $140,000, the best estimate of the cost at completion increases, the available contingency falls, and the budget remains the same, although this now corresponds to the 80% confidence limit rather than the 90% confidence limit.

Suppose now that the second activity is completed for a cost of $135,000. This is actually very slightly lower than the updated prediction (the best estimate for the step two cost after step one was completed was $135,960) but $35,000 higher than the original estimate. This additional evidence tends to confirm that the project costs are running higher than the estimates (or, that the estimates were low). The revised values for the best estimates of the incomplete steps are now given in Table 11.2.

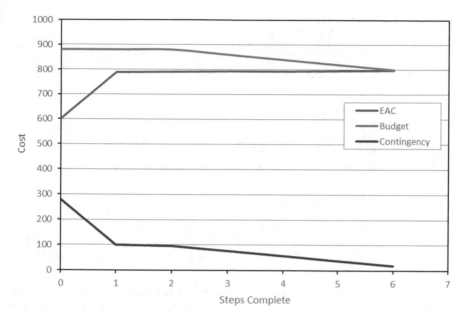

Fig. 11.2 Time history of revision

Table 11.2 Revised best estimate of cost

Work package	1	2	3	4	5	6	Total cost
Best estimate of cost	$140,000 Actual	$135,000 Actual	$131,500	$128,350	$125,520	$122,970	$783,340

Table 11.3 Revised best estimate of cost

Work package	1	2	3	4	5	6	Total cost
Best estimate of cost	$140,000 Actual	$135,000 Actual	$132,000 Actual	$128,800	$125,920	$122,970	$785,050

Even though step two was much higher than the original estimate, the best estimate of the cost at completion has fallen somewhat, from $787,240 to $783,340. After two of the six work packages have been completed, and assuming that the project budget has not been increased, the available contingency is now $96,140, slightly above the $92,240 after step one, but still below the value to meet the 90% confidence limit. In fact, the established budget with contingency corresponds to the 87% confidence value; there is a probability of 13% of exceeding the value $879,480.

Suppose now that the third activity is completed for a cost of $132,000. This is modestly higher than predicted (the best estimate for this cost after step two was completed was $128,350). The revised values for the best estimates of the incomplete steps are now given in Table 11.3.

Table 11.4 Revised best estimate of cost

Work package	1	2	3	4	5	6	Total cost
Best estimate of cost	$140,000 Actual	$135,000 Actual	$132,000 Actual	$130,000 Actual	$127,000	$124,300	$788,300

Table 11.5 Revised best estimate of cost

Work package	1	2	3	4	5	6	Total cost
Best estimate of cost	$140,000 Actual	$135,000 Actual	$132,000 Actual	$130,000 Actual	$129,000 Actual	$126,110	$792,110

Table 11.6 Revised best estimate of cost

Work package	1	2	3	4	5	6	Total cost
Best estimate	$140,000 Actual	$135,000 Actual	$132,000 Actual	$130,000 Actual	$129,000 Actual	$131,000 Actual	$797,000 Actual

After the third work package has been completed, the best estimate of the cost of completion has gone up slightly to $785,050, but the required contingency has gone down to $77,280 (it goes down even though the cost of activity three was higher than estimated, because there are fewer remaining risks), and the 90% confidence budget can now actually be decreased to $862,330, even though all the three work packages have been higher than the original estimates.

Suppose now that the fourth activity is completed for a cost of $130,000, slightly higher than predicted. The revised best estimates of the incomplete steps are now given in the Table 11.4.

After four of the six work packages have been completed, the best estimate of the cost of completion has gone up to $788,300, but the required contingency has gone down to $47,890 (it goes down even though the cost of activity four was high, because there are fewer remaining risks), and the 90% confidence budget can actually be decreased to $836,200.

Suppose now that the fifth activity is completed for a cost of $129,000, slightly higher than predicted. The revised best estimate of the single incomplete step is now given in Table 11.5.

After five of the six work packages have been completed, the best estimate of the cost of completion has crept up again, to $792,110, but the required contingency has gone down to $22,200, and the 90% confidence budget can be decreased to $814,310.

Suppose finally that the sixth activity is completed for a cost of $131,000. The actual costs are as given in Table 11.6.

The variations in the revised budgets and expected costs at completion, after each work package or step is completed, are shown in the figure given above. The objective of using the confidence limits is to keep the confidence band positioned so that it envelops the (unknown) actual cost at completion. That is, no one can predict the future (the actual cost at completion) with certainty, but we can try to define a confidence band that bounds where we expect to find it (with probability 90%, in this case). Although the lower confidence bound is not shown in the figure

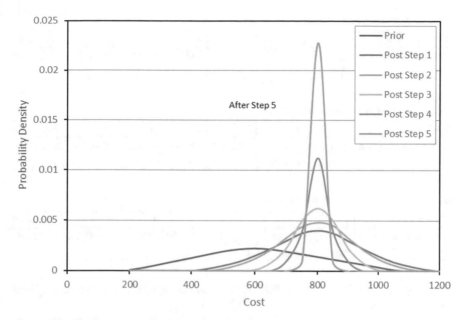

Fig. 11.3 Uncertainty on cost at completion

(it is the expected cost at completion minus the contingency), it is clear that, for this example at least, the method has achieved the goal of keeping the 90% confidence limit above the actual cost at completion (that is, above $797,000) for every step in the process up to project completion. Of course, even though the costs increased over the original estimates, this example was actually well behaved in that it never ran out of contingency (although contingency was being used up for some period of time).

Figure 11.3 shows the change in uncertainty regarding the final cost at completion as every work package is completed, providing additional information about where the project is going. Before any work is done, the prior estimate of the probability for the cost at completion is centered on $600,000, but has a very high variance (uncertainty), as shown in the figure below. After the completion of step one, the probability distribution shifts to the right, and is centered $787,240. As each subsequent work package is completed, the best estimate or most likely value for the cost at completion creeps up somewhat, depending on the latest actual cost reports, but the uncertainty (the width of the distribution) always lessens. The figure illustrates how the method starts with a high degree of uncertainty and zeros in on the target as more information becomes available about actual project performance.

Example 11.4

As another numerical example, the project is the same as in Example 11.3, but with different realizations. That is, all the work packages, estimates, correlations, etc., are identical to those in Example 11.3, but the actual project outcomes are different. Thus, as in Example 11.3, the expected value of the total project cost is

Table 11.7 Revised best estimate of cost

Work package	1	2	3	4	5	6	Total cost
Best estimate of cost	$90,000	$91,010	$91,910	$92,720	$93,450	$94,100	$553,190
	Actual						

Table 11.8 Revised best estimate of cost

Work package	1	2	3	4	5	6	Total cost
Best estimate of cost	$90,000	$95,000	$95,490	$95,940	$96,350	$96,710	$569,490
	Actual	Actual					

Table 11.9 Revised best estimate of cost

Work package	1	2	3	4	5	6	Total cost
Best estimate of cost	$90,000	$95,000	$105,000	$104,500	$104,050	$103,640	$602,190
	Actual	Actual	Actual				

$600,000, and the standard deviation of the total cost is $217,830, and the required budget at the 90% confidence value is $879,480.

The first work package is completed, this time at an actual cost of $90,000, lower than the expected value. The revised best estimates for the remaining work packages are shown in Table 11.7.

The fact that the first work package was completed for less than the best prior estimate of $100,000 means that there is some evidence that the cost estimates may be high on this project, and accordingly we should revise our estimates of the costs of the remaining work packages downward, given the correlations between the work package costs. That is, our best estimate of the total cost at completion is now $553,190, down from $600,000. The required contingency to cover the remaining costs at the 90% confidence level has gone down to $142,580, a significant reduction from the initial value of $217,830, freeing up $75,250 in contingency funds to be deployed elsewhere. The 90% confidence budget is now $695,770, down $183,710 from the original value of $879,480.

The second activity is completed for a cost of $95,000. This is lower than the original estimate but somewhat higher than the updated prediction (the best estimate for this cost after step one was completed was given in the table above as $91,010). The revised values for the best estimates of the work packages and cost at completion are now showing in Table 11.8.

After two of the six work packages have been completed, the best estimate of the cost of completion has gone up somewhat from $$553,190 to $569,490 (still less than the original estimate), but the required contingency has gone down to $109,110 and the 90% confidence budget has decreased to $678,600. The third activity is completed for a cost of $105,000. This is higher than predicted (the best estimate for this cost after step two was completed was $95,490). The best estimates of the costs of the incomplete steps are shown in Table 11.9:

After three work packages have been completed, the best estimate of the cost of completion has gone up to $602,190, but the required contingency has gone down

Table 11.10   Revised best estimate of cost

Work package	1	2	3	4	5	6	Total cost
Best estimate of cost	$90,000	$95,000	$105,000	$106,000	$105,400	$104,860	$606,260
	Actual	Actual	Actual	Actual			

Table 11.11   Revised best estimate of cost

Work package	1	2	3	4	5	6	Total cost
Best estimate of cost	$90,000	$95,000	$105,000	$106,000	$93,000	$97,250	$590,250
	Actual	Actual	Actual	Actual	Actual		

Table 11.12   Revised best estimate of cost

Work package	1	2	3	4	5	6	Total cost
Actual	$90,000	$95,000	$105,000	$106,000	$93,000	$103,000	$596,000

to $77,280. The 90% confidence budget has remained approximately steady at $679,470. The fourth activity is completed for a cost of $106,000, slightly higher than predicted. The revised best estimates of the incomplete steps are shown in Table 11.10.

The expected cost at completion goes up to $606,260, but the required contingency goes down to $47,890 and the 90% confidence budget decreases to $654,150. The fifth activity is completed for a cost of $93,000, lower than predicted. The revised best estimate of the single incomplete step is shown in Table 11.11:

After five of the six work packages have been completed, the best estimate of the cost of completion has gone down again, to $590,250, and the required contingency has gone down to $22,200, and the 90% confidence budget decreases to $612,450.

Finally, the sixth activity is completed for a cost of $103,000. The actual costs are shown in Table 11.12.

The variations in the revised budgets and expected costs at completion, after each work package or step is completed, are plotted in Fig. 11.4.

Figure 11.5 shows the change in uncertainty regarding the final cost at completion as every work package is completed. As each work package is completed, the estimated cost at completion goes down, then up, then down again, but the uncertainty in the cost at completion always decreases, zeroing in on the target as information becomes available.

11.4   Managing the Contingency: Cost to Complete

Suppose that a project consists of M work packages and that at some time N work packages remain to be done. (That is, $M-N$ work packages have been completed and reported.) Let z represent the remaining cost to complete the project, recalling that the work packages are numbered from M down to 1, the last:

$$z = \sum_{j=N}^{1} x_j$$

Then the expected cost to complete is:

$$E[z] = \overline{z} = \sum_{j=N}^{1} E[x_j] = \sum_{j=N}^{1} \overline{x}_j$$

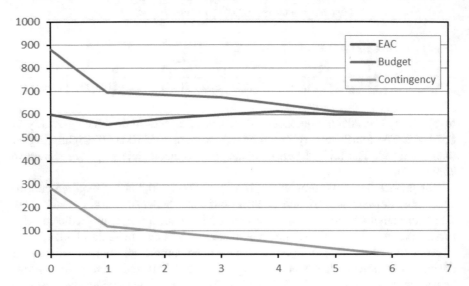

Fig. 11.4 Time history of cost revisions

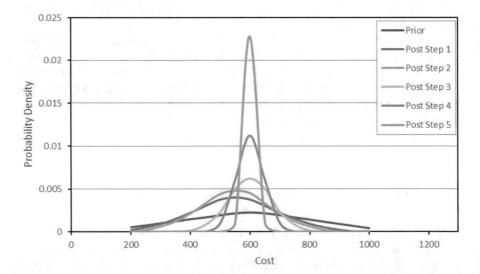

Fig. 11.5 Uncertainty on cost at completion

It is assumed that the expected values of the remaining work packages, along with their variances and the correlation matrix, have been updated at the completion of each work package by the method described earlier. Then,

$$E[z] = E[T] - \sum_{j=M}^{N+1} X_j$$

$$\mathrm{var}[z] = \mathrm{var}[T]$$

That is, the uncertainty in the remaining cost to complete is just the uncertainty in the total cost at completion, because completed work packages have no uncertainty. Let $f(z)$ be the probability density function for the remaining cost to complete, where $f(z)$ is conditional on the actual reported values $X_M, X_{M-1}, \ldots, X_{N+1}$. Let $B_{z,N}$ be the budget, including contingency, to cover the cost to complete the last N work packages (not the total cost at completion). The Expected Value-at-Risk is then:

$$EVaR_N = \int_{B_{z,N}}^{\infty} zf(z)\,dz$$

This represents the expected value of the budget overrun when N work packages remain and $B_{z,N}$ is the budget to complete. The expected value of the cost to complete is:

$$E[z] = \int_{-\infty}^{\infty} zf(z)\,dz$$

(Note that the lower limit of integration is merely for notational convenience; the cost to complete can never be negative.)

Now let's define the ratio R as:

$$R = \frac{EVaR_N}{E[z]} = \frac{\int_{B_{z,N}}^{\infty} zf(z)\,dz}{\int_{-\infty}^{\infty} zf(z)\,dz}$$

Then $100R$ is the expected budget overrun as a percentage of the expected cost to complete, when $B_{z,N}$ is the remaining budget to complete. In the previous examples, the contingency was adjusted in order to keep the probability of a budget overrun constant throughout the project. One alternate approach to contingency management might be to adjust the contingency (contained in $B_{z,N}$, the remaining budget to complete) so that R is a constant, where $100R$ is the expected budget overrun as a percentage of the expected cost to go. For example, one might set $R = 0.05$, which means that the budget to go should be set such that the expected budget overrun is always 5% of the cost to go.

In the current notation, α_N is the probability of exceeding the budget $B_{z,N}$ when there are N work packages yet to be done:

$$\alpha_N = \int_{B_{z,N}}^{\infty} f(z)\,dz$$

Given that $f(x)$ is Normal, which is the assumption here,

$$VaR = \sigma\phi(k) + \mu\left[1 - \Phi(k)\right]$$

Here $\phi(k)$ and $\Phi(k)$ are, respectively, the probability density function and the cumulative probability distribution for the Unit Normal. In the notation used here, this equation becomes:

$$VaR_N = \sqrt{\text{var}[z]}\,\phi(k) + E[z]\left[1 - \Phi(k)\right]$$

Dividing this by $E[z]$ gives:

$$R = \frac{VaR_N}{E[z]} = \left\{ \frac{\sqrt{\text{var}[z]}}{E[z]} \right\}\phi(k) + 1 - \Phi(k)$$

Note that $\dfrac{\sqrt{\text{var}[z]}}{E[z]}$ is the Coefficient of Variation of the cost-to-complete when N work packages remain, and $E[z]$ and $Var[z]$ are conditional on the reported costs for the work packages M through $N + 1$, computed by the method described above. Then, with R specified and $\dfrac{\sqrt{\text{var}[z]}}{E[z]}$ computed, the contingency to maintain a constant R can be found by solving the following equation for k:

$$R - 1 + \Phi(k) - \frac{\sqrt{\text{var}[z]}}{E[z]}\phi(k) = 0$$

Then the contingency to go is $k\sqrt{\text{var}[z]}$, the budget to go is $B_{z,N} = E[z] + k\sqrt{\text{var}[z]}$, and the probability that this budget will be overrun is $\alpha_N = 1 - \Phi(k)$.

Figure 11.6 shows the variation of α_N for various values of the Coefficient of Variation (COV) of the cost to go, when R is held constant (at $R = 0.05$ and $R = 0.10$). Note that higher values of the COV imply lower probabilities of overrunning the budget, when R is fixed. (Why?) As one might expect the COV of the cost to complete to decrease as the project moves forward, holding R constant implies that the budgets to go have an increasing probability of being overrun.

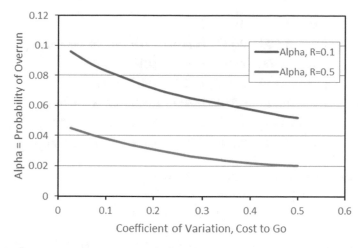

Fig. 11.6 Alpha (α) for various coefficients of variation

Example 11.5

The second moment method provides a simple, convenient way to adjust the risks, and hence the required contingencies to cover the risks, as a project proceeds and evidence is obtained on how well (or badly) it is going, compared to the initial estimates. The objective of this approach is to react as soon as possible to information on recent project performance that confirms or disconfirms the current estimates. The key parameter is the expected cost at completion (or, the expected time at completion). If the best estimate of the cost at completion, updated with the most recent progress information, is higher than the original estimate, then, assuming no scope changes, more contingency may be required or some program management corrective action may be needed to bring the project back on target. Conversely, if the updated best estimate of the cost at completion is the same as or lower than the original estimate, then the contingency required can be decreased and this contingency released to the program manager, as needed elsewhere. In the approach here, the estimates of all future units are updated as the actual cost for each completed unit becomes available.

Consider a project of 20 units. For ease of understanding, the units are considered to be identical. This is not required by the analysis, which can handle cases in which all units are different, but the simpler assumption is easier to follow. Then assume that the expected cost (the mean, the median, and the mode) for each unit is $100, and the uncertainty in each unit cost is given by the COV, which is assumed in this example to be 40% for every unit. As the COV is the standard deviation divided by the expected value (times 100 to give a percentage), the estimated values for the standard deviations for all units are $40 each. Then the best estimate of the total cost at completion is the sum of the expected values for the 20 units, or $2000.

The correlation matrix used in this example is as follows:

$$
\begin{bmatrix}
WP & 1 & 2 & 3 & 4 & 5 & 6 \\
1 & 1.000 & 0.900 & 0.810 & 0.729 & 0.656 & 0.590 \\
2 & 0.900 & 1.000 & 0.900 & 0.810 & 0.729 & 0.656 \\
3 & 0.810 & 0.900 & 1.000 & 0.900 & 0.810 & 0.729 \\
4 & 0.729 & 0.810 & 0.900 & 1.000 & 0.900 & 0.810 \\
5 & 0.656 & 0.729 & 0.810 & 0.900 & 1.000 & 0.900 \\
6 & 0.590 & 0.656 & 0.729 & 0.811 & 0.900 & 1.000
\end{bmatrix}
$$

These correlations were computed from the formula:

$$
\rho_{j,k} = \rho_o^{|k-j|}; \rho_o = 0.90
$$

Although the 20 by 20 covariance matrix is too large to show here, all the terms are computed by the equations given above, and the sum of all the covariances in the matrix is 355014.1. This is the *a priori* variance of the project total cost, and the square root of this, \$595.83, is the *a priori* standard deviation of the total project cost. Assume now that the project manager has some aversion to risk, and wants to be 90% confident of successful completion of the project within the budget. This level of risk aversion implies, using the one-sided Normal distribution, that the available budget must be \$2000 + 1.282(\$595.83) = \$2763.86. That is, the probability of a cost overrun at this project budget is 10% and this level of risk is acceptable to the project manager. As has been seen before, this value (\$2763.86 in this case) is called the Value-at-Risk (VaR). Note that if one were to set the budget at the expected total cost, \$2000, the probability of an over-run would be 50% and the project manager would have only a 50–50 chance of success. This would typically be considered an excessive probability of over-run. This does not imply that such risky budgets are never set; only that the high probability of failure should be recognized.

To summarize: before the project begins, the project manager concludes that the best estimate of the total project cost is \$2000, based on an average unit cost of \$100, and the project manager has access to an amount of \$2763.86 for 90% confidence that the cost will not exceed the available funding. This figure represents a contingency or markup of 38% over the expected cost, but it is not computed from the expected cost, it is computed from the risk in the unit costs. Note that \$2763.86 is not necessarily the bid price, because it does not include explicit allowance for overhead and profit.

Now suppose that the first unit (denoted here as $N = 20$) is completed for an actual cost of \$105, or 5% higher than the expected value of \$100. The fact that the first unit was completed for somewhat more than the best prior estimate of \$100 means that there is some evidence that the cost estimates on this project may be low, and accordingly we should revise our estimates of the costs of the remaining units upward. Given this information, the revised best estimates for the remaining 19 units, based on the method described earlier, are as given in the Table 11.13:

Table 11.13   Revised unit cost estimate

Unit	Prior expected cost of future units, $	Revised predicted cost of future units, $
20	105 actual cost	**105.00**
19	100	104.50
18	100	104.05
17	100	103.64
16	100	103.28
15	100	102.95
14	100	102.66
13	100	102.39
12	100	102.15
11	100	101.94
10	100	101.74
9	100	101.57
8	100	101.41
7	100	101.27
6	100	101.14
5	100	101.03
4	100	100.93
3	100	100.83
2	100	100.75
1	100	100.68
Total	2005	2043.92

This result is also shown in the following figures. Figure 11.7 below represents the prior situation, in which the predicted cost of each unit is $100.

Figure 11.8 just above shows the predicted values for all of the 19 remaining units, given that the first unit cost $105. These predicted values approach $100 for the later units, as the impacts of the $105 actual cost are diminished down the chain (diminished because the correlation coefficients are <1). The total job cost at completion, the sum of all the unit costs, actual and predicted, is forecast to be $2043.92, which is over the expected value but well under the allocated funding including management reserve of $2763.86.

The figure below shows the revised cost estimates for each remaining unit after the second unit has been completed at cost $111 (and the first unit at $105). The effect of these costs above the prior is to increase the estimate for the costs for each unit in the future, as shown graphically in Fig. 11.9.

Suppose now that the third unit is completed for a cost of $97. This is actually slightly lower than the prior estimate ($100) and the updated prediction (the best estimate for the third unit after the completion of the second unit is $109.90). This additional evidence tends to indicate that the unit costs are running both above and below the estimates. The revised values for the best estimates of the costs of the 17 incomplete units, using the same algorithm as before, are now given in the table below.

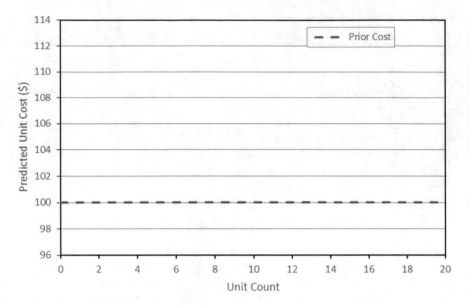

Fig. 11.7 Predicted unit cost with no units complete

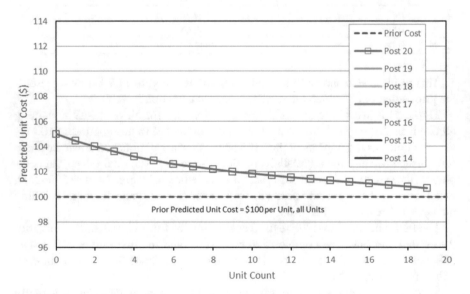

Fig. 11.8 Predicted unit costs after one unit complete

Figure 11.10 show the revised forecasts for the cost for each unit after the third unit has been shown to cost $97 (the curve with the large box symbols). The combined effect of these actual costs both above and below the previous forecasts is to give a new estimate close to the original estimate – in fact, somewhat below the original estimate ($1990.50 vs. $2000.00).

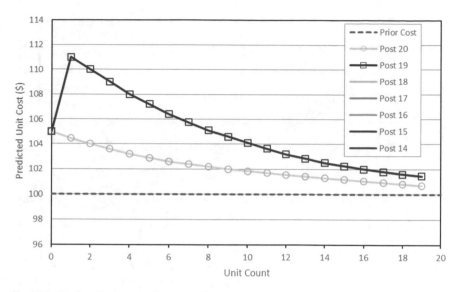

Fig. 11.9 Predicted unit costs after two units complete

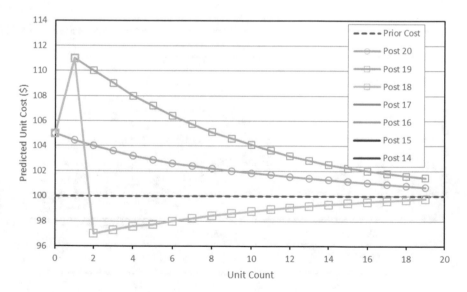

Fig. 11.10 Predicted unit costs after three units complete

Continuing on with this example, the fourth unit when complete is found to cost $108, again above the prior expected value and also above the latest prediction (see Fig. 11.11).

The fifth unit is completed at cost $112, above the prior expected value and above the latest prediction (see Fig. 11.12).

The sixth unit is found to cost $101. (See Fig. 11.13).

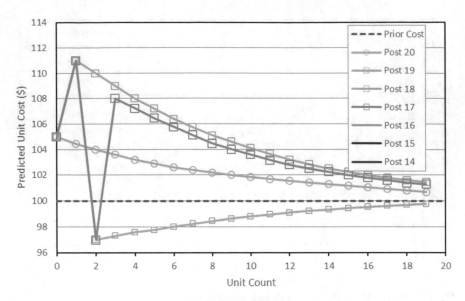

Fig. 11.11 Predicted unit costs after four units complete

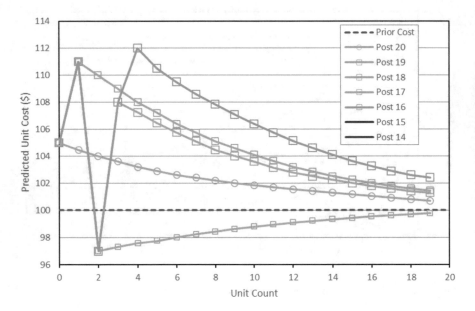

Fig. 11.12 Predicted unit costs after five units complete

Figure 11.14 shows the revised predicted costs per unit after the seventh unit completed has been shown to cost $112 (the curve with the large box symbols).

This series goes on as shown above, until all the units have been completed. Table 11.14 shows the actual costs for all units, the revised predictions of the cost at completion after each unit is completed, and the total project cost at completion.

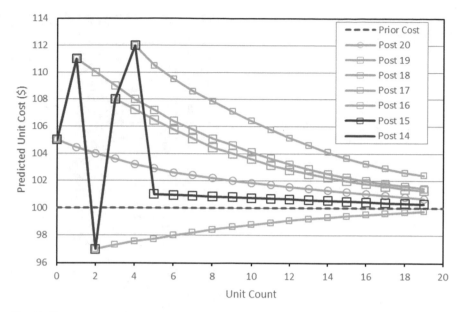

Fig. 11.13 Predicted unit costs after six units complete

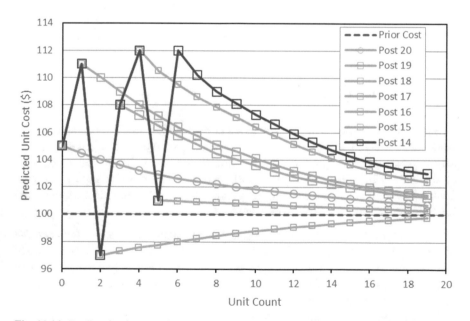

Fig. 11.14 Predicted unit costs after seven units complete

The actual cost at completion is $2586 or 29% more than the *a priori* estimated cost of $2000. However, this actual cost is less than the *a priori* 90% confidence budget, which was $2763.86. Therefore, the project ended up over the estimate but under the budget.

Table 11.14 Revised unit
cost estimate

Unit	Actual unit costs, $	Predicted cost at completion, $
20	105	2000.00
19	111	2043.92
18	97	2100.14
17	108	1990.50
16	112	2079.66
15	101	2118.76
14	112	2040.94
13	115	2126.55
12	124	2157.87
11	123	2233.22
10	145	2242.82
9	160	2401.09
8	155	2520.55
7	160	2526.24
6	158	2581.02
5	150	2599.76
4	155	2590.76
3	142	2625.15
2	136	2604.82
1	117	2601.40
Total	2586	2586.00

11.5   Practice Problems

Problem 11.5.1   Carlos is the project manager for a project that has two activities or work packages (so that this assignment can be done on a pocket calculator). Using a three-point estimation process, Carlos estimates the 5th, 50th, and 95th percentiles of the probability distributions for work package 1 and work package 2 (see Table 11.15). Then he uses the Pearson-Tukey equations to estimate the means and standard deviations of the two work packages.

Based on experience, Archie estimates the correlation coefficient between the two work packages to be 0.50. He also determines that he wants to be 90% confident that the total actual cost for the project will not exceed his budget, which is to be determined. Using $\alpha = 0.10$ and a table of the Normal distribution, what is the risk multiplier k Archie needs to use?

The *a priori* expected value of the total project cost is:

$$E[T] = \mu_1 + \mu_2$$

The covariance matrix **V** is given by:

Table 11.15 Problem data

Work package	X_{05}	X_{50}	X_{95}
1	$120K	$200K	$280K
2	$ 60K	$100K	$140K

$$V = \begin{pmatrix} \sigma_1^{\,2} & \rho_{12}\sigma_1\sigma_2 \\ \rho_{12}\sigma_1\sigma_2 & \sigma_2^{\,2} \end{pmatrix}$$

The *a priori* variance of the total cost is the sum of all the terms in the covariance matrix, or:

$$\mathrm{var}[T] \equiv \sigma_T^2 = \sigma_1^2 + 2\rho_{12}\sigma_1\sigma_2 + \sigma_2^2$$

What is the a priori expectation, variance, and standard deviation of the total cost? What is the a priori contingency $k\sigma_T$ (that is, the contingency before any work is done)? What is the a priori project budget?

Problem 11.5.2 Now, suppose that the two work packages are completed in sequence, first work package 2 and then work package 1. Carlos is informed that work package 2 has been completed, and the cost is now known for certain to be X_2 = $150K, which is well over the original estimated cost. Because the two work packages are correlated, this means the re-estimated cost of work package 1 will be higher than the original estimate too. As a result, Archie is going to request a budget increase in order to keep his risk of overrunning the project budget at the prior value $\alpha = 0.10$.

The conditional probability distribution on cost, given that x_2 is the known reported value X_2, is $f_{1|2}(x_1|x_2)$, which is determined from the joint probability density function and the marginal distribution by Bayes's Law. Then the conditional probability density function $f_{1|2}(x_1|x_2 = X_2)$ is Normal with parameters

$$\mathrm{mean} = \mathrm{E}\left[x_1 \middle| x_2 = X_2\right] = \mu_1 + \rho_{12}\left(\frac{\sigma_1}{\sigma_2}\right)(X_2 - \mu_2)$$

$$\mathrm{variance} = \mathrm{var}\left[x_1 \middle| x_2 = X_2\right] = \sigma_1^2\left(1 - \rho_{12}^2\right)$$

$$B_2 = X_2 + \mu_1 + \rho_{12}\left(\frac{\sigma_1}{\sigma_2}\right)(X_2 - \mu_2) + k\sigma_1\sqrt{1 - \rho_{12}^2}$$

with $x_2 = X_2 = 150K$ known, the cost to complete the project is x_1.

What is the variance of x_1 given $x_2 = X_2$ from the above equation? What is the contingency $k\sigma_T$? And What should be the proposed revised budget, after work package 2 is completed? Is it higher or lower than his original a priori budget? What is the expected value of x_1, the *cost to complete*?

References

Gelman A, Carlin J, Stern H, Dunson D, Vehtari A, Rubin D (2013) Bayesian data analysis. Chapman and Hall/CRC, New York

Hald A (1952) Statistical theory with engineering applications. Wiley, New York

Jorion P (2001) Value at risk: the new benchmark for managing financial risk, 2nd edn. McGraw-Hill, New York

Chapter 12
Statistical Project Control

Abstract In this chapter we introduce statistical project control methods. We focus on the problems of determining whether project-generated data fall within or outside specification limits. The examples included in this chapter address construction quality problems, earned-value management, and project performance prediction.

Keywords Statistical project control · Control charts · Range charts · Specification limits

12.1 Introduction

As has been discussed so far, all quantities and processes associated with project are subject to natural variation. These variations are due to *common causes*, which may be internal causes, external causes, or simply inability to predict the future. One of the central questions that derives from the areas of Statistical Quality Control (SQC) or Statistical Process Control (SPC) is to determine whether the observed variation in a process lies within its natural variability or is outside it.

If the process lies within its natural or inherent variability, it is said to be *in statistical control*. A process that is in statistical control may or may not be satisfactory from the viewpoint of the project requirements or specification (Thompson and Koronaki 2002). A process that is in statistical control and meets the specifications for variability is said to be *capable* of achieving the specifications; otherwise it is *not capable*. A process that does not stay within its natural variability is said to be *out of statistical control*. Therefore, any project process may be in one of three states:

1. The output of the process is controlled by *common causes* and lies within its natural process variability and meets the requirements and specifications for variability; it is capable and in statistical control.

© Springer Nature Switzerland AG 2020
I. Damnjanovic, K. Reinschmidt, *Data Analytics for Engineering and Construction Project Risk Management*, Risk, Systems and Decisions,
https://doi.org/10.1007/978-3-030-14251-3_12

2. The output of the process is controlled by *common causes* and lies within its natural process variability but does not meet the requirements and specifications for variability; it is in statistical control but not meeting the requirements.
3. The output of the process is controlled by some *assignable cause* or causes; it lies outside its natural process variability; it is not in statistical control.

Of these three states, one is acceptable and two are not acceptable. The response of project management to theses states is as follows:

1. Do nothing; the process is working.
2. The natural variability of the process as defined is excessive; it is necessary to find a process with lower variability. For example, the process could be stainless steel pipe welding, performed manually by welders. The variation in product quality (percentage of welds accepted or rejected after radiographic examination) may be within the natural variability of the manual welding process, but may be excessive compared to the requirements of the project. Hence, a switch to automatic pipe welding may be required to reduce weld quality variability. As another example, the process could be writing code in a software development project. The variation in product quality (number of software bugs) could be within the natural variability of software coding, but may be excessive to meet the requirements for completing the project on time. Hence, some change in the coding process may be required.
3. The process, which may originally have been capable and in statistical control, has changed in some (perhaps unknown) way. The *assignable cause* of this change in process must be found and corrected, or the entire process (and project) is at risk.

The project manager is faced with determining, from the available evidence, which of these three states the project is in and whether or not the process is in control. In this decision, the project manager is susceptible to two kinds of errors:

- Type I (error by obliviousness). The project manager determines that the process (or project) is in statistical control when it is actually out of control. Hence, no changes are made to the process when change would be beneficial, and the growing risks are not mitigated. The cost of this error is that the process may degrade until it is uncorrectable, leading to large risks, potential failure of the project, and failure on the part of the project manager.
- Type II (error by panic). The project manager determines that the process is out of control when it is in fact in control. Hence, changes are made when no changes are needed. The cost of this error is that the project is disrupted, the changes may be in fact detrimental, the credibility of project management is lost, and the project manager does not succeed.

Unfortunately, an attempt to avoid Type I errors may only lead to increasing the likelihood of Type II errors, and conversely. The best way to reduce the likelihood of each type of error is to learn how to distinguish the three states reliably.

12.2   Earned Value Management

Quality Control (QC) and Earned Value Management (EVM) are two project areas in which statistical process control may be applied. In EVM, project progress reports typically track Earned Value data by tabulations or graphs of the plan – the Budgeted Cost of Work Scheduled ($BCWS$), and the performance - the Budgeted Cost of Work Performed ($BCWP$) and the Actual Cost of Work Performed ($ACWP$), versus time (reporting period) (Solomon and Young 2007). Also, one may track the *variance*, or cumulative deviation of the actual performance compared to the plan. [The term *variance* as used in Earned Value and in this chapter has no statistical meaning or connection to σ^2. In Earned Value parlance, *variance* simply means *diference*.

However, these conventional forms of presentation, which track the cumulative $BCWP$ and $ACWP$ over time, may obscure short-term effects. That is, when well into the job, the cumulative $BCWP$ and $ACWP$ numbers are largely determined by history and little impacted by recent events. It takes a substantial change in any single reporting period to have any visible effect on the accumulated $BCWP$ and $ACWP$. Moreover, differences in rates from period to period are shown as changes in slope, and it may be difficult for an observer to judge derivatives. It is easy to discern long-term trends after the fact, but difficult to see what is happening currently, due to the necessary scale of the cumulative plot and the inertial effect of the past data.

Also, commonly tracked are the two dimensionless EV indices, the Schedule Performance Index (SPI) and the Cost Performance Index (CPI), where, by convention,

$$SPI = \frac{BCWP}{BCWS}$$

$$CPI = \frac{BCWP}{ACWP}$$

However, these dimensionless ratios, although useful, also suffer from the same problem, that after some time they are largely determined by the inertia of history. Therefore, they cannot serve the function of *leading indicators*. The question addressed here, as discussed above, is, how can a potentially adverse trend in the process be distinguished from mere random fluctuations in progress reporting? That is, how should a project engineer decide when some intervention is necessary, wishing to avoid the error of taking action when no action is needed, and to avoid the error of not taking action when action is needed?

Useful *leading indicators* may be obtained by applying the methods of Statistical Process Control (SPC). Statistical Process Control (or Statistical Quality Control, SQC) has been used in the manufacturing industries for over 70 years (e.g., Shewhart

1931). The SPC control charts also form an essential part of Total Quality Management (TQM) and Six-Sigma (Eckes 2001).

The standard EVMS reporting quantities are defined as:

- $BCWS(t)$= cumulative Budgeted Cost of Work Scheduled through reporting period t;
- $BCWP(t)$ = cumulative Budgeted Cost of Work Performed through reporting period t;
- $ACWP(t)$ = cumulative Actual Cost of Work Performed through reporting period t;

- $SPI(t) = \dfrac{BCWP(t)}{BCWS(t)}$ = Schedule Performance Index cumulative through reporting period t

- $CPI(t) = \dfrac{BCWP(t)}{ACWP(t)}$ = Cost Performance Index cumulative through reporting period t

To apply control charting methods, it is necessary to track metrics that are stationary throughout the life of the activity. $BCWP(t)$ varies over job duration as the logistic or S-curve, whereas $CPI(t)$ and $SPI(t)$ should be constant over a job. The earned value quantities for each reporting period t (such as week or month) may be written as follows (in which upper case denotes cumulative, lower case denotes incremental or period-by-period):

- $bcws(t)$ = incremental budgeted cost of work scheduled in reporting period t;
- $bcwp(t)$ = incremental budgeted cost of work performed in reporting period t;
- $acwp(t)$ = incremental actual cost of work performed in reporting period t;

- $spi(t) = \dfrac{bcwp(t)}{bcws(t)}$ = incremental schedule performance index

- $cpi(t) = \dfrac{bcwp(t)}{acwp(t)}$ = incremental cost performance Index

The cumulative and incremental definitions are linked by:

- $bcws(t) = BCWS(t) - BCWS(t - 1)$ or $BCWS(t) = BCWS(t - 1) + bcws(t)$
- $bcwp(t) = BCWP(t) - BCWP(t - 1)$ or $BCWP(t) = BCWP(t - 1) + bcwp(t)$
- $acwp(t) = ACWP(t) - ACWP(t - 1)$ or $ACWP(t) = ACWP(t - 1) + acwp(t)$

That is, $acwp(t)$ is the actual cost of work performed in the time period t, whereas $ACWP(t)$ is the cumulative cost of the work performed from the start through time t. Note that $CPI(t)$ is *not* equal to $CPI(t - 1) + cpi(t)$.

Due to random fluctuations in project conditions, the dimensionless indices $spi(t)$ and $cpi(t)$ will vary with reporting date. If the project is in a state of statistical control, the sample statistics will be characteristic of the underlying population and hence will be stable, varying around their constant central values. The mean values of $spi(t)$ and $cpi(t)$ should be 1.0 and the variances (that is, period-by-period variations) of both should be within the inherent limits of the process. If a job has gone out of statistical control, either the mean of $spi(t)$ or $cpi(t)$ is changing or the variance is changing, or both. We accept random variations as representing the effects of the *common causes* acting on the project process, as long as the variation is random and stable. If the $spi(t)$ or $cpi(t)$ plots are not random, but exhibit some pattern, then there is some *assignable cause* operating that is changing the process. If the process was in control to begin with, it may be going out of control due to the assignable cause. And this may mean that the job will go over schedule or over budget.

To evaluate whether a change is occurring in the mean or variance, one should first establish statistics based, if possible, on historical data on jobs that are considered to have been good performers. Then Upper and Lower Natural Process Limits, which are conventionally three standard deviations above and below the mean, can then be derived from experience. Then, the probability that the measured $spi(t)$ will be below the three sigma Lower Natural Process Limit (based on the Normal distribution), due to statistical fluctuations alone, is 0.0013, and the probability that $spi(t)$ would be above the Upper Natural Process Limit is also 0.0013.

Formally, we say that if the value of $spi(t)$ falls outside the natural process limits, we reject the null hypothesis that this is a random draw from a probability distribution describing a stable population. This is an indicator that the process may be going out of control, as the probability that this value would occur with the process in control (stable) is only about 1/400. More specifically, if project management were to follow up on every value of $spi(t)$ outside the Natural Process Limits to investigate a possible change in the process, management would be wasting its effort only once in 400 times.

As an indicator of variability, control charting methods often use the period-to-period range, which is the absolute magnitude of the difference between the current period value and that in the previous period, e.g.,:

- $spirange(t) = spi(t) - spi(t-1)|$ for the range of the schedule performance index.
- $cpirange(t) = cpi(t) - cpi(t-1)|$ for the range of the cost performance index.

The mean and the variance for the range can be determined by statistical methods, and the Upper Control Limit and the Lower Control Limit for the range established. Note that the mean of the process, $E[cpi(t)]$, could be changing with no change in the variance, or vice versa. Also, some changes will appear to be beneficial: a decrease in $E[cpi(t)]$ may indicate that the job is running over budget but an increase in $E[cpi(t)]$ may mean finishing under budget. Similarly, a reduction in the standard deviation of $cpi(t)$ appears to be favorable, whereas an increase in the standard deviation may be indicative of future problems. People are naturally more interested in adverse trends than favorable trends, but *both* kinds indicate that some assignable cause we don't know about is apparently changing a project process that we thought was stable, and the cause of this instability should be investigated and understood. That is, if $cpi(t)$ is decreasing, the project may go over budget, and so the project manager needs to identify the assignable cause and correct it before the problem becomes intractable. If, however, $cpi(t)$ is increasing, the project is trending under budget, and so the project manager wants to identify the assignable cause so that he can take advantage of it on other jobs.

Of course, the ±3 sigma process limits are simply points on a Normal probability distribution, and by themselves say nothing about quality. To use the reported progress data for control, one must know what acceptable performance is. That is, one must establish the *Upper and Lower Specification Limits* (USL and LSL), which define the band of acceptable performance; that is, the band in which the values should lie under acceptable project practices and specifications. Then, if $LSL < LNPL < UNPL < USL$, the process lies within the specification requirements; or if $LNPL < LSL < USL < UNPL$, the process lies outside the specification requirements, and management should be taking action.

A comparable metric is the capability index, Cp, which may be defined as

$$Cp = \frac{(USL - LSL)}{6\sigma}$$ if $Cp < 1$, the process is not capable; that is, it cannot produce

acceptable quality performance. As an obvious example, in the case of $spi(t)$ and $cpi(t)$ variables, the Natural Process Limits should include the target value 1.0: $\mu - 3\sigma \leq 1 \leq \mu + 3\sigma$. If not, the process is incapable of being on time or on budget.

Then, in the statistical analysis of a project process, we are interested in two separate issues:

- Is the process inherently capable of meeting the specification limits?
- Is the process consistent (stays within the Natural Process Limits)? Note that a process may be consistently within its own limits and still not meet the specification limits.

Generally, *three sigma quality* is regarded as excellent quality. It means that there is only one chance in 400 that any progress observation will lie outside the (two-sided) Specification Limits (which, of course, must be outside the Natural Process Limits). If the project process is highly variable, then the ±3 sigma process limits may be too wide. Suppose that the Specification limits lie *inside* the ±3 sigma control limits, $\mu - 3\sigma < LSL < USL < \mu + 3\sigma$ but lie outside the *four-sigma* limits $LSL \leq \mu - 2\sigma < \mu + 2\sigma \leq USL$.

Then one might say that management has achieved *two sigma* but not *three sigma* quality. This is not as good as three sigma, but it might be all that is achievable.

One of the difficulties in applying statistical process control to project is, how to set the specification limits. Appropriate specification limits for variability in concrete breaking strengths, for example, may be available from engineering considerations and long histories of concrete tests, but management may have little experience in specifying variability limits for the *process* of placing concrete. It is, of course, difficult to achieve high quality in anything if one doesn't know what quality is.

Because $spi(t)$ and $cpi(t)$ are normalized dimensionless ratios, they are not influenced by job size or duration. The spread between the upper and lower process limits can be considered a measure of the quality of job management. If this band gets smaller over a number of jobs, then management is improving. If reasonable, acceptable specification limits are set, and the ±3 sigma process limits consistently lie inside the specification limits, then one could say that project management has achieved three sigma quality.

12.3   Creating Control Charts for Project-Related Processes

If data are available from other projects or activities which are similar to the activity at hand and considered to be jobs under control, then the Control Limits should be set from these data. It is preferable that these data be within the specifications for the process, if the specifications are known, but this is not essential. It is essential that these baseline projects should be consistent; that is, in statistical control.

If comparable baseline projects are not available, then the initial period of the job at hand can be used to set the baseline control limits, if it appears from the data that the process starts off in control. Perhaps the only practical way to really determine this is to use the initial period to derive the Natural Process Limits and then to check if the initial baseline data are consistent with these limits. If they are not consistent, then the job has gotten off to a bad start and management attention is needed immediately. The discussion here assumes that there is an initial period in which the process starts in statistical control. If the subsequent data (after the baseline period) also appear to be consistent with the derived Control Limits, then the baseline may be extended and the Natural Process Limits recomputed with the additional data. Do not, however, use inconsistent (out of control) data to derive Control Limits.

Let Xj represent either the period j schedule performance index $spi(j)$ or the period j cost performance index $cpi(j)$, or some other measure of productivity at the project or activity level that is expected to be stationary over the period of the job. Assume that we have data on m periods in the baseline, which we will use to determine the Natural Process Limits for Xj (that is, for $spi(j)$ and $cpi(j)$). The best estimate of the population mean (in the baseline period, it is assumed that all observations are drawn from the same population) is the mean over the m samples:

$$\bar{X} = \frac{1}{m} \sum_{j=1}^{m} X_j$$

In a typical statistical quality control application, there are multiple (for example, m) observations in each sample, and these were averaged. In a process control application, there is only one observation per sample (that is, one $cpi(t)$ per reporting period), so that observation is the sample average.

The range is used to estimate the standard deviation, but in this application, with only one observation per period, the Range, R_j, is defined as the absolute difference between successive observations; that is,

$$R_j = \left| X_j - X_{j-1} \right|$$

Table 12.1 Control charts parameters

n	d_2	E_2	A_2	C_2	D_4	D_3
2	1.128	2.660	1.880	0.853	3.267	0.000
3	1.693	1.772	1.023	0.888	2.574	0.000
4	2.059	1.457	0.729	0.880	2.282	0.000
5	2.326	1.290	0.577	0.864	2.114	0.000
6	2.534	1.184	0.483	0.848	2.004	0.000
7	2.704	1.109	0.419	0.833	1.924	0.076
8	2.847	1.054	0.373	0.820	1.864	0.136
9	2.970	1.010	0.337	0.808	1.816	0.184
10	3.078	0.975	0.308	0.797	1.777	0.223

Note that there are only $m - 1$ ranges if we start counting with $j = 1$. Then the *Mean Range* is the average:

$$\bar{R} = \frac{1}{m-1}\sum_{j=2}^{m}R_j = \frac{1}{m-1}\sum_{j=2}^{m}\left|X_j - X_{j-1}\right|$$

We can get an estimate for the standard deviation of the population σ_X by dividing the mean range by a number called d_2 (this nomenclature was established by Shewhart 70 years ago). Values of d_2 are given in Table 12.1 for various values of n, the number of observations in each sample:

In the case of the moving range, use the value of d_2 for $n = 2$, that is, $d_2 = 1.128$.

The natural process limits for the observations are then defined by the mean plus and minus three standard deviations (for ±3 sigma limits):

$$LNPL = \bar{X} - 3\hat{\sigma}_X = \bar{X} - 3\bar{R}/d_2 = \bar{X} - E_2\bar{R}$$
$$UNPL = \bar{X} + 3\hat{\sigma}_X = \bar{X} + 3\bar{R}/d_2 = \bar{X} + E_2\bar{R}$$

where E_2 is given in the table above for both ±3 sigma process limits. [Note: $\hat{\sigma}_M = \hat{\sigma}_X$, the estimated standard deviation of the mean is equal to the process mean when there is only one observation per sample.] This defines the control chart for X [$spi(t)$ and $cpi(t)$].

To obtain the control limits for the range plot, we observe that the standard deviation of the range, σ_R, is a multiple of σ_X, $\sigma_R = c_2\sigma_X$, where c_2 is given in Table 12.1. Then,

$\sigma_R = c_2\sigma_X = c_2\bar{R}/d_2$
$UNPL = \bar{R} + 3\sigma_R = \bar{R} + 3\bar{R}c_2/d_2 = \bar{R}\left[1 + 3c_2/d_2\right]$
$UNPL = D_4\bar{R}$ where $D_4 = 1 + 3c_2/d_2$
$LNPL = D_3\bar{R}$ where $D_3 = \max\{0, 1 - 3c_2/d_2\}$

The factors D_4 and D_3 for the ±3 sigma limits are also given in Table 12.1. For the moving range, $n = 2$, $D_4 = 3.267$, and $D_3 = 0.0$ for ±3 sigma. For two sigma, replace 3 by 2 in the equations for UNPL and LNPL.

12.3.1   *Control Charts for Quality Control and Process Control*

To compare Quality Control and Process Control expressions, sees Table 12.2. Use the column at the left for Quality Control (the mean of *n* observations per sample) and the column on the right for Process Control (one observation per reporting period) (Breyfogle 1999).

Figures 12.1 and 12.2, respectively, show quality and process control charts. Figure 12.1 shows a control chart for the mean of three concrete cylinder tests, with *two* sigma upper and lower natural process limits, plotted against pour number.

Table 12.2 Quality control and process control comparison

Quality control	Process control
n observations for sample *j*: $X_{j,1}, X_{j,2}, X_{j,k}, \ldots, X_{j,n}$	1 observation for period *j*: X_j
	Mean for period *j*: X_j
Mean for sample *j*: $\bar{X}_j = \dfrac{1}{n}\displaystyle\sum_{k=1}^{n} X_{j,k}$	
Estimated process mean: $$\bar{X} = \frac{1}{m}\sum_{j=1}^{m}\bar{X}_j = \frac{1}{m}\sum_{j=1}^{m}\left[\frac{1}{n}\sum_{k=1}^{n}X_{j,k}\right]$$	Estimated process mean: $\bar{X} = \dfrac{1}{m}\displaystyle\sum_{j=1}^{m} X_j$
Range for sample *j*: $R_j = \max_k\{X_{j,k}\} - \min_k\{X_{j,k}\}$	Range for period *j*: $$R_j = \left|X_j - X_{j-1}\right| \quad for \;\; j \geq 2$$
Average range: $\bar{R} = \dfrac{1}{m}\displaystyle\sum_{j=1}^{m} R_j$	Average range: $$\bar{R} = \frac{1}{m-1}\sum_{j=2}^{m}R_j = \frac{1}{m-1}\sum_{j=2}^{m}\left|X_j - X_{j-1}\right|$$
Estimate of population standard deviation: $$\hat{\sigma}_X = \frac{\bar{R}}{d_2}$$	Estimate of population standard deviation: $$\hat{\sigma}_X = \frac{\bar{R}}{d_2}$$
In which d_2 is taken from the table for *n* = number of observations used in the computation of range	In which d_2 is taken from the table for *n* = 2 (number of observations used in the computation of range)
Estimate of the standard deviation of the mean: $$\hat{\sigma}_M = \frac{\hat{\sigma}_X}{\sqrt{n}}$$	Estimate of the standard deviation of the mean: $$\hat{\sigma}_M = \frac{\hat{\sigma}_X}{\sqrt{1}}$$
Center line of the control chart for X: $CL = \bar{X}$	Center line of the control chart for X: $CL = \bar{X}$
Upper natural process limit (± 3 *sigma*) for the control chart for X:	Upper natural process limit (± 3 *sigma*) for the control chart for X:
$$UNPL = \bar{X} + 3\hat{\sigma}_M = \bar{X} + 3\hat{\sigma}_X / \sqrt{n}$$ $$UNPL = \bar{X} + 3\bar{R} / \left(d_2\sqrt{n}\right) = \bar{X} + A_2\bar{R}$$	$$UNPL = \bar{X} + 3\hat{\sigma}_M = \bar{X} + 3\hat{\sigma}_X$$ $$UNPL = \bar{X} + 3\bar{R} / d_2 = \bar{X} + E_2\bar{R}$$

(continued)

Table 12.2 (continued)

Quality control	Process control
In which	In which
	$E_2 = 3/d_2$
$A_2 = 3/\left(d_2\sqrt{n}\right)$	
See table for values of A_2 vs. n	See table for values of E_2 (using n = 2)
Lower natural process limit (±3 *sigma*) for the control chart for X: $LNPL = \bar{X} - A_2\bar{R}$	Lower natural process limit (±3 *sigma*) for the control chart for X: $LNPL = \bar{X} - E_2\bar{R}$
Center line of the control chart for range: $CL = \bar{R}$	Center line of the control chart for range: $CL = \bar{R}$
Upper natural process limit (±3 *sigma*) for the control chart for range:	Upper natural process limit (±3 *sigma*) for the control chart for range:
$\hat{\sigma}_R = c_2\hat{\sigma}_X = c_2\bar{R}/d_2$	$\hat{\sigma}_R = c_2\hat{\sigma}_X = c_2\bar{R}/d_2$
$UNPL = \bar{R} + 3\hat{\sigma}_R = \bar{R} + 3c_2\hat{\sigma}_X$	$UNPL = \bar{R} + 3\hat{\sigma}_R = \bar{R} + 3c_2\hat{\sigma}_X$
$UNPL = \bar{R} + 3c_2\bar{R}/d_2 = \bar{R}\left(1 + 3c_2/d_2\right)$	$UNPL = \bar{R} + 3c_2\bar{R}/d_2 = \bar{R}\left(1 + 3c_2/d_2\right)$
$UNPL = \bar{R}D_4$ in which $D_4 = 1 + 3c_2/d_2$	$UNPL = \bar{R}D_4$ in which $D_4 = 1 + 3c_2/d_2$
Lower natural process limit (±3 *sigma*) for the control chart for range:	Lower natural process limit (±3 *sigma*) for the control chart for range:
$LNPL = \bar{R}D_3$	$LNPL = \bar{R}D_3$
in which $D_3 = $ max {0, $1 - 3c_2/d_2$}	in which $D_3 = $ max {0, $1 - 3c_2/d_2$}
See table for values of D_3 and D_4 vs. n	See table for values of D_3 and D_4 (recall that n = 2 for range)

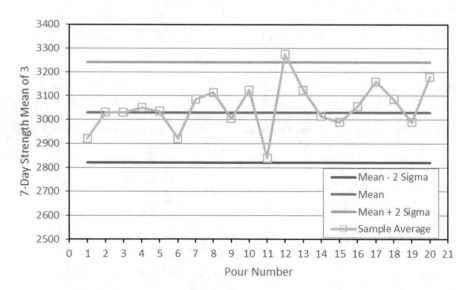

Fig. 12.1 Control chart for mean 7-day strength

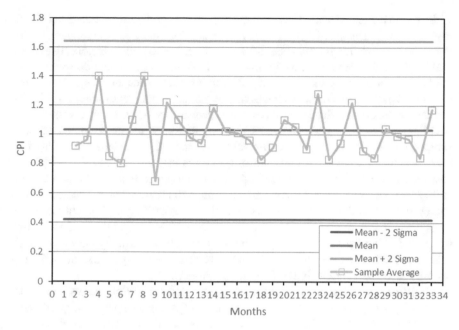

Fig. 12.2 Project cpi series

Figure 12.2 shows a plot of the month-to-month incremental cost performance index for a major (over $1,000,000,000) project with ±3 sigma upper and lower natural process limits. Is this process in statistical control? Is this process *capable*?

12.4 Statistical Quality Control

The underlying issue in statistical quality control is variability. Suppose an engineer is performing receipt inspection at a site, by examining one item of a shipment of parts or equipment to see if it meets the engineer's specification. If there were no variability, examination of one item would be conclusive: if that part is good, then the whole shipment is good; if that part is unacceptable, then so are all the rest. In this ideal case, statistical quality control is irrelevant. But, of course, there is variability, and so inspection of a single item in a shipment is not conclusive evidence of the remaining items.

At the other extreme, the engineer could perform 100% inspection. All the good parts would be identified and accepted; all the bad parts would be identified and rejected. There would never be any doubt about the quality of the uninspected parts because there would be no uninspected parts. In this ideal case, statistical quality control would be irrelevant.

In the presence of variability, inspection of one part is inadequate but 100% inspection is expensive. (And probably even fallible; there is some evidence that even 100% inspection is only about 80% effective, due to human factors (Breyfogle 1999).) Therefore, it is desirable to inspect some number or some proportion of the total shipment of parts, and to base a decision whether to accept or reject the entire shipment based on this sample. However, this sample is subject to small sample size limitations, such that there is always some probability of an error, insofar as the small sample is not representative of the population of all items. If, based on a sample, we decide to accept the entire shipment, then there is some probability that there is actually an excessive number of bad parts in the shipment. Conversely, if, based on a sample, we decide to reject the entire shipment, then there is some probability that the parts in the shipment actually meet the specification. The first case is sometimes called the consumer's risk (the risk of accepting a shipment when the parts are actually no good; the second case is sometimes called the producer's risk (the risk that the purchaser will reject the material when it actually meets his specification). Both risks decrease as the sample size gets larger (and theoretically become zero if the sample is 100% of the shipment).

Statistics is then just a way of estimating the risk (probability) that we will make an error by accepting a shipment of bad parts or rejecting a shipment of good parts. Either error costs money, wastes time, and makes us look bad. However, there is no way of avoiding these risks, short of inspecting every part.

Quality may be measured by the proportion of bad parts in a shipment, or in other ways, and quality-related decisions may go further than simply accepting or rejecting a shipment. Consider another type of quality that is determined by measurements: cast in place concrete. As a quality control measure, typically three test cylinders are cast along with the concrete placement, for every batch of concrete. These cylinders are tested 7 days after placement, and the compressive strength measured. The values for the three sample cylinders are averaged. If the measurements do not meet the specification requirements, then either the engineers have to be called in, to determine if the low strength concrete can be accepted, or the 7-day old concrete has to be jack hammered out. These things are expensive. But if the concrete strength is much larger than necessary, this costs money too. Because of variability, we cannot design the concrete mix to be exactly the strength desired; we must make it stronger. The cost of variability can be seen from a simple example.

Suppose we want concrete with ultimate strength of, say, 3000 at 7 days. If $\overline{X}_j$ represents the average for the j-th sample of n cylinders broken at 7 days, then we may require that $\overline{X}_j \geq 3000$. But, due to variability in the test specimens, we must restate this requirement in a probabilistic sense: $P\{\overline{X}_j < 3000\} \leq \alpha$, where α is the probability of an error in accepting that the entire concrete batch has strength >3000 when it actually has strength <3000. Suppose that the true distribution of the measured values has mean value μ and variance σ^2. That is, each cylinder test is a random draw from a population (assumed Normal) with mean value μ and variance σ^2. Of course, we don't know what these values are. Because of our uncertainty (or ignorance) about the strength parameters, we have to break the test cylinders to make some estimates of these parameters. From the tests, we compute the sample mean $\bar{X}_j$ and use it as an estimate of the population mean μ. Since, by definition, if $X_{j,k}$ is the breaking strength for cylinder k, in sample j, then

$$\bar{X}_j = \frac{1}{n}\sum_{k=1}^{n} X_{j,k}$$

If we have tested m samples, then we can compute the average over all samples as:

$$\bar{X} = \frac{1}{m}\sum_{j=1}^{m}\bar{X}_j = \frac{1}{m}\sum_{j=1}^{m}\left[\frac{1}{n}\sum_{k=1}^{n} X_{j,k}\right]$$

Here $\bar{X}$ is a better estimate of the process mean than the single sample mean. We know from elementary statistics that the variance of the sample mean is:

$$\sigma_M^2 = Var\left[\bar{X}\right] = \frac{1}{n}Var\left[X\right] = \frac{1}{n}\sigma^2$$

$$\sigma_M = \frac{\sigma}{\sqrt{n}}$$

In this case, we will call the value 3000 the Lower Specification Limit (LSL), that is, the lowest value for the sample mean that is acceptable by the specification and let $\alpha = 0.025$. That is, we will accept a process in which 1 out of 40 sample means is below the specification limit and 39 out of 40 are above the specification limit. This value of α corresponds approximately to the 4σ level. That is, using the standard tables for the Normal distribution, the probability that $\bar{X}_j$ is less than the mean minus $2\sigma_M$ (or greater than the mean plus $2\sigma_M$) is 0.023 (close enough for this work). Let $n = 3$. If we know the population standard deviation, say $\sigma = 125$, then $\sigma_M = \frac{\sigma}{\sqrt{3}} = \frac{175}{\sqrt{3}} \cong 100$. Then we must design the concrete mix to have a mean strength of $3000 + 2\sigma_M = 3200$ (see Fig. 12.3).

Suppose, however, that the concrete mix process was more variable, such that the population standard deviation was, say, $\sigma = 865$. Then $\sigma_M = 865/\sqrt{3} \cong 500$. To overcome the variability and assure that the Lower Specification Limit is violated not more than $\alpha\%$ of the time, we have to raise the mean strength. In this case, the mix design would require a mean of $3000 + 2\sigma_M = 4000$. Clearly the higher variability is going to increase the requirement for cement and hence the cost of the concrete.

The quality of the concrete is represented by the probability that a sample will be accepted as satisfying the specification. It is conventional to refer to this quality in terms of numbers of standard deviations. That is, in the concrete case we set the Lower Specification Limit to be $\mu - t\sigma_M$, where t is an integer, 1, 2, 3, etc. We consider only the one-sided or one-tailed case, because in general we don't reject con-

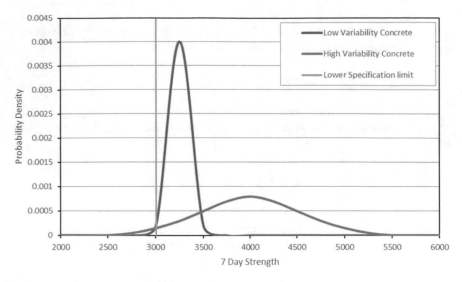

Fig. 12.3 Concrete mix – low and high variability

Table 12.3 Probability of rejecting concrete batch

LSL	% Batches rejected		Design μ if $\sigma = 865$	Design μ if $\sigma = 175$
$\mu - \sigma_M$	15.87%	~1/6	3500	3100
$\mu - 2\sigma_M$	2.28%	~1/44	4000	3200
$\mu - 3\sigma_M$	0.135%	~1/740	4500	3300
$\mu - 4\sigma_M$	0.00317%	~1/32,000	5000	3400
$\mu - 5\sigma_M$	0.0000287%	~1/3,500,000	5500	3500
$\mu - 6\sigma_M$	0.0000000987%	~1/1,000,000,000	6000	3600

crete for being too strong, only for being too week. Table 12.3 shows the probability of rejecting a batch of concrete for different values of t, and the mean design strength we must use to meet this level of quality for the cases in which $\sigma = 175$ and $\sigma = 865$.

Therefore, if we aspired to 6σ quality, we would reject only one batch in a billion, but we would have to design for a mean strength of 6000 to achieve this, if $\sigma = 865$. It is unlikely that anyone pouring concrete would require 6σ quality, as one would never see a batch rejected in a lifetime. However, setting the Lower Specification Limit (LSL) at $\mu - 3\sigma_M$ is not unreasonable, and would produce concrete such that about one in 740 batches would be rejected. In some cases, $LSL = \mu - 4\sigma_M$ might be appropriate, but this level of quality would require raising the mean design strength to 5000.

The above example illustrates why many people say that quality in costs a lot of money. It does, if quality is obtained by over design, as in this case. If $LSL = \mu - 3\sigma_M$, for example, about 98% of all batches have mean strength over 3500, when only 3000 is required, so a lot of cement is being wasted. The objective of the Six Sigma process is to meet the quality objectives by process improvement rather than gross over-design. In this example, if the process can be improved to the point that $\sigma = 175$, then the $LSL = \mu - 3\sigma_M$ criterion can be met by designing the concrete to achieve a mean sample strength of 3300 instead of 4500. Reduction in process variation can actually save money. For example, increasing quality by lowering the LSL from $\mu - 2\sigma_M$ to $\mu - 3\sigma_M$ reduces the frequency of rejects from 1 in 44 to 1 in 740. In other words, for every batch rejected at the 3σ quality level, 17 batches would be rejected at the 2σ level. As, by assumption, the concrete has been in the forms for 7 days prior to the tests, removing this rejected concrete could cost a lot of money, even if it were possible.

How can the process variation be reduced? That is what the engineer is paid to determine.

Suppose that instead of (or in addition to) specifying a constraint on the average breaking strength of n specimens per batch, the specification writer put a constraint on the minimum value of the n specimens.

In each batch, one computes the sample mean strength $\overline{X}_j$ from the three (or n) test specimens. It is also necessary to compute the sample variance. It is conventional in Statistical Quality Control to compute the sample standard deviation from the Range of the sample, where Range is defined as

$$R_j = \max_k [X_k] - \min_k [X_k] \text{ for sample } j$$

This usage may be in part because, when SQC was developed (by Shewhart and others), computing the squares and square roots in the usual equation for sample standard deviation was difficult, without computers or even pocket calculators, so simpler formulations were preferred. Also, the sample sizes are very small, so corrections were applied to these calculations to compensate for small sample size. These corrections are tabulated and are so widely used that they are essentially part of the method. Therefore, they will be used here.

If there are many samples in each batch, then one can approximate the sample standard deviation by:

$$\hat{\sigma} = \frac{R_j}{6}$$

Here, $\hat{\sigma}$ is an approximation to the true, but unknown, standard deviation, σ. This approximation is based on the fact that, for the Normal distribution, the area (probability) between the mean minus three standard deviations and the mean plus three standard deviations is 0.9974, which is close to 1.00. Then, the Range is very close to being six standard deviations ($6 - \sigma$), and one can estimate the standard deviation as one-sixth of the observed Range. However, this approximation cannot be used for small sample sizes.

Let $\overline{R}$ be the average range over some number of samples (that is, concrete batches). That is, there are m samples (or batches), indexed by j, and n specimens in a sample, indexed by k, then:

$$R_j = \max_k \left[X_{j,k} \right] - \min_k \left[X_{j,k} \right] \text{ for sample j}$$

$$\bar{R} = \frac{1}{m} \sum_{j=1}^{m} R_j$$

Then, an estimate for the population standard deviation that is valid for small samples is:

$$\hat{\sigma}_X = \frac{\bar{R}}{d_2}$$

Here d_2 is a function of n, the sample size, and is tabulated in Table 12.1 for n from 2 to 10. For values of n greater than 10, use $d_2 \cong \sqrt{n}$.

From the above discussion, an estimate for the standard deviation of the mean strength is:

$$\hat{\sigma}_M = \frac{\hat{\sigma}_X}{\sqrt{n}}$$

In using Statistical Quality Control, one often displays the information on a Control Chart. On a control chart for the mean sample strength $\bar{X}$ are shown the following, plotted for the number of samples (see figure below):

- The desired mean, μ, as given in the specification
- The Center Line, or CL, determined by the sample mean, $\bar{X}$
- The Upper Specification Limit, USL, usually given by $\mu + 3\sigma_M$
- The Lower Specification Limit, LSL, usually given by $\mu - 3\sigma_M$, defining the Six Sigma specification USL − LSL = $(\mu + 3\sigma_M) - (\mu - 3\sigma_M) = 6\sigma_M$. (However, in the example above we defined the LSL as $\mu - 2\sigma_M$)
- The Upper Natural Process Limit, UNPL, given by the estimated process param-

  eters, $\text{UNPL} = \bar{X} + 3\hat{\sigma}_M = \bar{X} + \dfrac{3\hat{\sigma}_X}{\sqrt{n}} = \bar{X} + \dfrac{3}{\sqrt{n}} \left(\dfrac{\bar{R}}{d_2} \right) = \bar{X} + A_2\bar{R}$. The table in

  Chap. 20 gives values for A_2
- The Lower Natural Process Limit, LNPL, using the estimated process parameters,

$$\text{LNPL} = \bar{X} - 3\hat{\sigma}_M = \bar{X} - \frac{3\hat{\sigma}_X}{\sqrt{n}} = \bar{X} - \frac{3}{\sqrt{n}} \left(\frac{\bar{R}}{d_2} \right) = \bar{X} - A_2\bar{R}.$$

- The observed sample means $\bar{X}_j$

Figure 12.4 shows a control chart for sample means with the Lower Specification Limit (LSL) indicated at the $\mu - 3\sigma_M = 3300 - 3(100) = 3000$ level. Also shown are the computed values for the process mean and the Lower Natural Process Limit

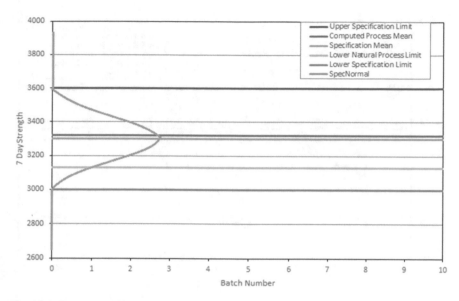

Fig. 12.4 Process capable

(LNPL). Here it is clear that the LNPL lies inside of (above) the LSL, at about the 3100 level. That is, the process is producing results that are acceptable or even better than the specification. If the process has LNPL > LSL, and, if appropriate, UNPL < USL, then the process is said to be *capable* of meeting the quality specification.

Figure 12.5 shows a control chart for sample means with the Lower Specification Limit (LSL) indicate at the $\mu - 3\sigma_M = 3300 - 3(100) = 3000$ level. Also shown are the computed values for the process mean and the Lower Natural Process Limit. Here it is clear that the LNPL lies outside of (below) the LSL, at about the 2850 level. That is, the process is actually producing results that, statistically, do not meet the specification because a higher number than permitted are below the LSL due to higher variation than the baseline. If the process has LNPL < LSL, and, if appropriate, UNPL > USL, then the process is said to be *not capable* of meeting the quality specification.

The values for the horizontal lines in the control chart are determined based on a process that is in statistical control. That is, there is variation, but this variation is considered an acceptable part of the process. If the process is under statistical control, the variations in the actual observations are random. Random variations are generated by some common cause or causes. If the variations are not random, then the variations are due to some assignable cause, and the process is not under statistical control. This means that the engineer must find the assignable causes and eliminate than before the process goes off track. The function of the control charts is to help the engineer identify whether the process is under control, in order to do something about an assignable cause before it becomes a major problem.

Are the actual sample averages in the figure just above random? Why or why not?

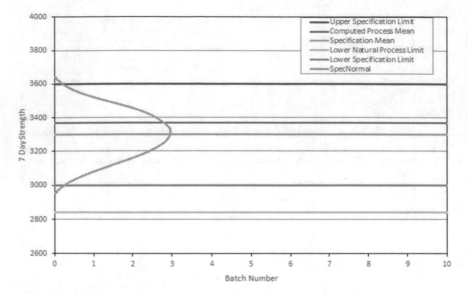

Fig. 12.5 Process not capable

Suppose, in the example above, we wanted a process with the LSL set as $\mu - 3\sigma_M$ = 3000. Then, assuming we consider a standard deviation of 500 to be acceptable variation, we must design the concrete mix for a mean strength of 4500 (that is, the Coefficient of Variation is 500 / 4500 = σ/μ = 11%). If this process as observed is performing under statistical control, the likelihood that a sample mean strength would be less than 3000 is 0.0013, or about 1 in 1000. Suppose that some sample average is in fact less than 3000. Then there are two possible hypotheses:

1. The process is in statistical control but just happened to generate a random event with probability 1 in 1000.
2. The process is not in statistical control.

If we consider 1 in 1000 to be a small probability, then we can reject hypothesis 1 and go looking for the assignable cause. Still, there is 1 chance in 1000 that we are wrong, that hypothesis 1 is correct, and the process is still in control.

12.4.1   Range Charts

In addition to the control charts for $\overline{X}$, the sample mean, one can define a number of other control charts. The chart that will be discussed here is the control chart for Range. Range was defined above. The control chart for range has a center line at the mean Range,

$$\overline{R} = \frac{1}{m}\sum_{j=1}^{m} R_j$$

The UNPL, also called the Upper Control Limit (UCL), and LNPL, also called the Lower Control Limit (LCL), are given by:

$$UCL = D_4 \overline{R}$$
$$LCL = D_3 \overline{R}$$

The coefficients D_3 and D_4 for various values of n, the sample size, are given in the Table 12.1.

Example 12.1

Consider project data in Table 12.4 (this example is adopted from Breyfogle (1999), pages 165–166).

Is $\overline{X}$ process in statistical control? How about Range process? Is it in statistical control? Figure 12.6 shows the data and natural process limits and process mean.

Note the three points indicated by arrows that lie outside the 6σ band. This process is not in statistical control and the engineer should seek to identify the assignable cause or causes for these three points. Figure 12.7 shows Range Control Chart for the same data.

Table 12.4 Project data

Sample no.	X1	X2	X3	X4	X5	Mean X	Range R
1	36	35	34	33	32	34	4
2	31	31	34	32	30	31.6	4
3	30	30	32	30	32	30.8	2
4	32	33	33	32	35	33	3
5	32	34	37	37	35	35	5
6	32	32	31	33	33	32.2	2
7	33	33	36	32	31	33	5
8	23	33	36	35	36	32.6	13
9	43	36	35	24	31	33.8	19
10	36	35	36	41	41	37.8	6
11	34	38	35	34	38	35.8	4
12	36	38	39	39	40	38.4	4
13	36	40	35	26	33	34	14
14	36	35	37	34	33	35	4
15	30	37	33	34	35	33.8	7
16	28	31	33	33	33	31.6	5
17	33	30	34	33	35	33	5
18	27	28	29	27	30	28.2	3
19	35	36	29	27	32	31.8	9
20	33	35	35	39	36	35.6	6
Mean						33.55	6.2

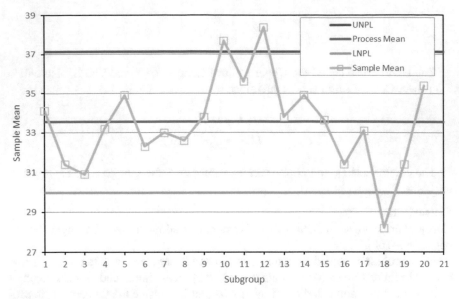

Fig. 12.6 Sample mean control chart

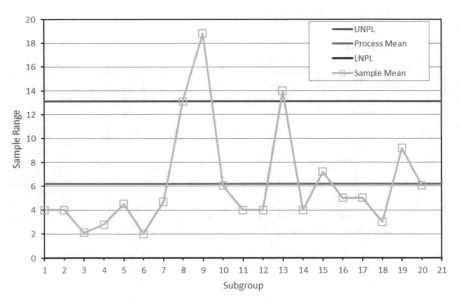

Fig. 12.7 Sample range control chart

Note the two points indicated by arrows that lie outside the 6σ band. The ranges for two of the samples are significantly greater than the ranges for the other 18 samples. This process is not in statistical control and the engineer should seek to identify the assignable cause or causes for these two points.

12.5   Control Charts and Statistical Project Control

Suppose the project manager wishes to use Project Control Charts for a project. A simple form of run chart could be obtained by plotting the cost of each sequential unit against unit number as completed. This is shown in Fig. 12.8. The mean cost is of course 100, and the standard deviation of each cost is 40, so the Upper Natural Process Limits are at mean+σ = 140, mean+2σ = 180, and mean+3σ = 220; only mean+σ = 140 and mean+2σ = 180 are shown. The run chart indicates that this process, consisting of the construction of identical units, is in statistical control; no points break through the mean+2σ line, much less the mean+3σ line. However, there is some question about a long run, with 14 consecutive values above the mean line, a highly improbable event, perhaps indicating that the project is not in statistical control. However, by the time the project manager could detect such a long run the project is nearly over.

Where on this run chart would one place the Specification Limit? One could put it at the mean line, or 100, implying the specification that unit costs should not exceed the estimated \$100 per unit, but this would result in, on the average, 50% of the units being declared out-of-spec. Is the process shown in the figure below in-spec or out-of-spec? Where should the Specification Limit be placed?

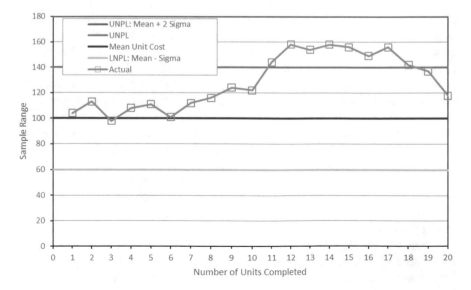

Fig. 12.8 Run chart for unit cost

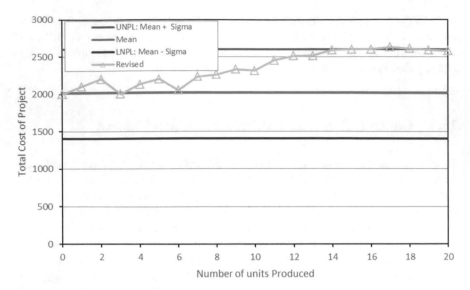

Fig. 12.9   Control chart for total project cost

An alternate approach could be to generate a run chart for the expected cost at completion, as shown in Fig. 12.9, rather than the individual unit costs. The a priori mean cost at completion is $2000 and the standard deviation of the cost at completion was determined previously to be $595.83, so mean+1$\sigma$ is $2595.83 and mean+2$\sigma$ is $3191.66. Again, the process appears at first to be in statistical control, with no value making it to the mean+2σ limit, let alone a mean+3σ limit. Still, this chart has 17 values in a row all above the mean line, a statistically highly improbable event, so the process may be declared out of statistical control. However, this run length criterion may be suitable for manufacturing operations, which are generally unlimited in time, but projects are limited in duration and detecting out-of-control situations based on run lengths may not be very helpful.

But, where is the Specification Limit? Again, it might be placed at the mean, but this gives a 50% probability of being out-of-spec, too large to be helpful. It might be placed at the mean+1σ line, where the probability of being out-of-spec is about 16%, but this is rather arbitrary. In fact, the location of the Specification Limit has nothing to do with the location of the Natural Process Limits; the two concepts measure different things.

The control chart in Fig. 12.10 adopts a statistical approach. It was stated earlier that the project manager wanted a budget such that the probability of success would be 90% or more and the probability of failure (exceeding the budget) would be 10% or less. That budget was determined to be $2763.86. Therefore, the project manager's clear specification for this project is that the probability of over-running the $2763.86 limit should be less than 10%. This is reflected in the chart below. The Specification Limit is the horizontal line representing a 10% probability of overrun-

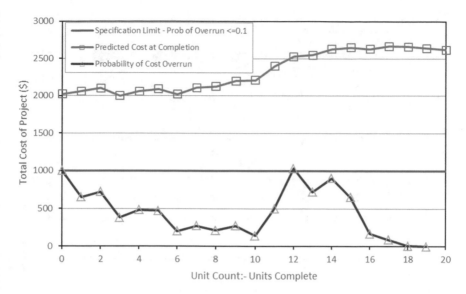

Fig. 12.10 Probability of cost overrun

ning. The probability of a cost overrun is computed by the approach given above and plotted in the control chart. Here it is seen that the probability of overrun exceeds 10% when 13 of the 20 units have been completed, and so the process is out-of-spec at that point.

Computing the probability of an overrun is not difficult is not difficult if a spreadsheet is used for the control chart. After seven units have been completed, the predicted cost at completion is $2126.55. The standard deviation of this prediction is calculated by the equations above to be $329.12. Then the probability that the cost at completion will under run the available funding, namely $2763.86, is 0.9736, so the probability of overrunning is 1.0–0.9736 = 0.0264. Therefore, based on the results from the first seven units completed, the project manager has only a probability of about 2.6% of failing to meet the available funding limitation, as shown in Fig. 12.10.

An alternate metric for tracking through a control chart is the Value-at-Risk. Before the project began, the project manager (or his client) was willing to accept a VaR at the 90% confidence level of $2763.86. As each unit is complete, the posterior VaR can be computed from the newly-acquired cost information. For example, after the completion of seven units, the revised expected value of the cost at completion is computed to be $2126.55. The computed standard deviation of the cost at completion is $329.12. The VaR at this time may be 2548.33. Hence the posterior VaR is less than the initial VaR by the amount $215.53, and the project is in-spec with regard to cost. Figure 12.11 shows the Specification Limit as the horizontal line at the initial or prior value of the VaR, at $2763.86. Any VaR below this line is in-spec; any VaR above it is out-of-spec with regard to cost.

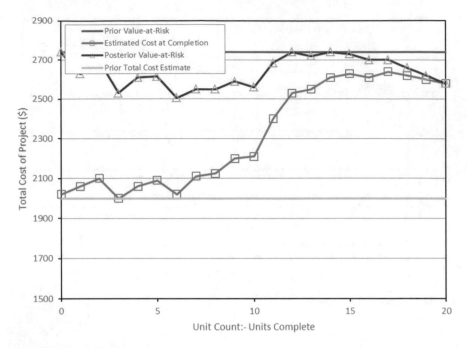

Fig. 12.11   Probability of cost overrun

The variations in the revised budgets and expected costs at completion, after each work package, unit, is completed, are shown in the figures given above. The objective of using the confidence limits is to keep the confidence band positioned so that it envelops the (unknown) actual cost at completion. That is, no one can predict the future (the actual cost at completion) with certainty, but we can try to define a confidence band that bounds where we expect to find it (with probability 90%, in this case). Although the lower confidence bound is not shown in the figure (it is the expected cost at completion minus the contingency), it is clear that, for this example at least, the method has achieved the goal of keeping the 90% confidence limit above the actual cost at completion for every step in the process up to project completion. Of course, even though the costs increased over the original estimates, this example was actually well behaved in that it never ran out of contingency (although contingency was being used up for some period of time).

Figure 12.12 shows the change in uncertainty regarding the final cost at completion as each unit is completed, providing additional information about where the project is going. (Some units have been omitted from the figure for clarity; plotting all 20 units results in a graph too difficult to read.) Before any work is done, the prior estimate of the probability for the cost at completion is centered on $2000, but has a very high standard deviation (uncertainty), as shown in the figure below. After the completion of the first unit, the probability distribution becomes narrower and shifts to the right. As each subsequent unit is completed, the best estimate or most likely value for the cost at completion increases somewhat, depending on the latest

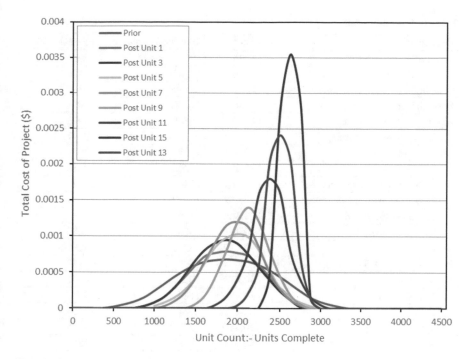

Fig. 12.12 Posterior probability density function

actual reported cost, but the uncertainty (the width of the distribution) always decreases. The figure illustrates how the method starts with a high degree of uncertainty and zeros in on the target as more information becomes available about actual project cost performance.

12.6   Practice Problems

Problem 12.6.1 You are a Construction Engineer working for an asphalt paving contractor. The general superintendent is concerned about the allowances for small tools and supplies that are included in bids for jobs. Obviously, a bid allowance that is too low leads to loss of profits, whereas an allowance that is too high may lead to loss of jobs to competitors. The allowance for small tools and supplies currently used in bid development is 10 cents per ton of asphalt placed, and the superintendent wants to know if this factor is still valid, if any particular jobs have significantly overrun the allowance, and if there is any time trend in the small tools expense. Data on 19 actual jobs, showing the actual tons of asphalt placed and the actual expenditures on small tools and supplies, are shown in Table 12.5, in chronological order.

Table 12.5 Problem data

Description	Job #	Tons asphalt placed	$ small tools
Guilford road resurfacing	1035	8961	2080
Mebane Oaks road and highway 119	1036	17,406	2726
State highway 49 at Trollingwood road	1037	32,048	3380
State highway 49 at Orange street	1038	12,377	1954
Alamance at Guilford road	1040	16,921	5973
Davidson county resurfacing	20,452	56,836	9217
City of Reidsville	20,454	2632	704
City of Winston	20,459	15,885	2882
US 52 Northbound lanes shoulder	20,461	9524	2626
US 52 Southbound lanes shoulder	20,462	10,196	2467
US 220 at Guilford road	20,466	6189	379
City of Thomasville	20,469	2905	1083
WBS 37193	20,472	4833	1899
Business 40 and US 431 ramps	20,474	3599	1158
WBS 36788	20,475	10,116	4641
Bluff school road	20,476	1684	442
Kivett drive	20,477	2561	1397
Orange and Caswell	20,480	13,725	2275
Davidson county	20,484	20,998	1560

Show a control chart for dollars expended on small tools per ton of asphalt, as a function of time. Is this process homogeneous (in statistical control)? Explain why or why not. Show upper and lower process limits. Is it possible to define specification limits? If so, what should they be? Do any jobs stand out, high or low? Is the actual expense for small tools a linear function of the job size, as is implied by the 10 cents per ton factor, or is some other relationship with job size better?

Problem 12.6.2 When the Wehner Building was under construction, project engineers made observations on the time it took to drill in foundations. The raw data for eight sampled holes are shown in Table 12.6, in order of drilling. The total drilling time for each hole is the sum of the shaft drilling time and the bell drilling time. Movement of the drilling equipment, etc., from one hole to another is not included. Setting of rebar cages is not included. Many other holes were drilled but not recorded, but there is no reason to believe that they differed statistically from this limited sample.

- Compute the means and standard deviations for the bell, the shaft, and the total drilling time per hole.
- Compute the Three Sigma Upper Natural Process Limit and the Lower Natural Process Limit for the bell and the shaft.
- Plot the run chart with LNPL and UNPL for the bell and the shaft (both on the same plot).

Table 12.6 Problem data

Hole (Number)	Shaft (Hours)	Drilling (Minutes)	Time (Seconds)	Bell (Hours)	Drilling (Minutes)	Time (Seconds)
P48K	0	33	41	0	55	50
P48H	0	30	57	0	52	12
P48G	0	47	9	1	15	47
P48A	0	25	34	0	48	23
P48U	0	31	46	1	6	52
P48R	0	42	49	1	3	43
P48M	0	40	52	1	10	24
P48L	0	38	57	1	3	38

- By examination of the run plots, are the bell times and the shaft times correlated? Compute the correlation coefficient between the bell time and the shaft time.
- Compute the mean and the standard deviation for the total time, and then plot the run chart with LNPL and UNPL for the total time.
- Repeat the above for the Range Charts for the bell, the shaft, and the total time each hole.

Is this process under statistical control? Why or why not? What is your recommendation about this process? Do you think there should be specification limits? If so, what should they be?

References

Breyfogle FW (1999) Implementing six sigma: smarter solutions using statistical methods. Wiley, New York

Eckes G (2001) The six sigma revolution: how general electric and others turned process into profits. Wiley, New York

Shewhart WA (1931) Economic control of quality of manufactured product. ASQ Quality Press, Milwaukee

Solomon PJ, Young RR (2007) Performance-based earned value. Wiley, Hoboken

Thompson JR, Koronaki J (2002) Statistical process control: the Deming paradigm and beyond, 2nd edn. Chapman & Hall, Boca Raton

Chapter 13
Forecasting Project Completion

Abstract In this chapter we discuss methods for forecasting future job progress. More specifically we focus on forecasting two important project performance criteria – completion time and cost-at-completion, on the basis of past progress data. We introduce a class of S-curves that is suitable for representing job progress as well as discuss how to develop the confidence intervals around the forecasts. In addition we show how Bayesian methods can be used to update the parameters of the S-curve models.

Keywords Earned value · Forecasting · S-curves

13.1 Introduction

One might say that project management is all about forecasting, because a project manager constantly needs to forecast future job progress, and in particular to forecast completion time and cost at completion, on the basis of past progress. If the predicted cost is close to the project budget, and the predicted completion date is close to the project schedule, then the project manager may not need to do anything. On the other hand, if the predicted cost at completion is much greater than the budget, or the predicted time at completion is much greater than the schedule, the project manager may need to do something immediately.

One might also like to place confidence intervals around the forecasts, in order to assess the reliability of the prediction, so that the project manager does not err by taking action when no action is needed, or by taking no action when action is required. Clearly, this approach has its limitations, one of which is the lack of data to work with, especially early in the job. Nevertheless, forecasting project completion is essential to effective project management, and often the lack of precision in all of the forecasting approaches means that there is value to be gained from using multiple, independent forecasting methods.

© Springer Nature Switzerland AG 2020
I. Damnjanovic, K. Reinschmidt, *Data Analytics for Engineering and Construction Project Risk Management*, Risk, Systems and Decisions,
https://doi.org/10.1007/978-3-030-14251-3_13

13.2   Prediction Using Earned Value Management

At some time t, it is desired to estimate the cost at completion, EAC or $ECAC$, from the progress to date. Then,

- BAC is the Budget at Completion, or the total budgeted cost.
- $EAC(t)$ is the estimated cost-at-completion based on information at time t, to be determined.
- $BCWS(t)$ is the plan, the Budgeted Cost of the Work Scheduled at time t.
- $ACWP(t)$ is the reported actual cost of construction performed up to time t.
- $BCWP(t)$ is the earned value reported at time t.
- $CPI(t)$ is the Cost Performance Index at time t, $CPI(t) = \dfrac{BCWP(t)}{ACWP(t)}$

Clearly, the estimated cost at completion has to exceed the cost of the project to date, if the project is incomplete, or $EAC(t) > ACWP(t)$. The remaining work, or value, at the budgeted rates, is $BAC - BCWP(t)$. Based on the project to date, the average ratio of actual cost to budgeted cost is $\dfrac{ACWP(t)}{BCWP(t)} = \dfrac{1.0}{CPI(t)}$ Then, based on the assumption that the average CPI observed for the work done to date will be true of the work to be done,

$$EAC(t) \cong ACWP(t) + \left[BAC - BCWP(t) \right] \left[\frac{ACWP(t)}{BCWP(t)} \right]$$

$$EAC(t) \cong ACWP(t) + BAC \left[\frac{ACWP(t)}{BCWP(t)} \right] - ACWP(t)$$

$$EAC(t) \cong \frac{BAC}{CPI(t)}$$

Therefore, to estimate the cost at completion at any time, one simply scales the original budget by the inverse of the current CPI.

Predicting the duration of the project is not so straightforward. Define

- $SDAC$ = the original Scheduled Duration at Completion, or the earliest time T at which $BCWP(t) = BAC$.
- $EDAC(t)$ = the Estimated Duration at Completion, made at time t.

By analogy with the cost estimate, one might say that

$$EDAC_1(t) \cong \frac{SDAC}{SPI(t)}$$

This estimate is here called $EDAC_1(t)$ because it is only one possible estimate. To arrive at another estimate, the average rate of accomplishing work, that is, earning value, is $\dfrac{BCWP(t)}{t}$ (given that the project started at time 0). The remaining work, or value to go, at the budgeted rates, is, as above, $BAC - BCWP(t)$. The time left until the estimated completion is $EDAC(t) - t$. That is, the remaining value, $BAC - BCWP(t)$, has to be earned in the remaining time $EDAC(t) - t$. Assuming that the average rate of doing work (value earned per unit time) is the same in the future as it was to date,

$$\frac{BAC - BCWP(t)}{BCWP(t)/t} = EDAC_2(t) - t$$

$$\frac{BAC}{BCWP(t)}t - t = EDAC_2(t) - t$$

$$EDAC_2(t) = t\left(\frac{BAC}{BCWP(t)}\right) = t\left(\frac{BAC}{BCWS(t)SPI(t)}\right)$$

Therefore, by the second method, to estimate the project duration, or the date at completion, one simply divides the current project time by the proportion of the value earned to date.

Unfortunately, the two methods do not usually give the same predictions. Consider a project where BAC = 100 and SDAC = 66.

Figure 13.1 shows $BCWS(t)$ (dotted line) and $BCWP(t)$ (solid line), while Fig. 13.2 shows $SPI(t)$. The actual earned value starts slower than the plan but then

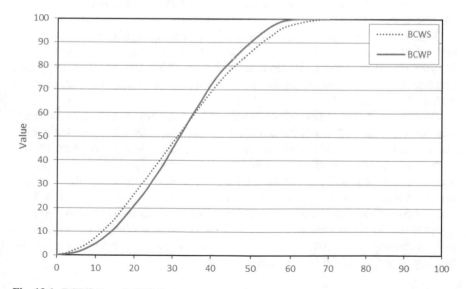

Fig. 13.1 BCWS(t) vs. BCWP(t)

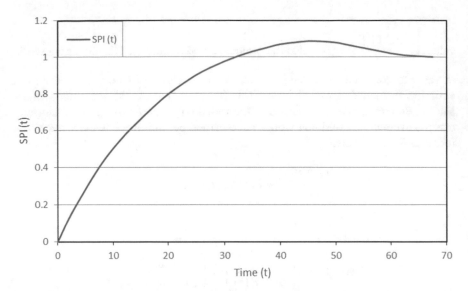

Fig. 13.2 Schedule Performance Index SPI(t)

catches up, so that the actual project duration is exactly the scheduled duration, 66 weeks.

Figure 13.2 shows the Schedule Performance Index $SPI(t)$ determined at each time t.

Based on the SPI, the project manager would conclude that the project shown here is behind schedule up to the half-way point ($t = 33$ weeks), and ahead of schedule after that.

Figure 13.3 shows the estimated dates of completion given by the two methods defined above. Method 1 (EDAC1) is the dotted line and Method 2 (EDAC2) gives the solid line. Both substantially overestimate the duration, up to the half-way point, and then both underestimate the duration. The second method gives more extreme deviations than the first.

Other methods for prediction have been proposed, but none is generally accepted. The two methods discussed here have the common characteristic of overestimating the project duration based upon early results, and underestimating the duration late in the project. If these prediction methods are actually used by project managers, as seen in the example, the forecasts (and the SPI) might encourage them to add resources early on, even though the project would have completed on time with no intervention. Then, later in the project, both methods underestimate the duration, possibly leading project managers to release resources prematurely, in the optimistic expectation of early completion. Thus, erroneous forecasts could lead to stagnation late in projects and hence lead to overruns.

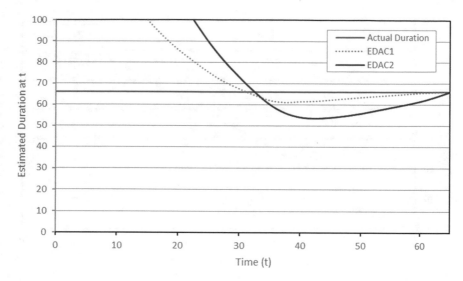

Fig. 13.3 Estimated duration at completion

Table 13.1 Percentage job complete, first ten reporting periods	Period	Progress %
	0	0
	1	1
	2	2.4
	3	3.2
	4	4.7
	5	6.1
	6	7.8
	7	12.3
	8	14.1
	9	25.1
	10	27.3

13.3   Linear Regression

Perhaps the first thing that comes to mind regarding forecasting is to plot the actual reported progress curve, *BCWP*, against time, *t*, fit a straight line to the past data points, and then extrapolate this line to obtain the Estimated Date at Completion (*EDAC*), when the *BCWP* equals the Budget at Completion (*BAC*). However, very often tasks, jobs, or projects start slowly and the rate of progress increases over time, rather than remaining constant, as the linear extrapolation would assume. Consider, as an example, the reported progress (in percentage of the total Budget at Completion) for the first ten reporting periods shown in Table 13.1.

Figure 13.4 shows the result of a linear fit to the reported data set. Clearly the fit is poor, and using this function to extrapolate to the date at which the job would be complete would not be credible.

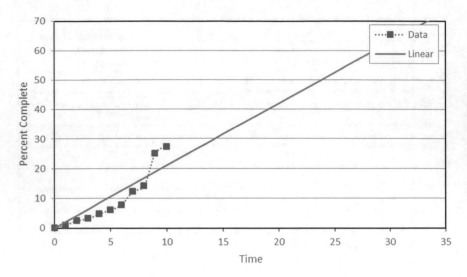

Fig. 13.4 Linear fit to first ten periods in Table 13.1, and linear extrapolation

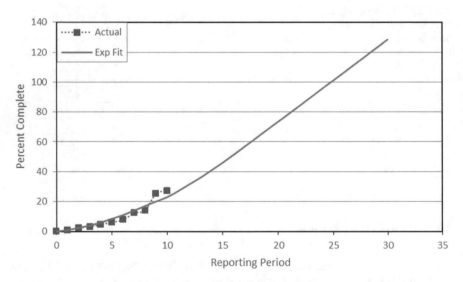

Fig. 13.5 Exponential fit to first ten periods in Table 13.1, and exponential extrapolation

If the linear fit is not very attractive, the next thought, based on the convex curvature of the plot of the data, might be that a quadratic or exponential fit would be appropriate (Pindyck and Rubinfeld 1976). The following plot in Fig. 13.5 shows an exponential fit to the same data points. Clearly, the fit is somewhat better, but one might observe that real projects tend to slow down toward the end, and the exponential fit indicates that the rate of progress (the slope) is still accelerating just as the project finishes (becomes 100% complete). A quadratic fit gives similar results. These results lead us to try sigmoidal, or S-shaped curves, as a better representation of how projects actually go through their life cycles.

13.4 Sigmoidal Curves

The S-shaped, or logistic, curve is ubiquitous in projects. In fact, this curve might be said to be at the heart of all projects, because it has a beginning, a middle, and an end. Consider, as a simple model, that there are two pools of work: *Work-done* and *Work-to-be-done*, and the job consists of the process of transferring work units from *Work-to-be-done* to *Work-done*. When all the work has been transferred from *Work-to-be-done* to *Work-done*, then the job is finished. Because the project is finite (indefinite work without a beginning or an end is not a project), there is a rising S-shaped curve that shows the cumulative *Work-done* as a function of time. There is also a falling, or reverse S-curve, which shows the decline of *Work-to-be-done* over time. The sum of *Work-done* plus *Work-to-be-done* is, of course, a constant (barring scope changes and rebaselining). The fundamental nature of the logistic in projects can hardly be overemphasized. Many projects are actually managed by the S-curves. Some people misunderstand the S-curve as the result of the changing levels of resources used during the successive stages of the life cycle. However, this statement has it backwards, by confusing cause and effect: the changing levels of resources are the result, not the cause, of the inherent S-shape of progress on projects.

To generate a simple model, we first assume that we have some reliable metric of work performed on a project (*Work-done*). For example, in a construction field activity, this might be the number of units (such as piping spool pieces) installed to date. We assume that all the units are equal, or that each has been assigned some weighting factor to make them equivalent, so they can be added together. One typical weighting factor is the estimated number of man-hours for each type and size of spool piece, and progress is the cumulative sum of the product of spool pieces installed times man-hours per spool piece, expressed as equivalent man-hours. Here we assume that we know the total work at completion (the initial value of *Work-to-be-done*) and the cumulative work accomplished (*Work-done*) at each reporting period.

13.4.1 The Pearl Curve

To generate a simple model with enough generality for many types of jobs, let us consider the rate of doing work (the rate of flow of completed work out of *Work-to-be-done* and into *Work-done*). Let $\Delta y(t)$ be the amount of work accomplished in a reporting period Δt (for example, a week or a month). Let $y(t)$ be the cumulative

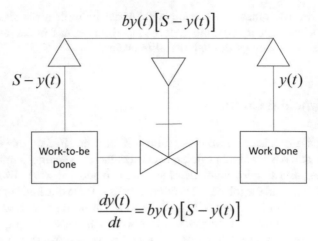

$$\frac{dy(t)}{dt} = by(t)[S - y(t)]$$

Fig. 13.6 Work-to-be-done schematic diagram

Work-done accomplished up to time t, and let S be the initial amount of *Work-to-be-done* (and the final amount of *Work-done*). Then $S - y(t)$ is the amount of work remaining to be done at time t (see Fig. 13.6).

What might we say about the flow between these variables? One might reason, based on observation of projects, that the rate of doing work $\dfrac{\Delta y(t)}{\Delta t}$ is related to the *Work-done*, $y(t)$. This seems reasonable because, on any project, some work must be accomplished in order to enable other work to be done. Work tends to *fan out* from the project inception, as shown by typical project networks. The work that can be done, based on the work already done, is sometimes called the *work face*. When little work has yet been accomplished, the work face is very limited, little can efficiently be built on it or from it, and so progress initially is slow.

So, we may say, an approximate relation might be:

$$\frac{\Delta y(t)}{y(t)} \propto \Delta t = \text{constant}$$

$$\frac{dy(t)}{dt} \propto y(t)$$

Conversely, we might also say that the rate of doing work $\dfrac{\Delta y(t)}{\Delta t}$ is related to the *Work-to-be-done*, $S - y(t)$, the amount of work remaining at time t. Work tends to *fan in* toward the project completion, as shown by typical project networks. When little work remains to be done, the work face is again limited, less work can be accomplished in a given time period, and progress slows down. Then,

$$\frac{\Delta y(t)}{S - y(t)} \propto \Delta t = \text{constant}$$

$$\frac{dy(t)}{dt} \propto S - y(t)$$

Thus, we may combine these concepts into an approximate relation as:

$$\frac{\Delta y(t)}{\Delta t} \propto y(t)\left[S - y(t)\right]$$

$$\frac{\Delta y(t)}{\Delta t} = by(t)\left[S - y(t)\right]$$

In this equation, b is a rate constant dependent on the nature of the job and the amount of resources applied to it. In this formulation, the rate of work $\dfrac{\Delta y(t)}{\Delta t}$ is zero when $y(t) = 0$, and approaches zero again as $y(t) \rightarrow S$. In the limit this gives the differential equation:

$$\lim \Delta t \rightarrow 0, \frac{\Delta y}{\Delta t} \rightarrow \frac{dy}{dt} = by(S - y)$$

The solution to this differential equation is the well-known *logistic equation* (also called *the Pearl curve*):

$$y(t) = \frac{S}{1 + ae^{-bt}}$$

In this equation, the parameter b controls the slope, whereas the parameter a is a constant of integration that shifts the curve along the time axis. You may verify this solution by differentiating $y(t)$ in the equation and comparing this with the derivative given above. This equation has the familiar S-shape, and is sometimes called the *Pearl curve* (Nelder 1961). Figure 13.7 shows the general shape of the Pearl curve, in which $S = 100$ and the duration of the job is 25-time units. Two additional S-shaped curves are also shown, which will be discussed later.

Note that the parameter b may depend upon the level of resources applied. If the resource level is increased, then b is increased, and the S-curve is steeper, but it is always S-shaped. The logic of the above derivation is, no matter how many resources are thrown at the project, the curve of *Work-done* is *always* S-shaped, due to the fan out – fan in characteristics discussed above. As applying too many resources would be inefficient and costly, the actual number of resources used is tailored to the amount of work $\Delta y(t)$ that can be done in each time period. That is, the changing level of resources in a project is the result of the inherent *S*-shape, not the cause of it. And this simple model produces the classic S-shape. It has only two parameters,

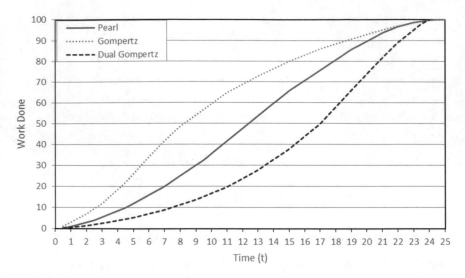

Fig. 13.7 Three sigmoidal curves (Pearl, Gompertz, and Dual Gompertz)

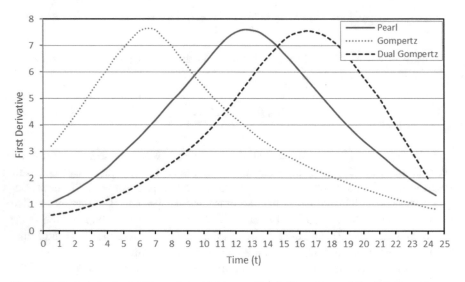

Fig. 13.8 First derivatives of three sigmoidal curves (Pearl, Gompertz, and Dual Gompertz)

so we don't need a lot of data to define the parameters of the curve. However, this also means that we don't have a lot of control over the shape of the curve.

The first derivatives of these curves are given in Fig. 13.8.

13.4.2   The Gompertz Curve

In deriving the Pearl curve, we assumed that the rate of doing work $\dfrac{\Delta y(t)}{\Delta t}$ was proportional to *Work-done*, $y(t)$, multiplied by *Work-to-be-done*, $S - y(t)$, the difference between the total project scope and the *Work-done* to date; that is:

$$\frac{\Delta y(t)}{\Delta t} \propto y(t)\big[S - y(t)\big]$$

Suppose instead we assume that the rate of work is proportional to the difference between the *logarithms* of the total project work and the *Work-done* to date, or

$$\frac{\Delta y(t)}{\Delta t} \propto y(t)\big\{\ln(S) - \ln\big[y(t)\big]\big\}$$

$$\frac{dy}{dt} = b\,y(t)\ln\frac{S}{y(t)}$$

That is, we assume that the rate at which work is accomplished is proportional to the work done to date, $y(t)$, multiplied by the logarithm of the ratio of S to $y(t)$ rather than the difference of S and $y(t)$. The rate is zero when $y(t) = 0$, and approaches zero as $y(t) \to S$. The resulting curve is sigmoidal, as with the Pearl formulation, but the rate is different, as the derivative approaches zero as $\ln\left[\dfrac{S}{y(t)}\right] \to 0$ rather than as $S - y(t) \to 0$. The solution to this differential equation is:

$$y(t) = S \times e^{\left[-ae^{(-bt)}\right]}$$

in which b is the rate coefficient and a is a constant of integration (shift constant). You may verify this solution by differentiating $y(t)$ in the equation and comparing this to the derivative given above.

This is sometimes called the *Gompertz curve*. The Pearl curve given earlier is anti-symmetric about the median (the first derivative is symmetric), but the Gompertz curve is not anti-symmetric, and the inflection point is below the mid-point. The inflection point can be seen in Fig. 13.7, but is most easily seen in Fig. 13.8, as the inflection point occurs when the first derivative goes through a maximum. The Gompertz curve is *skewed to the right*, because the curve, and its first derivative, have a short tail to the left and a long tail to the right, i.e., the job starts with a fairly rapid rate but takes a relatively long time to accomplish the last units of work. Compared to the Pearl curve, the Gompertz curve represents a job that is a fast starter but a slow finisher. See Figs. 13.7 and 13.8 above for the Gompertz curve and its first derivative, compared to the Pearl equation.

13.4.3   The Dual (Reverse or Complementary) Gompertz

Previously we assumed that the rate of work was proportional to the product of *Work-done* and *Work-to-be-done*, $y(t)[S - y(t)]$, to derive the equation for the Pearl curve. Alternately, we assumed that the rate of work was proportional to the product of the work done to date and the logarithm of the ratio of the total work to the work done to date, $y(t)$ $y(t) \times \ln\left[\dfrac{S}{y(t)}\right]$, to derive the equation for the Gompertz curve.

Suppose now that we assume that the rate of work is proportional to the product of the *Work-to-be-done* and the logarithm of the ratio of total work to work remaining, or:

$$\frac{dy}{dt} = b(S - y)\ln\left[\frac{S}{S - y}\right]$$

Notice that this is similar to the differential equation that led to the derivation of the Gompertz function, except that $S - y$ appears here where y appears in the Gompertz formulation. In this formulation, the derivative $\dfrac{dy}{dt} \to 0$ as $y(t) \to S$, and is zero for $y(t) = 0$, because $\ln\left[\dfrac{S}{S - y(t)}\right] \to 0$ as $y(t) \to 0$. Therefore, the behavior is sigmoidal, similar to that obtained before, but the rates are somewhat different. This differential equation has the solution:

$$y(t) = S \times \left(1 - e^{\left[-ae^{(+bt)}\right]}\right)$$

This S-curve doesn't seem to have a standard name, but because it is sort of the complement or reverse of the Gompertz, we may call it here the *Dual Gompertz* curve. It is in a way the mirror image of the Gompertz, *skewed to the left*, that is, it has its inflection point above the midpoint, with a shorter tail to the right and a longer one to the left, as shown in Figs. 13.7 and 13.8. A job described by this function is a slow starter but a fast finisher, compared to the Pearl curve.

13.4.4   Fitting the Logistic Curves

As mentioned above, one practical problem with the Pearl equation and the dual Gompertz equations may be immediately observed: with the initial condition $y(0) = 0$, then $\dfrac{dy}{dt}(0) = 0$ and work never gets started, never mind completed. We can never plot the point $y(t) = 0$ for it occurs at $t = -\infty$. However, we can make an approximation, just as we use the Normal distribution, which is defined on the range $-\infty \leq x \leq +\infty$, but apply it to a finite range.

One engineering approximation is to define an upper asymptote, say $A_U > S$, and a lower asymptote, say $A_L < 0$, such that the Pearl and Gompertz curves fit inside the asymptotic limits $A_L < y(t) < A_U$. Then, for example, the Pearl curve is given by:

$$y(t) = A_L + \frac{A_U - A_L}{1 + ae^{-bt}}$$

$$\frac{dy(t)}{dt} = b\left[y(t) - A_L\right]\left[\frac{A_U - y(t)}{A_U - A_L}\right]$$

Then, we can use as a boundary condition the finite time at which the project starts, that is, the time t_0 at which $y(t_0) = 0$ using the offset equation. We use this boundary condition to solve for the shift factor, a, in the modified Pearl equation. Substituting in the modified equation above:

$$y(t_0) = 0 = A_L + \left[\frac{A_U - A_L}{1 + ae^{(-bt_0)}}\right]$$

With algebraic manipulation gives:

$$-A_L = \left[\frac{A_U - A_L}{1 + ae^{(-bt_0)}}\right]$$

$$1 + ae^{(-bt_0)} = -\left[\frac{A_U - A_L}{A_L}\right]$$

$$ae^{(-bt_0)} = -\left[\frac{A_U - A_L}{A_L}\right] - 1 = -\frac{A_U}{A_L}$$

$$a = -\frac{A_U}{A_L}e^{(-bt_0)}$$

So, given the start date t_0, we can solve for the shift parameter a. If the time axis for the project is scaled such that $t_0 = 0$, then this equation reduces to $a = -\frac{A_U}{A_L}$. Note that $a > 0$ because $A_L < 0$ by definition.

If we specify the finite time at which the job finishes as t_s, when $y(t_s) = S$, then we have two points on the curve: $y(t_0)$ at t_0 and $y(t_s)$ at t_s. With these two endpoints, we can solve for the rate constant b and the shift coefficient a that fit the Pearl curve to the two time points t_0 and t_s:

$$y(t_s) = A_L + \left[\frac{A_U - A_L}{1 + ae^{(-bt_s)}}\right] = S$$

Some algebraic manipulation leads to:

$$S - A_L = \left[\frac{A_U - A_L}{1 + ae^{(-bt_s)}} \right]$$

$$1 + ae^{(-bt_s)} = \left[\frac{A_U - A_L}{S - A_L} \right]$$

$$\frac{a}{e^{(-bt_s)}} = \left[\frac{A_U - A_L}{S - A_L} \right] - 1 = \left[\frac{A_U - S}{S - A_L} \right]$$

$$e^{(-bt_s)} = a \left(\frac{S - A_L}{A_U - S} \right)$$

$$b = \frac{1}{t_s} \ln \left[a \left(\frac{S - A_L}{A_U - S} \right) \right]$$

If the job starts at $t_0 = 0$, then

$$a = -\frac{A_U}{A_L}$$

$$b = \frac{1}{t_s} \ln \left[-a \left(\frac{A_U}{A_L} \right) \left(\frac{S - A_L}{A_U - S} \right) \right]$$

The simple Pearl and Gompertz equations are highly generic, as they assume that tasks typically follow the S-shape of the logistic. By adjusting the asymptotes A_U and A_L one can get tails as long or short as one likes. There are only two undetermined coefficients, a and b, and one of these (a) is used up just defining the start time. The symmetric Pearl function may be used in default of any better information, for example, if we don't have good information about the work schedule (i.e., the $BCWS(t)$ as a function of time) or if we just want a simple approximation. If we have data, then we use whichever equation fits the data best.

13.5   Finding the Best Fit Curve

We can find the parameters of a sigmoid curve from the values at the endpoints t_0 and t_s, but that gives no help in forecasting, as the finish date t_s is what we are looking for. Of course, we could use any two points to define the curve, because all of the functions above have only two undetermined parameters, but if there are three or more reporting dates it is not obvious which two to use. To generate the most reliable forecasts with any of these equations, we want to use all the reported data points. So, we find the best fit of the parameters to the reported progress data using some curve-fitting technique and then use the equation to extrapolate from the present to the future; that is, to job completion (when $y(t) = S$).

Suppose that earned value y_j is reported at the end of reporting period j (that is, at time t_j), for $1 \leq j \leq n$. Then we would like to find values for a and b (solving for S is considered later) that minimize the sum of the squares of the deviations of the reported values from the postulated curve. That is, for the Pearl approximation, we want to find a and b that minimize

$$SSD = \sum_{j=1}^{n} \left[y_j - \frac{S}{1 + ae^{(-bt_j)}} \right]^2$$

The deviations are squared so that negative deviations count as much as positive deviations, and the square is used because there is a whole body of statistical development and regression based on the sum of squares. The sigmoid functions are nonlinear, so we could solve for the optimal values of a and b that minimize SSD using some nonlinear optimization procedure, such as the Solver function in Excel. Here, however, we will use a more simple method, linear regression analysis, because it is familiar to most engineers and managers, and it provides the confidence intervals we would like to have about the precision of the forecast.

To use standard linear regression with ordinary least squares, we need to linearize the problem. One way this can be done is to rearrange, for example, the Pearl equation $y(t) = \dfrac{S}{1 + ae^{-bt}}$ into the form:

$$1 + ae^{(-bt)} = \frac{S}{y(t)}$$

$$ae^{(-bt)} = \frac{S}{y(t)} - 1 = \frac{S - y(t)}{y(t)}$$

and then take logarithms:

$$\ln\left[\frac{S - y(t)}{y(t)}\right] = \ln a - bt$$

This is equivalent to the linear equation $Y = A + Bt$ if one makes the substitutions:

$$Y = \ln\left[\frac{S - y(t)}{y(t)}\right]$$

$$A = \ln a$$

$$B = -b$$

This is a linear equation in time t and the transformed dependent variable $\ln\left[\dfrac{S-y(t)}{y(t)}\right]$ and basically the procedure is to plot reported values of the transformed variable $Y(t) = \ln\left[\dfrac{S-y(t)}{y(t)}\right]$ against t, draw a straight line that best fits the points, and determine the equation of the line. Using the modified Pearl equation with the asymptotes A_U and A_L gives the equation derived earlier:

$$y(t) = A_L + \left[\frac{A_U - A_L}{1 + ae^{(-bt)}}\right]$$

Some algebraic manipulation leads to:

$$y(t) - A_L = \left[\frac{A_U - A_L}{1 + ae^{(-bt)}}\right]$$

$$1 + ae^{(-bt)} = \left[\frac{A_U - A_L}{y(t) - A_L}\right]$$

$$ae^{(-bt)} = \left[\frac{A_U - A_L}{y(t) - A_L}\right] - 1 = \left[\frac{A_U - y(t)}{y(t) - A_L}\right]$$

$$\ln\left[\frac{A_U - y(t)}{y(t) - A_L}\right] = Y(t) = \ln\ a - bt = A + Bt$$

That is, we plot $Y(t) = \ln\left[\dfrac{A_U - y(t)}{y(t) - A_L}\right]$ against t and find the best-fit straight line through these points. Following that we can determine the intercept A and slope B and finally compute $a = e^A$ and $b = -B$.

To forecast, we can either:

1. Determine the Pearl parameters $a = e^A$ and $b = -B$ from the intercept and slope of the best fit straight line, and then forecast values of $y(t)$ for some future time t using the Pearl equation;
2. Forecast values of $Y(t)$ for some future time t using the linear equation $Y(t) = A + Bt$ and then transform these values back into the original dependent variable $y(t)$ by

$$y(t) = \frac{S}{1 + e^{Y(t)}}$$

The same approach may be taken to linearize the Gompertz equation for linear fitting to the data. From above, the Gompertz equation was derived as:

$$y(t) = S \times \left(e^{\left[-ae^{(-bt)} \right]} \right)$$

Performing some algebraic manipulations leads to:

$$\frac{y_t}{S} = e^{\left[-ae^{(-bt)} \right]} = \frac{1}{e^{\left[ae^{(-bt)} \right]}}$$

$$e^{\left[ae^{(-bt)} \right]} = \frac{S}{y_t}$$

$$a\, e^{-bt} = \ln\left(\frac{S}{y_t} \right)$$

$$\ln(a - bt) = \ln\left[\ln\left(\frac{S}{y_t} \right) \right]$$

This is the form of a linear equation,

$$Y_t = A + Bt$$

$$\text{with} \quad Y_t = \ln\left[\ln\left(\frac{S}{y_t} \right) \right] \text{ for } 0 < y_t < S$$

$$A = \ln(a)$$

$$B = -b$$

Extrapolated forecasts Y_t can be made with the linear equation for larger values of t and may be inverted to obtain forecasts in the original variables by reversing the above derivation:

$$a = e^A$$

$$b = -B$$

$$Y_t = \ln\left[\ln\left(\frac{S}{y_t} \right) \right]$$

$$e^{(Y_t)} = \ln\left(\frac{S}{y_t} \right)$$

$$e^{\left[e^{(Y_t)} \right]} = \frac{S}{y_t}$$

$$y_t = \frac{S}{e^{\left[e^{(Y_t)} \right]}}$$

Thus, by transforming the original reported progress data $y(t)$ into the new variable $Y(t)$, we can fit a straight-line $Y(t) = A + Bt$, and then determine the parameters $a = e^A$ and $b = -B$, from the intercept and slope of the linear equation. We can then either forecast $y(t)$ directly, using these parameters in the Gompertz equation, or forecast $Y(t)$) using the linear equation $Y(t) = A + Bt$ and transform back to the original variables using $y(t) = \dfrac{S}{e^{e^{[Y(t)]}}}$.

Finally, the same approach may be taken to linearize the Dual Gompertz equation for linear fitting to the data. From above, the Dual Gompertz equation was derived as:

$$y_t = S \times \left(1 - e^{\left[-ae^{(+bt)} \right]} \right)$$

We can rewrite this as:

$$y_t = S - S e^{\left[-ae^{(+bt)} \right]}$$

$$S - y_t = S e^{\left[-ae^{(+bt)} \right]} = \frac{S}{e^{\left[-ae^{(+bt)} \right]}}$$

$$e^{\left[-ae^{(+bt)} \right]} = \frac{S}{S - y_t}$$

Take the logarithm of both sides of the above equation:

$$ae^{(bt)} = \ln\left(\frac{S}{S - y_t} \right)$$

Take logarithms again:

$$\ln a + bt = \ln\left[\ln\left(\frac{S}{S - y_t} \right) \right]$$

This is the form of a linear equation,

$$Y_t = A + Bt$$

$$\text{with} \quad Y_t = \ln\left[\ln\left(\frac{S}{S - y_t} \right) \right] \text{for } 0 < y_t < S$$

$$A = \ln(a)$$

$$B = b$$

Extrapolated forecasts Y_t can be made with the linear equation for larger values of t and may be inverted to obtain forecasts in the original variables by reversing the above derivation:

$$a = e^A$$

$$b = B$$

$$Y_t = \ln\left[\ln\left(\frac{S}{S - y_t}\right)\right]$$

$$e^{(Y_t)} = \ln\left(\frac{S}{S - y_t}\right)$$

$$e^{\left[e^{(Y_t)}\right]} = \frac{S}{S - y_t}$$

$$S - y_t = \frac{S}{e^{\left[e^{(Y_t)}\right]}}$$

$$y_t = S\left\{1 - \frac{1}{e^{\left[e^{(Y_t)}\right]}}\right\}$$

Thus, by transforming the original reported progress data $y(t)$ into the new variable $Y(t)$, we can fit a straight-line $Y(t) = A + Bt$, and then determine the parameters $a = e^A$ and $b = -B$, from the intercept and slope of the linear equation. We can then either forecast $y(t)$ by using these parameters in the Dual Gompertz equation directly, or forecast $Y(t)$ using the linear equation $Y(t) = A + Bt$ and transform back to the original variables using $y_t = S\left\{1 - \frac{1}{e^{\left[e^{(Y_t)}\right]}}\right\}$.

We can use the entire record for completed jobs of the same type, to determine the equation that best fits the progress curve for the past work. Then, this equation can be used to plan and track future jobs of the same kind. For example, if the progress data for completed jobs of the same type are best fit by Gompertz functions, then this is evidence that we should use a Gompertz function to fit the data on a job in progress that we are trying to forecast. To eliminate the effect of job size, we can normalize the equation by dividing by the total number of units installed (or *BAC*) for each job, so that in the equation $S = 1.0$ (or $S = 100\%$) and $y(t)$ represents the fraction (or percentage) of the total work done at any time t, and at completion $y(t_s) = S = 1$ (or 100%).

Then, for a job in progress, we can use the progress record up to the current date to find the best fitting line of the type selected in Part 1, and extrapolate the line to completion. For example, if

$S = $ Budget at Completion (*BAC*), and
$y = $ Budgeted Cost of the Work Performed [*BCWP(t)*],

then we compute the transformed variable $Y(t) = \ln\left[\dfrac{BAC}{BCWP(t)-1}\right]$ and plot

$Y(t)$ versus t. If the reported data points plot as (approximately) a straight line, then $BCWP(t)$ is (approximately) following the Pearl curve. If the plotted points do not follow a straight line, even approximately, then some other equation is needed to fit the data.

Linear regression eliminates the guesswork and subjectivity from the forecasting process and at the same time determines the confidence band on the extrapolated values. Linear regression finds the values of A and B that minimize the sum of the squares of the deviations of the reported points from the straight line:

$$SSD = \sum_{j=1}^{n}\left[Y_j - (A + Bt)\right]^2$$

in which n is the number of progress reporting periods prior to the time of the forecast. Having found A and B for the first n reporting periods, we use the equation

$Y(t) = A + Bt$ to forecast $Y(t)$ (and hence $y(t) = \dfrac{S}{\left(1+e^Y\right)}$) for any time t in the future.

Note that these regressions against the independent variable *time* often give very large values for R^2 ($R^2 > 0.99$) is not uncommon). This is because the amount of work completed is obviously highly correlated with the passage of time. Note also that, if the data are not well fit by the Pearl or Gompertz equations (which only have two unknown coefficients, a and b), then the regression residuals will not be randomly distributed and independent, but will often show long runs of positive or negative values. The *Durbin-Watson test* may be used to evaluate whether the residuals are autocorrelated.

Example 13.1

As an example, we use the $n = 10$ data points in Table 13.1 given earlier. In this example, we set $S = 100$, so $y(t)$ represents the percentage of the total job completed at time t. Using these ten points, the linear regression method using the Dual or Reverse Gompertz equation gives the values $A = 4.616$ and $B = 0.360$. These values give the parameters $a = e^A = 0.00989$, $b = 0.360$ in the Dual Gompertz equation, with $R^2 = 0.977$. The closeness of the fitted values is indicated in the Table 13.2.

Figure 13.9 shows some results for these data. The reported progress percentages and the fitted Dual Gompertz curve are shown up to time period 10 to indicate the closeness of the fit. Then the fitted function is extrapolated forward to indicate completion in about time period 19. Of course, due to uncertainty, there is a likelihood of about 50% that the job would actually take longer. In forecasting, we use the best-fit equation plus some statistics output from the linear regression program to give the confidence intervals on the forecasts. In this case, an 80% confidence interval was used; that is, we expect from the variance computed from the historical data

Table 13.2 Fitted values for data from regression equation, $n = 10$

Time	Actual value	Fitted value
1	1.0	1.41
2	2.4	2.01
3	3.2	2.87
4	4.7	4.09
5	6.1	5.81
6	7.8	8.23
7	12.3	11.58
8	14.1	16.17
9	25.1	22.34
10	27.3	30.41

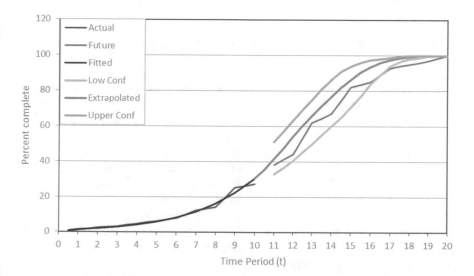

Fig. 13.9 Dual Gompertz fit to reported data and extrapolation

that the interval will cover the future progress, when it is reported, 80% of the time, with 10% above the upper limit, 10% below. We could also use 50%, 90%, or 95% confidence intervals. Figure 13.9 shows the 80% confidence band for the forecast from period 11 through period 20. Also shown in this figure for comparison are the future data points for periods 11 through 20, when the job actually completed. There is a 10% chance that the actual results would fall below the lower confidence line generated at time 10, and we see that in this case the results did fall outside this limit starting in period 16.

Is this a good forecast? The answer to that depends on what the forecast is used for. Clearly, this job took somewhat longer than the best forecast, and fell in a range

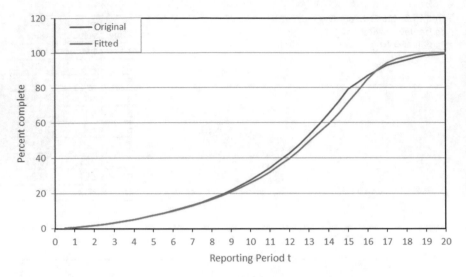

Fig. 13.10   Reverse Gompertz function fitted to entire job sequence

with probability less than 10%. However, the actual results clearly fall inside the
six-sigma band around the forecast, so we would probably say that this job was
under statistical control, although the variance may be greater than we would like to
see.

Was the Dual Gompertz a reasonable function to use for forecasting? After this
job is over, we can fit the Dual Gompertz to the full set of data points. This *ex post*
fit is shown in Fig. 13.10, below. Clearly this job is a slow starter and a fast finisher,
although perhaps not as fast as the Dual Gompertz function would indicate. Still, the
fit is close enough that it would be difficult to find another mathematical function
that would fit better. Based on these results, we would probably feel justified in
using the Dual Gompertz function to forecast future jobs of this same type – although
you are welcome to try to find a better one.

At this point, one may well ask what is the probability distribution of the forecasts. The answer is the Normal, which in this case is justified by an argument from
statistical independence. The linear regression equation can be written:

$$Y_j = A + Bt_j + \varepsilon_j$$

where

Y_j is the observed (transformed) variable and
ε_j is a random error term. The *classical Normal linear regression model* assumes
   that:

$E[\varepsilon_j] = 0, \quad \forall\, j$; the error terms ε_j all have zero mean;
$E[\varepsilon_j^2] = \sigma^2, \quad \forall\, j$; the error terms ε_j all have constant variance
$E[\varepsilon_j \varepsilon_k] = 0, \quad \forall\, j, \forall\, k$ such that $j \neq k$; the ε_j are all statistically independent

ε_j is Normally distributed, $N[0, \sigma^2]$

These assumptions imply that, if the classical Normal linear regression model applies, then the dependent variables Y are also Normally distributed, by the reproductive property of the Normal, with expected value $E[Y_j] = A + Bt_j$ and constant variance σ^2. Also, the Y_j are assumed to be statistically independent, because the covariance between the Y_j and Y_k at different times is, by the above assumptions:

$$E\left\{\left(Y_j - E\left[Y_j\right]\right)\left(Y_k - E\left[Y_k\right]\right)\right\} = E\left\{\left[A + Bt_j + \varepsilon_j - \left(A + Bt_j\right)\right]\left[A + Bt_k + \varepsilon_k - \left(A + Bt_k\right)\right]\right\}$$

$$\text{Covariance}\left(Y_j, Y_k\right) = E\left[\varepsilon_j \varepsilon_k\right] = 0 \quad for \ \forall j, \ \forall k \text{ such that } j \neq k$$

It turns out, however, that by this set of assumptions the differences are not statistically independent. Whether or not these assumptions are met in any given case should be checked. Also, these are the assumed properties of the transformed variables Y_j, where $Y_j = \ln\left[\dfrac{(S - y_j)}{y_j}\right]$, *not* the original data. As the values of Y_j are assumed to be Normally distributed, then $\dfrac{(S - y_j)}{y_j}$ must be lognormally distributed, but the original variables $y_j = \dfrac{S}{\left[1 + e^{Y_j}\right]}$ are not lognormally distributed, as the lognormal is defined on the range $[0, \infty]$, but y_t is non zero only over the finite range $0 < y_j < S$.

For prediction, however, we determine the confidence limits for the forecasts of the transformed variable Y and then convert these to confidence limits for the original variables y, rather than determining the confidence limits for y directly. The confidence limits are actually set by the t probability distribution, not the Normal, because in practice we don't know the error variance but have to estimate it from the data. However, the t-distribution is very close to the Normal distribution when the number of data points is large (say 30 or so). This procedure, although approximate, is much easier because it permits the use of standard linear regression codes, and the error introduced by using the transformed confidence limits for Y as the confidence limits for y is generally negligible in comparison to the likely accuracy in which y can be measured, much less forecasted.

Applying this approach generates the entries in Table 13.3, below, which repeats the actual and fitted points for the first 10 reporting periods from Table 13.2, and forecasts the job progress forward, for time periods 11 through 20. The table shows the best predicted value Corresponding to a 50% likelihood that the job will take longer), the lower confidence limit (corresponding to a 10% likelihood that the job will take longer), and the upper confidence limit (corresponding to a 90% likelihood that the job will take longer). The entries in this table are the values plotted in Fig. 13.10.

As soon as the reported progress is received from the next reporting period, the entire calculation is repeated, now with 11 data points, to generate a new regression equation, new values for A and B, and a new set of forecasts for time 12 through

Table 13.3 Forecast with 80% confidence bounds, reverse Gompertz curve

Reporting (Period)	Reported (Percent complete)	Fitted value Reverse Gompertz	Forecast Lower 10% bound	Forecast Best fit mean	Forecast Upper 10% bound
0	0.0				
1	1.0	1.41			
2	2.4	2.01			
3	3.2	2.87			
4	4.7	4.09			
5	6.1	5.81			
6	7.8	8.23			
7	12.3	11.58			
8	14.1	16.17			
9	25.1	22.34			
10	27.3	30.41			
11			28.91	40.53	54.68
12			38.08	52.52	68.59
13			48.91	65.63	81.69
14			60.91	78.36	91.75
15			73.06	88.86	97.46
16			83.94	95.70	99.55
17			92.17	98.90	99.97
18			97.11	99.84	100.00
19			99.27	99.99	100.00
20			99.89	100.00	100.00

time 20 (or whatever seems appropriate). This rolling procedure continues as the job advances.

13.5.1   Inverting the Variables

In the previous example of predicting project duration, time, that is, reporting period, was the independent variable, and earned value, or *BCWP*, was the dependent variable. This is consistent with common usage, in which the abscissa is usually time and the ordinate is whatever variable is dependent on time. It is also consistent with normal regression using ordinary least squares, in which it is assumed that the independent variable is known perfectly and has no error, and all the errors are concentrated in the dependent variable. Certainly, the reporting period should be known for sure, although the observations or measurements on the BCWP are certainly subject to error. However, although the reporting period may have no error, it does have variance. If the reporting interval is 1 month, different months have different numbers of workdays. Even if the reporting interval is 1 week, weeks may have different length due to holidays. Even so, it makes sense that time is the independent variable, because nothing can be done to influence it, and although one can say that the passage of time is the cause of work being accomplished, one would not say that work causes time to pass. (If it did, that would be another good reason to stop working)

As we have seen, placing all the errors on the *BCWP* means that the probability distributions and confidence bands are on *BCWP*. This allows one to determine the probability that some value for *BCWP* or more (or less) will be earned by some fixed date. However, what we may really be concerned about is the confidence band on the project duration, so that one can state the probability that the project will be completed on or before some given critical date. This can be done, but it may require breaking all the rules given above.

Using the same example as given just before, we seek to predict the completion of this project based on (in the example) ten observations on the *BCWP*. However, here we treat *BCWP* as the independent variable and reporting time as the dependent variable. Using ordinary least squares, this is equivalent to the assumption that there are no errors in observing the earned value, and all the errors lie in the reporting date. The project is complete when $BCWP(t) = BAC$ (the budget at completion), so that one only has to enter the abscissa at $BCWP = BAC$ and read the completion time off the curve.

Figure 13.11 below shows the same observed data as before, which could have been plotted exactly as before, but to emphasize the difference the abscissa remains the independent variable, and now becomes the reported earned value, and the ordinate remains the dependent variable, and now becomes the time.

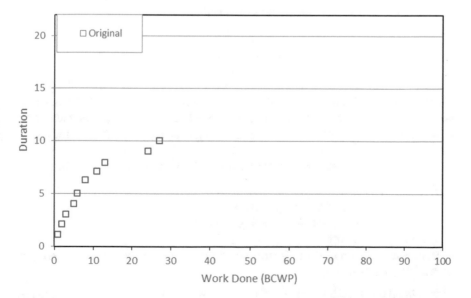

Fig. 13.11 Dual Gompertz data

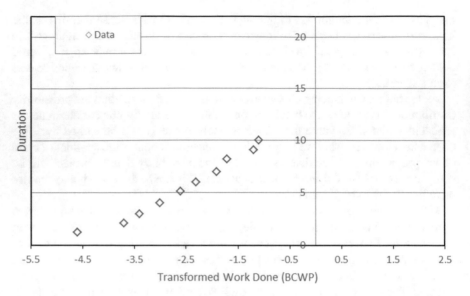

Fig. 13.12 Linearized data

We apply the linear transformation as used before, again assuming the Dual Gompertz function. That is, for the Dual Gompertz,

$$BCWP(t) = y_t = S\left\{1 - \exp\left[-a \exp(bt)\right]\right\}$$

This leads, as before, to the transformed variable

$$Y_t = \ln\left[\ln\left(\frac{S}{S - y_t}\right)\right]$$

Here $S = 100$ as before. The previous example plotted $Y_t = A + Bt$. Here, however, we plot $t = A' + B'Y_t$. This plot is shown in Fig. 13.12. The transformed data points seem to follow a straight line, an indication that the Dual Gompertz is giving a good fit.

Performing the regression on the transformed data set gives the following results:

- Adjusted $R^2 = 0.974$
- Intercept $= A' = 12.65139$; t-test $= 30.5$, $p < 1.5 \times 10^{-9}$
- Slope $= B' = 2.713743$; t-test $= 18.5$, $p < 7.4 \times 10^{-8}$
- $F = 344$; $p < 7 \times 10^{-8}$
- Standard deviation of the estimate $= 0.484$ (this will be used in determining the confidence band).

The fitted line is plotted in Fig. 13.13 in the transformed (linearized) coordinates. Also shown is the extension of this line, extrapolated out to project completion. The

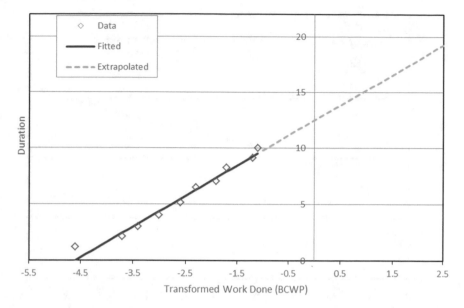

Fig. 13.13 Linearized data with fitted line

budget at completion (*BAC*) is $S = 100$; effective completion of 99.999 becomes, in the transformed coordinates, $Y = 2.44$. It is easily seen in the linear plot that the expected time of completion is about 19 months.

The fitted line, now transformed back to the original coordinates, is plotted in Fig. 13.14. Also shown is the fitted Dual Gompertz function extrapolated to project completion at about 19 months.

To obtain the confidence band on the times for various amounts of work accomplished, we determine the standard error of the forecast from the standard error of the estimate. Using a 80% confidence band as before (the probability of a duration more than the upper confidence limit is 0.10; the probability of a duration less than the lower confidence limit is 0.10), the Fig. 13.15 shows the confidence band added to the linear (transformed) plot. It is clear that there is about a 10% chance of finishing in less than 18 months, and about a 10% chance of taking longer than 20.5 months.

Figure 13.16 shows the confidence bounds transformed back into the original coordinates.

13.5.2   Determination of Confidence Bands on Forecasts of S-Curves

When making project forecasts, it is highly desirable to have some indication of the confidence in these forecasts. If we can compute some estimate of the confidence band, we can estimate the likelihood of overrunning and given EDAC (Estimated

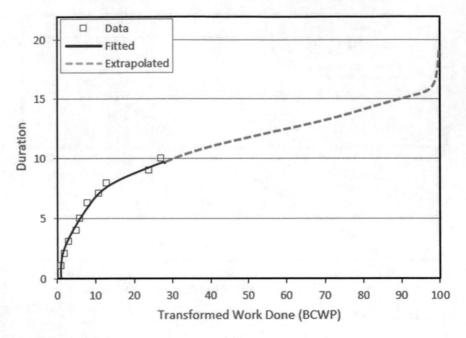

Fig. 13.14 Dual Gompertz data and fitted model

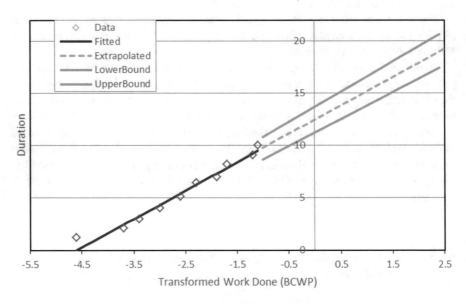

Fig. 13.15 Linearized data with model fit and confidence intervals

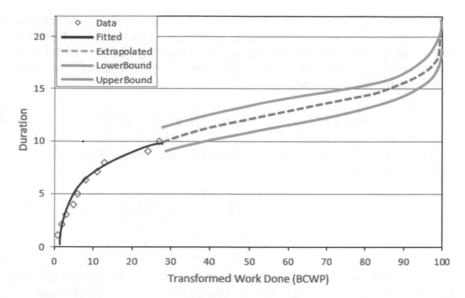

Fig. 13.16 Dual Gompertz data with model fit and confidence intervals

Date at Completion). Suppose that the true relationship between the (transformed) dependent variable and time (or reporting period) is given by:

$$Y = A + Bt$$

By regression through n time steps (reporting periods), we determine estimates for the regression coefficients $\hat{A}$, $\hat{B}$. We wish to forecast values for Y, k time steps into the future, by using the linear regression equation:

$$\hat{Y}_{n+k} = E[Y_{n+k}] = \hat{A} + \hat{B}t_{n+k}, \quad k = 1,2,3,\ldots$$

The forecast error at future period $n + k$ is:

$$\hat{e}_{n+k} = \hat{Y}_{n+k} - Y_{n+k} = \left(\hat{A} - A\right) + \left(\hat{B} - B\right)t_{n+k} + \varepsilon_{n+k}$$

Note that there are two sources of error in this forecast:

1. The random error or noise term ε_{n+k}
2. The errors due to the estimates of the regression coefficients, which are themselves random variables. Note in particular how any error in the slope coefficient $\left(\hat{B} - B\right)$ is magnified by the time ahead at which the forecast is desired, t_{n+k}.

   Qualitatively, it is obvious that the confidence band, whatever it is numerically, must widen with increasing forecast time, t_{n+k}.

The error of the linearized forecast must be Normally distributed, because the ordinary regression method assumes that $\hat{A}$, $\hat{B}$, ε_{n+k} are all Normally distributed, and any linear combination of Normal variates is Normally distributed. Moreover, the error of the forecast has zero mean, which we see by taking expectations (note that t_{n+k} is not a random variate):

$$E[\hat{e}_{n+k}] = E[\hat{A} - A] + E[\hat{B} - B]t_{n+k} + E[\varepsilon_{n+k}] = 0$$

Define the standard error of the forecast k steps ahead as $\sigma_f(k)$; then the variance of the forecast error is:

$$\sigma_f^2(k) = E\left[(\hat{e}_{n+k})^2\right] = E\left[(\hat{A} - A)^2\right] + E\left[(\hat{B} - B)^2\right]t_{n+k}^2$$
$$+ E\left[(\varepsilon_{n+k})^2\right] + E\left[(\hat{A} - A)(\hat{B} - B)\right]2t_{n+k}$$

In this, a number of the cross-product terms in the squares of the forecast errors, $\hat{e}_{n+k}^2$, are zero because it is assumed in deriving the regression equation that the random error terms are not serially correlated, and hence are not correlated with the regression estimates. Note that, for regressions involving fitting of the various sigmoidal curves to data, this may or may not be a good assumption. How good or bad this assumption is may be tested by visual inspection, by plotting the residuals, or numerically, by using the Durbin-Watson test for serial correlation. Proceeding nevertheless, we have then, from the equation above:

$$\sigma_f^2(k) = \text{var}(\hat{A}) + 2t_{n+k}\text{cov}(\hat{A}, \hat{B}) + t_{n+k}^2\text{var}(\hat{B}) + \sigma^2$$

From the basic equations for ordinary least squares fit to the data, we have (see a textbook on regression analysis for the derivations of the following results):

$$\text{var}(\hat{A}) = \frac{\sigma^2 \sum_{i=1}^{n} t_i^2}{n \sum_{i=1}^{n}(t_i - \bar{t})^2}$$

$$\text{var}(\hat{B}) = \frac{\sigma^2}{\sum_{i=1}^{n}(t_i - \bar{t})^2}$$

$$\text{cov}(\hat{A}, \hat{B}) = -\frac{\sigma^2 \bar{t}}{\sum_{i=1}^{n}(t_i - \bar{t})^2}$$

in which $\bar{t}$ is the mean value of the time periods 1 through n covered by the regression:

$$\bar{t} = \frac{1}{n}\sum_{i=1}^{n}t_i$$

If the reporting periods are equally spaced (each week, for example, so that $t_i = i(\Delta t)$), and there are data for every time period, then these equations can be simplified. The equations above allow for varying intervals between the observations. Substituting these three equations into the previous one gives:

$$\sigma_f^2 = \sigma^2\left\{\frac{\sum_{i=1}^{n}t_i^2}{n\sum_{i=1}^{n}(t_i-\bar{t})^2} - \frac{2t_{n+k}\bar{t}}{\sum_{i=1}^{n}(t_i-\bar{t})^2} + \frac{t_{n+k}^2}{\sum_{i=1}^{n}(t_i-\bar{t})^2} + 1\right\}$$

$$\sum_{i=1}^{n}t_i^2 = \sum_{i=1}^{n}(t_i-\bar{t}+\bar{t})^2$$
$$= \sum_{i=1}^{n}\left[(t_i-\bar{t})^2 + 2(t_i-\bar{t})\bar{t} + \bar{t}^2\right]$$

But,   therefore:

$$= \sum_{i=1}^{n}\left[(t_i-\bar{t})^2 + \bar{t}^2\right]$$

$$\frac{\sum_{i=1}^{n}t_i^2}{n\sum_{i=1}^{n}(t_i-\bar{t})^2} = \frac{\sum_{i=1}^{n}(t_i-\bar{t})^2 + n\bar{t}^2}{n\sum_{i=1}^{n}(t_i-\bar{t})^2} = \frac{1}{n} + \frac{\bar{t}^2}{\sum_{i=1}^{n}(t_i-\bar{t})^2}$$

which gives, after some algebraic simplification,

$$\sigma_f^2 = \sigma^2\left[1 + \frac{1}{n} + \frac{(t_{n+k}-\bar{t})^2}{\sum_{i=1}^{n}(t_i-\bar{t})^2}\right]$$

Recalling that $\bar{t}$ is fixed by the n observations available for the regression, note especially the term in $(t_{n+k}-\bar{t})^2$; as one tries to forecast further into the future, the error goes up as the square of the difference between that future time and the mean time of the past observations. Unfortunately, of course, we do not know σ^2; we can only estimate it from the regression, as s^2:

$$s^2 = \frac{1}{n-2} \sum_{i=1}^{n} \left(Y_i - \hat{Y}_i \right)^2$$

Then,

or

$$s_f^2 = s^2 \left[1 + \frac{1}{n} + \frac{\left(t_{n+k} - \bar{t} \right)^2}{\sum_{i=1}^{n} \left(t_i - \bar{t} \right)^2} \right]$$

With s_f, the standard error of the forecast, we can determine confidence intervals about the future forecasts of $\hat{Y}_{n+k}$ for all values of k of interest. These confidence intervals are, of course, with reference to the transformed variables used in the linear regression; we must now transform them back into the original variables, $\hat{y}_{n+k}$. That is, we reverse transform the confidence intervals in the same way that we reverse transform the forecasts. In the case of the Pearl function, the linearization derived above was:

$$Y_t = \ln \left[\frac{S - y_t}{y_t} \right]$$

The inverse transformation is then:

$$e^{Y_t} = \frac{S - y_t}{y_t}$$

$$y_t e^{Y_t} = S - y_t$$

$$y_t \left(1 + e^{Y_t} \right) = S$$

$$y_t = \frac{S}{1 + e^{Y_t}}$$

in general, and in particular,

$$\hat{y}_{n+k} = \frac{S}{1 + e^{\hat{Y}_{n+k}}}$$

13.6   Assessment of the Methods Discussed

Extensive bench marking and use of this approach to forecast actual construction projects has shown that the forecasts given by this method can often be more accurate than forecasts made by more sophisticated methods. However, they can sometimes be seriously in error, when the regression is based on only a few points. This problem is most severe when the data have high variability around the regression line and the progress to date is well below the inflection point of the curve (as The Pearl equation is symmetric, the inflection point occurs at the midway point, or when $y(t) = \dfrac{S}{2}$. It is very difficult to forecast accurately when this inflection point will occur, given only data below the inflection point. Given reported data above the inflection point, the method is more reliable. Of course, forecasts at early stages of work are much more useful than those made later. The user is advised to apply this method to historical data first, if possible to obtain a feel for how well the method fits the jobs of interest, before placing too much confidence in it. In any case, the forecasts made using this approach should be compared to other information.

Some of the indicators that may be symptomatic of problems:

1. The extrapolated completion date is much later than the project scheduled completion. This may indicate difficulties in calculating progress in the early stages of the job, but may also indicate that the job is starting slowly and the time lost may not be recovered later unless there is a significant learning effect in operation.
2. The confidence band is unusually large. This probably indicates erratic progress or inaccurate progress reporting, which should be investigated and corrected in order to get a valid picture of job progress. This is similar to the information presented in a control chart for the incremental schedule performance index spi.
3. The projected completion date or the width of the confidence band changes significantly from one reporting period to another. This isn't unusual in the early stages of the job, for less than 10% completion, but if it occurs later than this it may indicate that some major change has occurred with the job.

Several points should be kept in mind about this procedure:

1. The Sum of Squares that is minimized by the linear regression uses the deviations from the *linearized equation*, not the deviations from the *logistic equation*. Hence, the fitted values A and B which are optimal for the linearized equation do not necessarily convert to optimal values a and b for the nonlinear logistic equation. In this sense, the fitted curve is not necessarily the best fit to the logistic function. Both the regression coefficients and the confidence bands will be affected.
2. The method assumes that the entire job will follow a standard generic S-curve, and tries to find the parameters of this curve as a basis for extrapolation. It cannot account for later changes that affect the entire shape of the curve. It is necessary to keep in mind that the method inherently assumes that the underlying model of the project does not change. If the project changes, then the initial model may become invalid. Suppose, for example, that a forecast is made using one of the above methods, and this forecast indicates that the project will complete far in excess of the scheduled date. As a result of this prediction, the project manager puts the project on extended work weeks to recover the schedule. As a result of this change, the project finishes on the original scheduled date. Therefore, the

early forecast is now wrong – the project actually finished on time whereas the prediction was to finish late. However, the project finished on time because the underlying model changed (from regular time to overtime), and without the forecast the change might not have been made. Therefore, the forecast served its purpose. The forecast was conditional on model remaining constant, but the function of the project manager is to change the model when necessary.

3. The forecasts are extrapolated from the initial data. Early on, when there are few data points, inaccurate data will have a large effect on the fitted equation, leading to poor (in hindsight) forecasts, in which the confidence band does not cover the actual results. Difficulties can be diagnosed if (i) the confidence band is very broad, or (ii) the confidence band changes substantially when a new data point is added. Basically, the method is founded on the assumption that historical data incorporate information about the future, and if this assumption is violated the forecasts will unquestionably be bad. Unfortunately, bad data give bad forecasts. Sometimes even good data give bad forecasts.

4. The function used above is about the simplest possible, with one independent variable (time). It is hardly necessary to use a linear regression package to solve this small a problem, in one independent variable – the solution equations can easily be written out explicitly. However, it is certainly feasible to construct more realistic fitting functions with several or many independent variables, giving some function $y(t) = f(t, x_1, x_2, \ldots x_n)$, where x_1, $x_2, \ldots x_n$ are additional independent variables. To use standard linear regression, it must be possible to linearize this expression, as discussed earlier. Also, it is necessary to chose the independent variables such that they can be measured, and also such that they can be independently predicted. For example, the weather might be an independent variable that could help explain job progress, but to be useable in forecasting it must be capable of being predicted ahead to job completion, perhaps using National Weather Service long-range forecasts.

5. In computing confidence intervals, it is essential to use the *standard error of the forecast*, not the *standard error of the estimate* from the regression analysis. The standard error of the *estimate* reflects the scatter of the reported data points about the regression line, but the standard error of the *forecast* reflects both this scatter and the error in fitting the regression line from the data. As the forecasts represent future time, the variances of both these effects must be included in determining the confidence bounds.

The method described may appear simplistic, but it has been used for tracking many projects. It is, based on experience, often as accurate as much more expensive and time-consuming methods. Some of its advantages are:

1. It is automatic and completely objective. The whole forecast can be computer-generated. A spreadsheet will suffice. There are no subjective factors, guesses, hunches, judgment, or biases. The only thing the user has to do is to enter the reported progress every reporting period. The objectivity of the forecasts is often useful when having discussions with projects as to whether they are or are not on schedule.

2. It generates confidence intervals on the forecasts – 50%, 90% or whatever is desired.
3. It can easily be used by higher levels of management.

Quick and dirty forecasting is often useful, and construction engineers may wish to have multiple methods for evaluating project progress and forecasting completion, and should combine the results from several independent methods. (There is no point to using methods that are not independent; if several methods give highly correlated results, all but one can be dropped).

With all these methods to choose from, plus many more, how is one to select the best forecast? One useful exercise is to obtain feedback from real projects on the accuracy of cost estimates. When a project is complete, a *post mortem* analysis is needed to improve forecasting accuracy.

13.7   Methods Based on Bayesian Inference

Bayesian inferencing, or the revision of beliefs about future activities based on the information gained about past activities, offers a number of opportunities for making more reliable forecasts of future outcomes – and for making reliable forecasts sooner, when they are more valuable to project management (Gelman et al. 2013). Of course, as elsewhere in this book, we are concerned not with making point estimates of future outcomes, but with determining confidence bands on our estimates – and for making these confidence bands as narrow as possible, as soon as possible.

With a number of families of sigmoid curves available, it is assumed that one can find some functional relation that can be fitted to the available $BCWS(t)$ plan, either by linear regression, by minimizing the sum of the squares of the errors, or by eye. As stated earlier, the method assumes that the actual $BCWP(t)$ progress will follow the same family as the fitted BCWS(t) curve, but with different parameters. The method given here tries to determine the best estimate for these parameters using the $BCWP(t)$ progress reports as they arrive.

The focus of this method is not so much on high accuracy as on early warning. An accurate prediction of the actual task duration, b in the above functions, obtained late in the task, is worth very little. An approximate prediction of the task duration obtained very early in the task may be highly valuable to the project manager. This method seeks an early warning of deviations of the actual progress (*BCWP*) from the plan (*BCWS*). Although Bayes' rule is slightly more complicated, and probably less familiar, than linear regression, test results indicate that it may be able to give earlier warning that other methods.

We start with two parameter problems: b and n. Actually, one parameter would be simpler, but the major conceptual difficulty is in going from one parameter to two or more, so this discussion focuses on a two-parameter problem.

Assume that we have fitted a function from one of the families above to the planned $BCWS(t)$ data and have obtained the values $b*$ and $n*$ as the best fit parameters. Now, as progress reports on the $BCWP(t)$ are obtained, we wish to determine whether the actual progress data support the same values of b^* and n^*, in which case the project manager can believe that the task is on schedule, or whether the data support some other values of b and n, and the project manager should change his belief, especially if $b > b^{**}$. We apply Bayes' rule to advise the project manager on whether he should alter his belief.

The parameters b and n are continuous variables, and there are mathematical methods available to handle Bayesian inferencing problems using continuous variables. However, the presentation here will discretize the parameters. This greatly simplifies the presentation, and such discretization is a common practice in engineering. Suppose, to be specific, that we discretize both b and n into nine values each. Moreover, we center the range of the discrete values of b on the value b^*, which is our a priori best estimate for the task duration, obtained from the fit to the BCWS(t) curve; and similarly for n. Then, we have a nine by nine table of matrix for b and n:

$$
\begin{bmatrix}
 & b_{-4} & b_{-3} & b_{-2} & b_{-1} & b^* & b_1 & b_2 & b_3 & b_4 \\
n_{-4} & & & & & & & & & \\
n_{-3} & & & & & & & & & \\
n_{-2} & & & & & & & & & \\
n_{-1} & & & & & & & & & \\
n^* & & & & & b^*,n^* & & & & \\
n_1 & & & & & & & & & \\
n_2 & & & & & & & & & \\
n_3 & & & & & & & & & \\
n_4 & & & & & & & & &
\end{bmatrix}
$$

Each of these pairs is considered an *event*. That is, the event $[b = b_j, n = n_k]$ is the *event* that the true value of b is b_j and the true value of n is n_k, and there are $9 \times 9 = 81$ possible events. Values of b or n that fall outside the matrix are considered impossible (have probability zero), so the range of the table should be extend far enough to avoid excluding real possibilities. Of course, one can use a table with more entries than 9×9. Whatever the size of the matrix, the sum of the prior probability distributions over all the events must equal one:

$$\sum_j \sum_k P\left[b = b_j, n = n_k = 1.0\right]$$

The Bayesian approach requires a prior probability distribution over the set of all events. This prior distribution must be set by the decision maker, in this case the project manager. The prior distribution depends on the project manager's degree of belief in the various outcomes. For example, if the project manager very strongly believes that the actual task BCWP(t) is going to track the planned $BCWS(t)$ very closely, then he could set $P[b = b^*, n = n^*] = 0.90$ and assign all of the other events probability 1/800. This kind of strong prior would make it very difficult for the project manager's beliefs to be changed based on the actual $BCWP(t)$ data, no matter what they were.

.	b_{-4}	b_{-3}	b_{-2}	b_{-1}	b^*	b_1	b_2	b_3	b_4
	0.00125	0.00125	0.00125	0.00125	0.00125	0.00125	0.00125	0.00125	0.00125
	0.00125	0.00125	0.00125	0.00125	0.00125	0.00125	0.00125	0.00125	0.00125
	0.00125	0.00125	0.00125	0.00125	0.00125	0.00125	0.00125	0.00125	0.00125
	0.00125	0.00125	0.00125	0.00125	0.00125	0.00125	0.00125	0.00125	0.00125
n^*	0.00125	0.00125	0.00125	0.00125	0.90000	0.00125	0.00125	0.00125	0.00125
	0.00125	0.00125	0.00125	0.00125	0.00125	0.00125	0.00125	0.00125	0.00125
	0.00125	0.00125	0.00125	0.00125	0.00125	0.00125	0.00125	0.00125	0.00125
	0.00125	0.00125	0.00125	0.00125	0.00125	0.00125	0.00125	0.00125	0.00125
	0.00125	0.00125	0.00125	0.00125	0.00125	0.00125	0.00125	0.00125	0.00125

Another method would be to have the prior probability the largest for $\{b^*, n^*\}$, but with decreasing probability for $\{b, n\}$ further away from this value. An example might be the prior probability matrix just below, in which every entry is to be divided by 369.

.	b_{-4}	b_{-3}	b_{-2}	b_{-1}	b^*	b_1	b_2	b_3	b_4
n_{-4}	1	2	3	4	5	4	3	2	1
n_{-3}	2	3	4	5	6	5	4	3	2
n_{-2}	3	4	5	6	7	6	5	4	3
n_{-1}	4	5	6	7	8	7	6	5	4
n^*	5	6	7	8	9	8	7	6	5
n_1	4	5	6	7	8	7	6	5	4
n_2	3	4	5	6	7	6	5	4	3
n_3	2	3	4	5	6	5	4	3	2
n_4	1	2	3	4	5	4	3	2	1

One could use a bivariate normal distribution to set the prior probabilities, or some other probability distribution.

Another prior could be the uniform prior, $P[b = b_j, n = n_k] = 1/81, \quad \forall j, \quad \forall k$. In this case the prior probability matrix would be the following table, with each entry divided by 81.

$$
\begin{bmatrix}
\cdot & b_{-4} & b_{-3} & b_{-2} & b_{-1} & b^* & b_1 & b_2 & b_3 & b_4 \\
n_{-4} & 1 & 1 & 1 & 1 & 1 & 1 & 1 & 1 & 1 \\
n_{-3} & 1 & 1 & 1 & 1 & 1 & 1 & 1 & 1 & 1 \\
n_{-2} & 1 & 1 & 1 & 1 & 1 & 1 & 1 & 1 & 1 \\
n_{-1} & 1 & 1 & 1 & 1 & 1 & 1 & 1 & 1 & 1 \\
n^* & 1 & 1 & 1 & 1 & 1 & 1 & 1 & 1 & 1 \\
n_1 & 1 & 1 & 1 & 1 & 1 & 1 & 1 & 1 & 1 \\
n_2 & 1 & 1 & 1 & 1 & 1 & 1 & 1 & 1 & 1 \\
n_3 & 1 & 1 & 1 & 1 & 1 & 1 & 1 & 1 & 1 \\
n_4 & 1 & 1 & 1 & 1 & 1 & 1 & 1 & 1 & 1
\end{bmatrix}
$$

This is sometimes referred to as *the noninformative prior*, because it provides no information to discriminate between any events. That is, the project manager shows no preference for any particular values of b and n, and wishes to have these values chosen entirely by the data. In the examples following, the noninformative prior will be used in order to allow the solutions to be determined entirely by the data, not by any prior opinions. However, it must be emphasized that, in the Bayesian approach, the project manager always has some prior information or knowledge, and *there is no reason why the project manager should not allow his prior convictions to influence the results, through the prior probability matrix, if he so desires.*

The discussion in Chap. 10 identified Bayes' rule as the following:

- $D =$ the data set, which initially is the full $BCWS$ plan, and then is revised as each $BCWP(t)$ progress report is received.

- $\Theta = \{b, n\}$ the set of parameters for the $\dfrac{BCWS}{BCWP}$ curve.

- $P(D|\Theta) =$ the *conditional probability* that the particular outcomes D would be observed, given the probability parameters Θ;

- $P(D|\Theta) =$ the *conditional probability* that Θ is the value taken on by the set of parameters *given that* the outcomes D were observed.

In the Bayesian approach, both the observables $D = BCWP(t)$ and the model parameters $\Theta = \{b, n\}$ are considered random variates. We start then with a *prior distribution* $P\{\Theta | BCWS\}$ the initial distribution on the parameters $\Theta = \{b, n\}$, conditional on the fit to the $BCWS(t)$ curve. Then the joint distribution of the observations D, which are the values reported for the $BCWP(t)$, and the parameters Θ, is:

$$
P\{D \cap \Theta\} = P\{D|\Theta\} P\{\Theta\} = P\{\Theta|D\} P\{D\}
$$

where $P\{D\}$ is the marginal distribution of the observables D. Then, this expression can be rewritten as:

$$P\{\Theta | D\} = P\frac{\{D \cap \Theta\}}{P\{D\}} = \frac{P\{D | \Theta\}}{P\{D\}}$$

where $P\{\Theta | D\}$ is now the *posterior distribution* of the parameters Θ, *given that* the outcomes D were observed. Determining this posterior distribution is the objective of Bayesian analysis.

The last expression may be rewritten in terms of the two parameters of the sigmoid function as:

$$P\{b = b_j, n = n_k \mid BCWP(t)\} = P\frac{\{BCWP(t) \mid b = b_j, n = n_k\}}{P\{BCWP(t)\}}$$

Here, $P\{b = b_j, n = n_k | BCWP(t)\}$ is the posterior distribution on the parameters $\{b, n\}$ and $P\{b = b_j, n = n_k\}$ is the prior distribution on the parameters, which after every $BCWP(t)$ time step is reset equal to the previous posterior. That is, the posterior distribution computed at time t becomes the prior distribution used in the computation at time $t + 1$, and so on.

The term in the denominator of the above equation is just the probability of observing the data that were observed, over all possible values of the parameters:

$$P[BCWP(t)] = \sum_j \sum_k P[BCWP(t) \mid b = b_j, n = n_k]$$

The remaining term in the equation is $P\{BCWP(t) | b = b_j, n = n_k\}$. This is just the probability of observing the actual data, conditional on the parameters. This is sometimes called the likelihood function. Given that the parameters have values $b = b_j$ and $n = n_k$, the likelihood that the actual data BCWP(t) would be observed is $P\{BCWP(t) | b = b_j, n = n_k\}$. Suppose that we assume some values of the parameters, such that $b = b_j$ and $n = n_k$. This choice defines a sigmoid curve for *BCWP* versus time. The actual progress reported at time t, $BCWP(t)$, may not lie exactly on this curve. However, this could still be the true curve, as we must allow for some error in measuring the reported $BCWP(t)$. We assume that the measurement errors are distributed Normally, as $N[0, \sigma^2]$ around the sigmoid curve. That is, the mean error is zero, so the errors are not biased up or down, but have standard deviation σ. This is the same assumption made in regression analysis, in which it is also assumed that the error variance is constant (homoscedastic). In the Bayesian approach, we can also assume homoscedasticity, or we can assume that the variance is not constant. In the example here, it is assumed that the reporting error is small at the beginning of the task (there should be little error in reporting *BCWP* when very little work is being done) and increases proportionally to the amount of work accomplished in each time period (that is, the reporting error is largest when the most work is going on, when the rate of progress is greatest). The assumption that the measurement error is proportional to the rate of progress implies that the reporting error becomes small as the task nears completion; this may or may not be a realistic assumption. In any case, the user can make his own assumption about the reporting errors and is not limited to homoscedasticity.

Then, at every time period t, the reported progress is $BCWP(t)$. The progress if the task were actually following the sigmoid function with parameters $b = b_j$ and $n = n_k$ would be $y(t| b = b_j$ and $n = n_k)$. The deviation is:

$$d\left(t|b = b_j \text{ and } n = n_k\right) = y\left(t|b = b_j \text{ and } n = n_k\right) - BCWP\left(t\right).$$

Using the Normal distribution for the error, $N[0, \sigma^2]$, we can determine the likelihood that a deviation this large would be observed due to random reporting error, when the true parameters are $b = b_j$ and $n = n_k$. One simple way to determine this likelihood is to compute the Normal probability density function

$$\phi\left[\frac{d\left(t \mid b = b_j, n = n_k\right)}{\sigma}\right]$$

and then multiply this by some interval to convert from probability density to probability. The actual value used is not critical, as we normalize all the quantities after they have been computed. That is, the probability that the actual reported progress belonged to one of the sigmoid curves defined by the matrix is exactly one (values of the parameters not in the matrix are impossible), so we adjust all the computed values such that

$$P\left[BCWP(t)\right] = \sum_j \sum_k P\left[BCWP(t) \mid b = b_j, n = n_k = 1\right]$$

Thus, we have now identified all the terms in the Bayesian equation, and we can compute the posterior distribution over all the pairs of parameters at each reporting period t from:

$$P\left\{b = b_j, n = n_k \mid BCWP(t)\right\} = P\left\{BCWP(t) \mid b = b_j, n = n_k\right\} P\left\{b = b_j, n = n_k\right\}$$

Having the posterior distribution at each time t, we must infer the values of b and n. One method is to use the most likely values. That is, we select the pair $b = b_j$, $n = n_k$ that maximizes the posterior probability $P\{b = b_j, n = n_k| BCWP(t)\}$. However, it is possible that the most likely set changes from time to time, until the process settles down on some steady-state value. Therefore, it has proved more successful to compute the mean values of b and n based on the posterior distribution, and then to use these mean values to generate a sigmoid curve that represents the best prediction of the future $BCWP$ curve. Unlike the modes (the most likely values), the means are stable over multiple time periods.

As mentioned earlier, we are interested in the confidence band for the projects of BCWP. If a sigmoid function like one of those discussed above, in which the parameter b represents the duration of the task, one can obtain an estimate of the confidence bounds on b by either plotting a histogram of the distribution of b, or by computing the standard deviation of b from the posterior probabilities for each b_j, and then using the Normal distribution to assign confidence bands at the desired confidence levels.

Example 13.2

As a simple example, assume that a project has a planned BCWS curve that is reasonably approximated by a symmetric Pearl curve, as shown in the Fig. 13.17. The Budget at Completion = $950 (in thousands), and the upper asymptote of the Pearl curve is taken as S = 1000. The planned duration is 29 weeks, and progress [BCWP(t)] is reported every week. The fitted parameters of the Pearl curve are $a = 60$ and $b = 0.25$.

$$y(t) = \frac{S}{1 + ae^{-bt}} = BCWS(t)$$

In this example we will use a 9×9 matrix of possible pairs of values for a and b, with the noninformative prior as shown below (divide all cell values by 81):

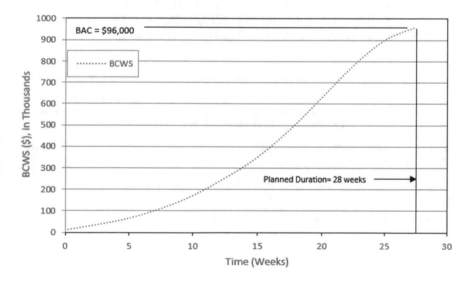

Fig. 13.17 BCWS vs. Time

$$\begin{bmatrix} \dfrac{a}{b} & 0.17 & 0.19 & 0.21 & 0.23 & 0.25 & 0.27 & 0.29 & 0.31 & 0.33 \\ 48 & 1 & 1 & 1 & 1 & 1 & 1 & 1 & 1 & 1 \\ 51 & 1 & 1 & 1 & 1 & 1 & 1 & 1 & 1 & 1 \\ 54 & 1 & 1 & 1 & 1 & 1 & 1 & 1 & 1 & 1 \\ 57 & 1 & 1 & 1 & 1 & 1 & 1 & 1 & 1 & 1 \\ 60 & 1 & 1 & 1 & 1 & 1 & 1 & 1 & 1 & 1 \\ 63 & 1 & 1 & 1 & 1 & 1 & 1 & 1 & 1 & 1 \\ 66 & 1 & 1 & 1 & 1 & 1 & 1 & 1 & 1 & 1 \\ 69 & 1 & 1 & 1 & 1 & 1 & 1 & 1 & 1 & 1 \\ 72 & 1 & 1 & 1 & 1 & 1 & 1 & 1 & 1 & 1 \end{bmatrix}$$

For each pair of values $\{a_j, b_k\}$ we can generate a Pearl curve $\{y(t)|a_j, b_k\}$. Then, as each weekly report of BCWP(t) is received, we compute the likelihood that this value would be obtained as a random error if the true curve that the *BCWP* is following were $\{y(t)|a_j, b_k\}$. With these values for all the possible pairs, we compute by Bayes' Theorem the posterior probability for each of the combinations $\{a_j, b_k\}$. From these probabilities, we compute the average values for the parameters a and b and use these to forecast. Figure 13.17 shows a snapshot after week 8 of the example. The actual reported $BCWP(t)$ are shown with triangular symbols.

It would be obvious at week 8, just by plotting the reported $BCWP(t)$ compared to the planned $BCWS(t)$, that the project is falling behind schedule. The more interesting question is, When is it going to finish? This is shown in Fig. 13.18 by the

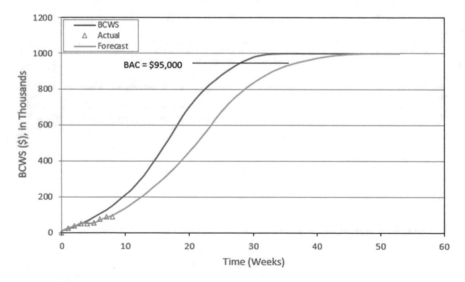

Fig. 13.18 Forecast at week 8

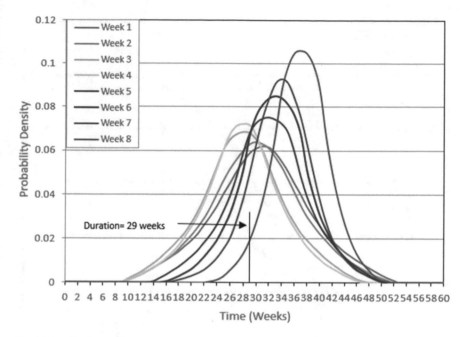

Fig. 13.19   Confidence bounds

forecast curve, which is a Pearl curve with the average parameters computed by the
Bayes process through week 8, extrapolated to the future. The curve plotted in the
figure represents the mean or expected value of the forecast $BCWP(t)$. Figure 13.19
shows the probability distributions for the estimated date at completion computed
from the reported data for weeks 1 through 8.

The probability distributions are shown here as Normal distributions. It can be
seen here how the probability distributions on the duration narrow as more information is obtained, and shift to the right. At week 8, the probability distribution is
centered around the mean value of 38 weeks. Moreover, it is clear from the week 8
probability distribution that the original estimate of the duration, 29 weeks, has less
than a 10% chance of happening (the area under the week 8 probability distribution
to the left of 29 weeks is less than 0.10).

Figure 13.20 shows the actual reported BCWP(t) for the entire project, the triangular symbols indicating the reported values. Also shown, as the curve labeled
$BCWP$, is the Pearl curve that best fits the actual performance across the complete
project. The differences between these two curves are the random reporting errors.

Figure 13.21 that plots the confidence band for the prediction of the project duration against the time at which the prediction was made. The upper or 90% confidence limit shows the values for the duration that would be exceed with probability
0.10, and would not be exceeded with probability 0.90. Because of the assumption
of symmetry, the mean line is also the median, or 50% confidence limit; it is equally
likely that the actual duration would be above or below this line. The lower confi-

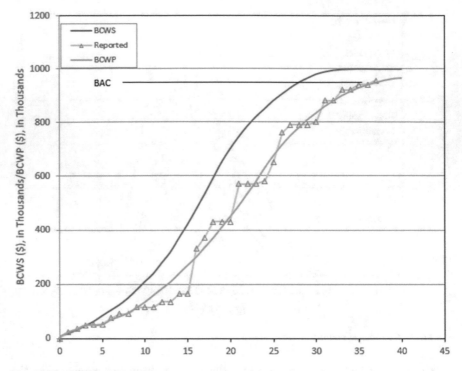

Fig. 13.20 BCWS and BCWP

dence limit is the value that would be exceeded with probability 0.90, and not exceeded with probability 0.10. Therefore, the actual values should lie between the upper and lower confidence bands about 80% of the time.

Of course, at any time, only the confidence band to the left of t (that is, before t) is known. By inspection of the confidence limits at week 8, it is apparent at this time that the planned duration of 29 weeks lies below the lower confidence limit, and so there is less than 10% probability that the project will finish in the planned time. At this same time, the expected value of the duration, the mean line, is about 37 weeks. That is, after week 8, meeting the original schedule is highly improbable, and project management should be working with the assumption of a median duration of 37 or 38 weeks instead of 29.

Comparing this figure with the previous one, it is possible to see how the fluctuations in the reported values for $BCWP(t)$ cause the confidence band to vary with time. However, the width of the confidence band narrows as more information is obtained on subsequent values for the actual $BCWP(t)$. Of course, the confidence limits shown here are actually too narrow, because they relate to the comparison of the actual reported values to the fitted Pearl curve. There are some errors in fitting the Pearl function to the planned $BCWS(t)$, namely the sum of the squares of the deviations between the fitted Pearl curve and the planned values, and neglecting this source of variance makes the confidence band look narrower than it should be. It is

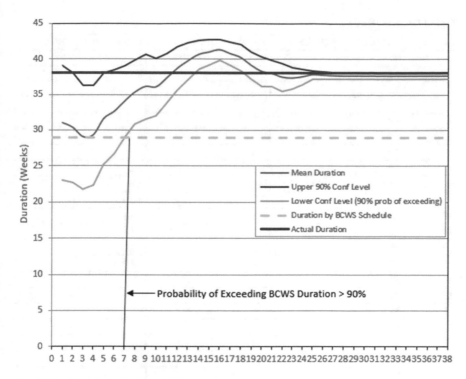

Fig. 13.21 Confidence bands on duration

possible that this variance could be included, but in general, if the fit between the *BCWS(t)* and the idealized function (Pearl or one of the others discussed) is close, the error is considered negligible.

As it turns out, the actual completion date is at week 38, but the point here is not that this value was accurately predicted, but that, with a few exceptions, it always was inside the 80% confidence band. Although we would like the confidence band to be as narrow as possible, the main objective is that the ultimate solution should lie within the confidence band from the beginning of the process to its completion. In this case, the Bayesian revision of probabilities has substantially met this objective.

13.8   Practice Problems

Problem 13.8.1 You are a Construction Engineer on a large coal-fired power plant jobsite, acting as supervisor for large pipe hanger installation. The schedule for completion of the large pipe hanger installation is 36 months after start of this phase. The Planned Value or Budgeted Cost of the Work Scheduled is shown in the second column of Table 13.4. Quantities have been normalized so that BAC = 100.

Table 13.4 Problem data

Month	Cumulative PV (BCWS)	Monthly earned value	Cumulative EV (BCWP)
0	0	0	0
1	0.2	0	0
2	0.9	0	0
3	2	1.1	1.1
4	3.4	1.1	2.2
5	5.3	2	4.2
6	7.4	1.6	5.8
7	9.9	1	6.8
8	12.6	0.1	6.9
9	15.6	3.5	10.3
10	18.9	2.7	13.1
11	22.3	3.2	16.3
12	25.9	1.5	17.8
13	29.7	1.2	18.9
14	33.6		
15	37.6		
16	41.7		
17	45.8		
18	50		
19	54.2		
20	58.3		
21	62.4		
22	66.4		
23	70.3		
24	74.1		
25	77.7		
26	81.1		
27	84.4		
28	87.4		
29	90.1		
30	92.6		
31	94.7		
32	96.6		
33	98		
34	99.1		
35	99.8		
36	100		

At the end of the 13th month of installation, the project-to-date values of the Earned Value per month and the cumulative EV or Budgeted Cost of the Work Performed are shown in the third and fourth columns of the table.

At this time, the Site Superintendent calls you to his trailer to tell you that the company VP for power plant construction, who reads the monthly EV reports, called him to say that he is very much concerned because the large bore pipe hanger SPI is only 0.6. At this rate, the hanger completion will be 24 months late. (Using the relation Estimated Duration at Completion = Planned Duration/SPI = 36/0.6 = 60). A schedule overrun in the large pipe hanger area will surely delay startup of the project and jeopardize the company's incentive fee for on time start of system tests (and eliminate your bonus as well).

The Site Superintendent also knows that, with 23 months to go and no current problems with on-time deliveries of large bore pipe hangers from the fabricator, approximately 3–4 months can be gained by rescheduling large bore pipe installation efforts to concentrate first on those piping systems that will go into system test first, thereby overlapping hanger installation and system tests by about 3 months. This rescheduling will increase some labor costs, due to installation out of optimal sequence, but this additional cost would be far less than the incentive fee at risk.

The Site Superintendent, to respond to the corporate VP, directs you to prepare a report to him giving a reliable prediction of the time at which large bore pipe hanger installation will be complete on this site.

• What is your forecast date for large bore pipe hanger completion? What curve would you use to make the forecast? State your assumptions.
• How confident are you that you will finish on or before this committed date?

Problem 13.8.2 Johnny Mize, Assistant Project Manager at the Odessa refinery upgrade project, was sitting at home watching the St. Louis Cardinals on TV when he got a call from Leo Durocher, Vice President for Construction at the home office. "I have heard rumors about the situation down at the Chocolate Bayou job, and I want you to be there tomorrow morning and report back to me about what is going on."

Johnny didn't know much about the Chocolate Bayou job and didn't want to walk into something blind, so he decided to call some friends of his in the company at various jobs to get some advance scoop on the situation. So Johnny placed a call to Paul Dean, Chief Electrical Supervisor at the Odessa job, who told him that "All I have heard about that site is that everything is OK except they are under-staffed and have been affected by a shortage of crafts."

Frankie Frisch, Chief Construction Engineer at the company's site in Abu Dhabi, told Johnny: "I have heard that, due to the owner's financial condition, that job is heavily cost-driven. The owner controls the cash flow expenditures every week down to the penny, and nobody can get an authorization to spend any money above the preset spending limits. Everybody is spending a lot of time and effort managing to the cash flow limits, which results in stop-start inefficiencies."

Pepper Martin, Welding Supervisor at the Odessa job, told Johnny that "Chocolate Bayou initially had a hard time hiring qualified pipe welders at that location, due to the volume of construction going on around there that sucked up all the welders in the vicinity, and pipe welding is on the critical path. So they decided to bring in travelers. They imported a bunch of pipewelder travelers and had to promise them a lot of overtime to get them there. But once they got them, they have been going like gangbusters, beating their numbers."

(Note: a traveler is a construction craftsman who does not work near his home but rather travels around the globe working at major industrial construction projects. Travelers are highly skilled and very productive, and demand lots of overtime to come to any job, from which they make lots of money).

Finally, Nub Kleinke, General Superintendent at the Guinea LNG job, told Johnny that "From what I hear, the job is highly schedule-driven. Due to changing economic conditions, the client decided he really wants the project done on time, so, shortly after they mobilized, the owner decided to set up some schedule milestones and to offer some large cash incentives for hitting these milestones, and so they are holding tight to the original schedule."

From this, Johnny wonders if they are all talking about the same project. So, he goes onto the company's Web site for the Chocolate Bayou project and finds the EVMS chart for this work, current up to the 30th week of the project. Johnny wants to reconcile the quantitative EVMS data with the anecdotal information and narratives he got in his phone calls. Help Johnny out by linking these verbal scenarios to the graphs in color shown below. Give the best match-up between the anecdotal remarks and the EVM plots. Note: You must fully explain your logic and reasoning as to why each dialogue best fits the chosen EVM chart.

For each of the four charts shown below (Figs. 13.22, 13.23, 13.24, and 13.25):

- Compute SPI and CPI at the end of 30 weeks and write these values on each chart.
- Forcaset completion for both the BCWP and ACWP curves. State your assumptions and explain your reasoning.
- The comments made by Paul Dean best fit the situation shown in which Chart Number?
- The comments made by Frankie Frisch best fit the situation shown in Chart Number?
- The comments made by Pepper Martin best fit the situation shown in Chart Number?
- The comments made by Nub Kleinke best fit the situation shown in Chart Number?

(Explain why you reached the conclusions).

Problem 13.8.3 Consider Example 13.1 from this Chapter. Using ten data points, the linear regression method using the Reverse Gompertz equation gives the values $A = -4.616$ and $B = 0.360$. These values give the estimates of the parameters $a = e^A = 0.00989$, $b = 0.360$ in the Dual Gompertz equation, with $R^2 = 0.977$. Table 13.5 shows the initial 10 data points used to estimate the parameters and 10 subsequent

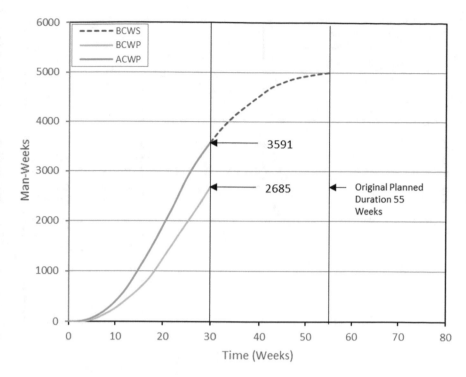

Fig. 13.22 Earned value chart 1

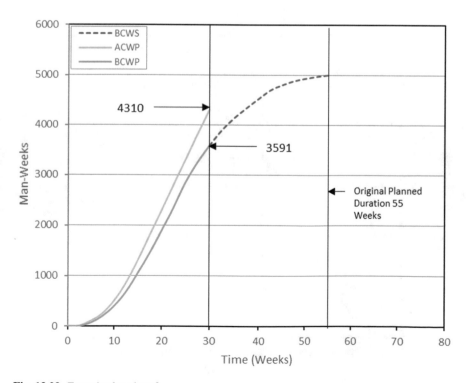

Fig. 13.23 Earned value chart 2

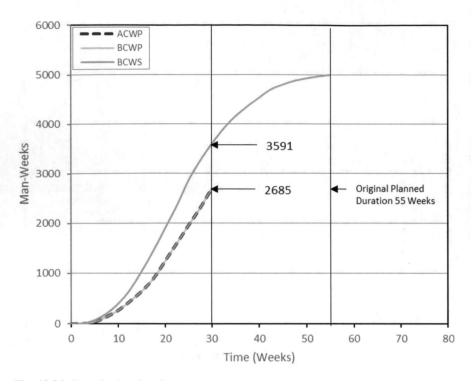

Fig. 13.24 Earned value chart 3

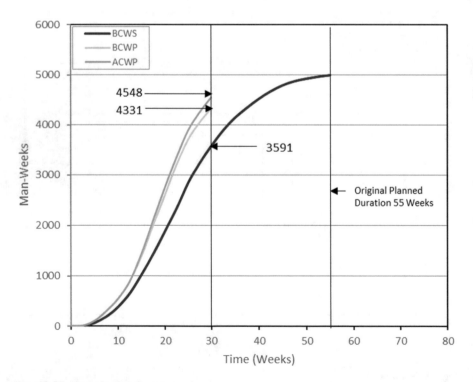

Fig. 13.25 Earned value chart 4

Table 13.5 Project data

Reporting period	Reported percent complete	Fitted value reverse Gompertz
0	0.0	
1	1.0	1.41
2	2.4	2.01
3	3.2	2.87
4	4.7	4.09
5	6.1	5.81
6	7.8	8.23
7	12.3	11.58
8	14.1	16.17
9	25.1	22.34
10	27.3	30.41
11	(36.2)	
12	(40.9)	
13	(49.5)	
14	(56.9)	
15	(73.2)	
16	(86.2)	
17	(92.0)	
18	(96.1)	
19	(98.2)	
20	(99.2)	

measurements, from Period 11 to 20) (shown in parenthesis). Use Bayesian method to show how the parameter estimates would change with each new observation. State your assumptions.

References

Gelman A, Stern HS, Carlin JB, Dunson DB, Vehtari A, Rubin DB (2013) Bayesian data analysis. Chapman and Hall/CRC, Boca Raton

Nelder JA (1961) The fitting of a generalization of the logistic curve. Biometrics 17(1):89–110

Pindyck RS, Rubinfeld DL (1976) Econometric models and economic forecasts. McGraw-Hill, New York

Chapter 14
Forecasting with Learning Curve Models

Abstract In this chapter we discuss the effect of learning on project efficiency and/or productivity. We introduce the concept of learning curves and provide the modeling approaches to forecast project completion time and cost. We use examples from projects characterized with repetitive tasks and where the learning effect is highly visible such as tunneling.

Keywords Learning model · Project learning · Forecasting

14.1   Introduction

Everyone is familiar with the situation, in everyday life, in which one becomes more proficient at some activity. One says, "Practice makes perfect." The same is true in industry: the larger the number of cumulative repetitions, the more efficient the process becomes. Projects may exhibit similar behavior, if they last long enough. The term *learning curve* refers to a situation in which the efficiency or productivity of an operation improves as the work progresses. This effect may be due to a variety of reasons, which are collectively called *learning*. These may include traditional learning (and its complement, *teaching*), but may also include such factors as increased capital investment in machinery, etc.

14.2   Learning Curve Model

Learning can be expressed as a reduction in the unit rate or the marginal cost of each unit with the number of units completed. The *Crawford learning curve model* expresses the marginal cost of the *n*-th unit, $mc(n)$ as a power function (Mosheiov and Sidney 2003):

$$mc(n) = T_1 n^b$$

© Springer Nature Switzerland AG 2020
I. Damnjanovic, K. Reinschmidt, *Data Analytics for Engineering and Construction Project Risk Management*, Risk, Systems and Decisions,
https://doi.org/10.1007/978-3-030-14251-3_14

in which T_1 and b are parameters to be determined. In this model, the ratio of marginal costs for different units in the sequence, say unit n and unit φn, where φ is some dimensionless proportionality factor, depends only on their relative position in the sequence and not on their absolute numbers:

$$\frac{mc(\varphi n)}{mc(n)} = \frac{T_1(\varphi n)^b}{T_1 n^b} = \varphi^b$$

In particular, the *learning slope* or *learning rate* is conventionally defined as the ratio of marginal costs when the number of units is doubled. That is, for $\varphi = 2$,

$$\rho = \frac{mc(2n)}{mc(n)} = \frac{T_1(2n)^b}{T_1 n^b} = 2^b$$

The plausible range for the learning rate is often taken to be:

$$\frac{1}{2} < \rho \leq 1$$

$$-1 < b \leq 0$$

For example, the 90% learning rate corresponds to $\rho = 0.90$, which corresponds, in the equation just above, to b = −0.152. Because of the doubling characteristic, the 64th unit costs 90% of the cost of the 32nd unit, which costs 90% of the cost of the 16th unit, which costs 90% of the cost of the 8th unit, which costs 90% of the cost of the 4th unit, which costs 90% of the cost of the 2nd unit, which costs 90% as much as the first unit. Or, summarizing the sequence, the 64th unit costs $(0.9)^6 = 53\%$, about one-half as much as the first unit. Figure 14.1 shows the marginal cost versus the unit number for learning rates 0.80, 0.85, 0.90, 0.95. Note that the larger the learning slope or learning rate, the smaller the reduction in marginal cost. That is, a learning slope of 1.00 or 100% means no reduction in marginal cost at all.

By taking logarithms of the basic Crawford power function,

$$mc(n) = T_1 n^b$$
$$\ln[mc(n)] = \ln[T_1] + b \ln[n]$$

Plotting $\ln[mc(n)]$ vs. $\ln[n]$ for various observed values of n, if this learning curve model is valid, should give a straight line, with intercept $\ln[T_1]$ and slope b. (But note that this b is negative, from the above discussion.)

Figure 14.2 shows the logarithm of the marginal cost versus the logarithm of the unit number, for learning rates 0.80, 0.85, 0.90, 0.95. Note the negative slopes in the log-log plots.

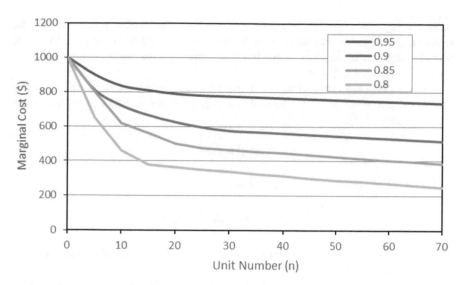

Fig. 14.1 Learning curves

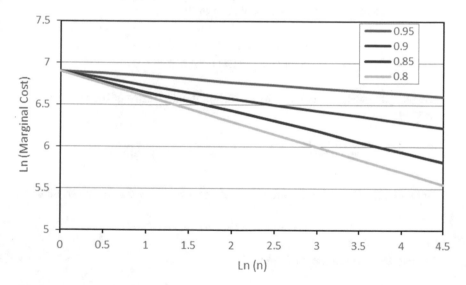

Fig. 14.2 Learning curves in log-log scales

14.3   Learning Curve for Projects

Although knowing marginal costs is desirable for management purposes, measuring marginal costs in a project is not necessarily easy. Typically, we have to deal with a reporting period, such as a week or month, for which we can observe the number of units accomplished (the incremental or period-by-period work performed) and the labor and other costs (the actual cost of the work performed during the reporting period). That is, the marginal cost is in practice the average cost of some number of units in some reporting period.

To avoid estimating marginal costs, the *Wright learning curve model* expresses the cumulative *average cost*, ac(n), of unit n as a power function of n (Anzanello and Fogliatto 2011):

$$ac(n) = A_1 n^\beta$$

By taking logarithms of this, one obtains:

$$\ln[ac(n)] = \ln[A_1] + \beta \ln[n]$$

Plotting ln[ac(n)]   vs. ln[n] should, if the model is appropriate, give a straight line with intercept ln[A_1] and slope β.

A relation may be established between the Crawford model and the Wright model. Using the Wright power function, the *total cost*, tc(n), after *n* units is the product of the average cost and the number of units:

$$ac(n) = A_1 n^\beta$$
$$tc(n) = n[ac(n)] = A_1 n^{1+\beta}$$

The marginal cost, for any *n*, is by definition the derivative of the total cost with respect to *n*:

$$mc(n) = \frac{d}{dn} tc(n) = \frac{d}{dn}\left[A_1 n^{1+\beta}\right] = A_1(1+\beta)n^\beta$$

Compare this expression for the marginal cost to that for the Crawford model:

$$mc(n) = T_1 n^b$$

The two expressions are equivalent if:

$$T_1 = (1+\beta)A_1$$
$$b = \beta$$

The definitions given above represent the classical approach to learning curves. However, these definitions are not based on any particular theory, and actual data may or may not show this behavior. For example, in these definitions, the greatest amount of learning occurs when going from the first unit to the second. The slope of the learning curve, that is, the change in the marginal cost, decreases thereafter. This behavior may be questioned for a number of reasons.

In order to determine the shape of the learning curve, one needs to address the question of why learning should occur at all on projects. In many cases, it probably does not, if the nature of the work changes before much learning can occur. Learning is presumed to occur only when there are enough repetitions of similar but not necessarily identical operations in a construction process. On the one hand, learning is desirable because it represents a reduction in the unit costs. On the other hand, learning may be undesirable because it indicates that the initial unit costs were high.

Whatever learning may be, it is not caused by the construction crafts learning their skills. It is not carpenters learning to drive nails or saw lumber. It is not iron-workers learning how to tighten bolts. Skilled craftsmen already know these things. If any person in the process is learning anything, it is the construction supervisor, who is learning how to staff, plan, control, and manage the work, in a form of on-the-job-training. If a construction process were well planned, if the plan were simulated to find the optimal methods, if the workers were trained in advance, then there should be little or no learning on the job – because it would have started close to the peak of productivity, far down the learning curve. Learning may occur because the process was not adequately planned, simulated, and optimized before the work started. Learning occurs when these activities that should have been done up-front take place only after the actual work starts.

We may then consider that learning consists of a process of trial and error, in which the supervisor and the work crew in general experiment with different approaches, sequences, allocations of resources, etc., retaining the best and discarding the worst. Consequently, learning may be considered to be a search process conducted in real time, on the job, by the crews actually doing the work. From an optimization viewpoint, it may be considered a form of hill-climbing or random search, in which one adjusts the direction of change according to the results obtained. Or, it may be considered a form of genetic algorithm, in which many work methods are proposed and tested, and new work methods are formed out of combinations of the previously tested ones, in an evolutionary process. Because construction work is never perfectly repetitious (as manufacturing may be), some adjustments to the process are always required. These changes go on until the job is complete. Then on the next job, the tasks are different, the personnel change, planning is again inadequate, and learning starts all over again.

Although the general learning model above is not specific, if it is valid in general outline, then we might expect some departures from the classical power function models given earlier. These might include:

- Learning (improvement in unit rates or marginal costs) would not be fastest immediately after starting the job, as it would take some time for the trial and error process described above to become effective. Therefore, the learning curve would not be convex everywhere, as with the power function, but would be concave to start and then become convex.
- Learning improvement would not continue forever, with the unit rate approaching zero, as with the power function, but would be asymptotic to some horizontal line or minimum value.

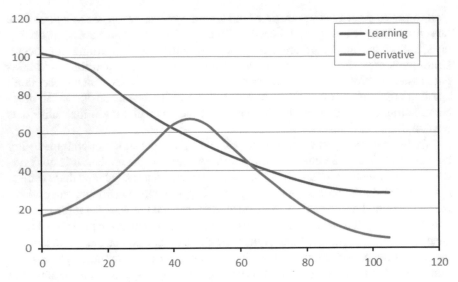

Fig. 14.3 Learning vs. derivative of learning

- Improved planning, simulation, optimization, and training would decrease the potential for learning (initial unit rates would be closer to the long term asymptote), and would increase the rate of learning (trial and error changes would be more effective).

As an example of what this might look like, consider Fig. 14.3.

To obtain an equation for a learning curve like that in Fig. 14.3, let $x(n) = x_n =$ unit rate after n units are complete; $x(\infty) = x_\infty =$ asymptotic lower bound on unit rate; $x(0) = x_0 =$ upper bound on unit rate; and $b =$ constant. Now suppose that the slope of the learning curve $dx(n)/dn$ is proportional to the amount of learning done so far, $x_0 - x_n$ and the potential for improvement: the amount of potential learning yet to do, $x_n - x_\infty$; therefore, $\dfrac{dx(n)}{dn} = b(x_n - x_\infty)(x_0 - x_n)$. By a change of variable, this expression can be transformed into the following equation. Let $y_n = x_n - x_0$ or $x_n = x_0 + y_n$ and $S = x_\infty - x_0$ then:

$$\frac{dx_n}{dn} = \frac{dy_n}{dn} = b(x_0 + y_n - x_\infty)(-y_n) = by_n(x_\infty - x_0 - y_n)$$

$$\frac{dy}{dn} = by(S - y)$$

The solution to this differential equation is given as:

$$y = \frac{S}{1 + ae^{-bn}}$$

in which a is a constant of integration.

Converting back to the current notation, this becomes:

$$x_n = x_0 + \frac{x_\infty - x_0}{1 + ae^{-bn}}$$

This is the function shown Fig. 14.3, in which $x(1) = 100$; $S = 30$; $a = 20$; $b = 0.075$. Also shown is the first derivative (not to the same scale). In this particular situation, the rate of improvement of the unit rate is a maximum at $n = 40$ units, and improvement has virtually ceased at $n = 100$ units.

14.4   Forecasting with Learning Curves

Consider a simple case of a repetitive construction operation, drilling a tunnel. The number of meters of advance is recorded for every day. Table 14.1 shows the situation for a certain actual tunnel after 35 days or 5 weeks of construction. Clearly the daily advance is highly random, but taken from a limited set of only six different values: 1.2, 1.5, 2.4, 3.0, 4.5, and 7.5 m/day.

Figure 14.4 shows a plot of the daily rate of advance for the first 35 days. The average advance for these 35 days is $97.5/35 = 2.7857$ m/day. The total length of the tunnel when completed will be 1035 m. The objective of the data analysis is to predict the completion of the tunnel given that the current methods and policies are continued.

Therefore, the best estimate of the completion time as a result of the first 35 days' experience is $1035/2.7857 = 372$ days. The question at this point is, is there a learning effect that could lead to an earlier completion, and what is the probability that the tunnel will be completed on any given date?

This example will be used to illustrate the learning curve power function model given at the beginning of this chapter. Here, the interest is in time to complete, as it is considered that the cost of the tunnel is directly proportional to the time it takes to drill it. Using the Crawford learning curve model, the marginal cost is the number of days per meter of advance, the reciprocal of the rate in meters per day.

Let m represent the number of units (meters) completed, and let $d(m)$ represent the marginal cost in days per meter when m meters have been done. Then the Crawford model is $d(m) = T_1 m^b$, in which T_1 and b are to be determined from the first 35 days' data. Data showing $d(m)$ vs. m for the 35 days' is presented in Fig. 14.5.

Taking natural logarithms, as suggested above, for $d(m)$ and m transforms the data as shown in Fig. 14.6.

A straight line fit to these 35 points gives $\ln[d(m)] = -0.0915 - 0.2509 \ln (m)$ with $R^2 = 0.41$. This relation would fit the equation as above

$$d(m) = T_1 m^b$$
$$d(m) = 0.91256 m^{-0.2509}$$
$$\rho = 2^{-0.2509} = 0.84$$

Table 14.1 Tunneling data

Day	Meters accomplished	Cumulative advance (m)
1	1.2	1.2
2	1.2	2.4
3	2.4	4.8
4	2.4	7.2
5	2.4	9.6
6	1.2	10.8
7	2.4	13.2
8	2.4	15.6
9	2.4	18
10	2.4	20.4
11	2.4	22.8
12	1.2	24
13	1.2	25.2
14	2.4	27.6
15	2.4	30
16	3	33
17	3	36
18	3	39
19	3	42
20	3	45
21	3	48
22	3	51
23	3	54
24	3	57
25	3	60
26	3	63
27	3	66
28	3	69
29	6	75
30	3	78
31	4.5	82.5
32	3	85.5
33	1.5	87
34	3	90
35	7.5	97.5

This would indicate a learning slope of 84%.

However, as the marginal costs (days/meter) vary widely, and as the stated objective of the analysis is to estimate the completion date, not the marginal costs, the example from this point will take an alternate path. Let $t(m)$ represent the total time to advance a total of m meters. Then, extending the Wright model given above to total cost (total time) we get $t(m) = A_1 m^{1+\beta}$. If we now take logarithms we get:

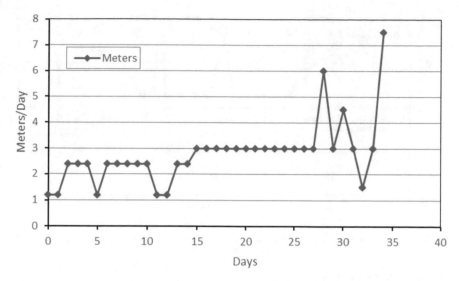

Fig. 14.4 Job progress (meters per day for first 35 days)

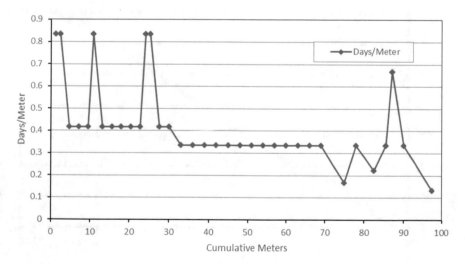

Fig. 14.5 Job progress data $d(m)$ vs. m

$$ln\left[t\left(m\right)\right] = ln\left[A_1\right] + \left(1 - \beta\right)ln\left[m\right]$$

The plot of $ln[t(m)]$ vs. $ln[m]$ for the known 35 days is shown in Fig. 14.7. The equation for the least squares fitted line shown in Fig. 14.7 also:

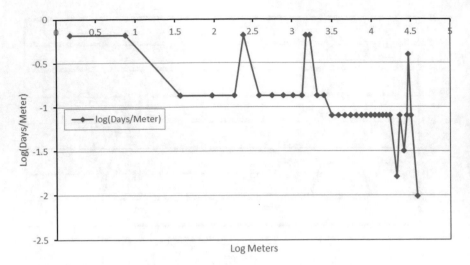

Fig. 14.6   Transformation of progress data

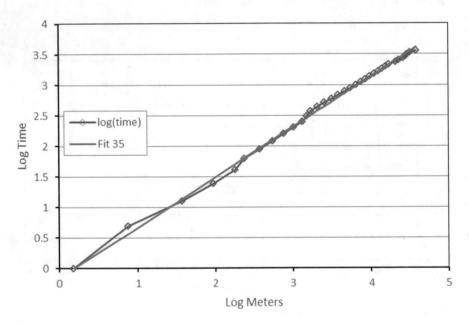

Fig. 14.7   Model fit

$$\ln\big[t(m)\big] = -0.13968 + 0.818147\ln\big[m\big]$$
$$t(m) = 0.869636 m^{0.818147}$$
$$\beta = 0.818147 - 1.000000 = -0.181853$$
$$\rho = 2^{-0.181853} = 0.88$$

The learning slope 0.88 is not far off from the estimate of 0.84 made with the marginal costs. The adjusted $R^2 = 0.997$ is much higher than before, but this is attributed to the high correlations between successive values of the logarithms of total time (0.999), compared to the relatively low correlations between successive values of the logarithms of the marginal costs (0.371). From the linear regression calculations, the residual sum of squares is 0.065045, so the residual variance is $0.065045/(35-2) = 0.001971$; and the residual standard deviation is then $\sqrt{0.001971} = 0.044396$.

This value is also known as the standard error of the estimate. However, the concern here is with the standard error of the forecast, given that the desired forecast for the completion date requires considerable extrapolation. Earlier, it was determined that a simple estimate of the time to drill 1035 m at the average of the first 35 days would be 372 days, meaning that the linear model derived from the known data must be extrapolated more than tenfold (372/35 = 10.6). To establish confidence bounds on the forecast to complete that far out, it is necessary to determine the standard error of the forecast when $m = 1035$ m.

The main results are given below, with a change of notation to correspond to the nomenclature used here.

$$\bar{m} = \frac{1}{35}\sum_{j=1}^{35} m_j$$

$$s_f^2(m) = s^2\left\{1 + \frac{1}{35} + \frac{(m-\bar{m})^2}{\sum_{j=1}^{35}(m_j - \bar{m})^2}\right\}$$

Here, m_j is the observed total progress through day j, $1 \leq j \leq 35$; s^2 is the variance of the residuals from the regression analysis; and m is the forecast number of meters completed, $m_{35} < m \leq 1035$.

The forecast process is then as follows:

- Forecast the mean of the logarithm of the time at the meters of advance m using the linear equation $ln[t(m)] = -0.13968 + 0.818147\, ln\,[m]$
- Forecast the confidence bounds for the linear equation using the relation $mean - ks_f(m) \leq confidence\ band \leq mean + ks_f(m)$. Here, the 80% confidence interval is used, with 10% probability that the tunnel completes in time less than the lower bound, and 90% probability that it completes in less time than the upper bound, so $k = 1.282$, on the assumption that the residuals are Normal.
- Convert the values for the mean and the two confidence bounds to the original variables by computing $exp(x)$.

The forecast for the linear model is shown in Fig. 14.8.

The dotted line shows the actual observations for the first 35 days. Note that the forecast period appears short compared to the 35 day period of the observations

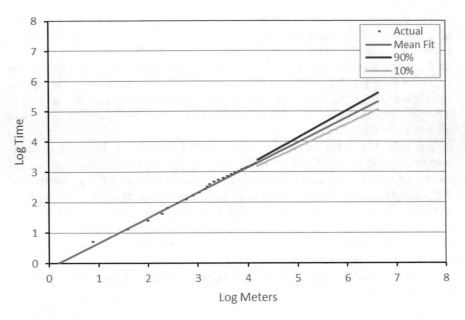

Fig. 14.8 Linear prediction of job progress

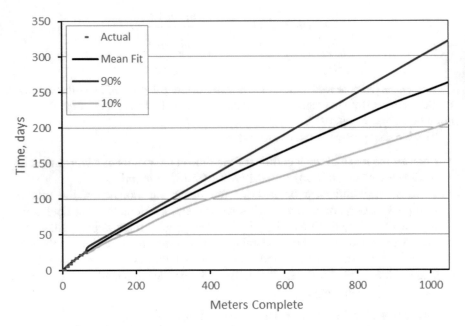

Fig. 14.9 Prediction with learning effect

because the axes are in logarithms. Figure 14.9 shows transforming the confidence bands back to the original variables and forecasting at day 35:

This plot clearly shows how the uncertainty expands with forecasts further out. By estimating the learning effect, the expected completion date (the time at 1035 m) is now 255 days, with a 10% probability if finishing in less than 206 days, and a 10% chance of taking more than 314 days. The curvature in the mean forecast reflects the learning effect. Note that in the linear (that is, logarithmic) plot, the confidence bounds are symmetric, but this is not the case in the graph above, in which the lower bound is (255–206) = 49 days below the median, and the upper bound is (314–255) = 59 days above the median. This indicates that the probability distribution of the time to complete the tunnel is somewhat shifted to the right (that is, to higher values). In this formulation, the probability distribution is on the time to reach any given distance of advance, not on the distance achieved in a given time.

As the tunnel advances, more data are obtained and the analysis above can be repeated again and again to generate revised forecasts for the completion date. As the tunnel advances, the time to completion should decrease and the confidence band should get narrower. If the confidence band widens, this would indicate an increase in variability. If the expected completion date gets further away, this would indicate some decline in the rate of advance. In this approach, one manages the job by the forecast of the time at completion, including both the both the expected completion time and the confidence band on the completion time.

14.5 Practice Problems

Problem 14.5.1 You are a Construction Engineer on a jobsite, acting as the supervisor in charge of two crews on a certain construction activity. This is your first job with real supervisory responsibility and you want to make a good impression on the site resident manager.

On Tuesday of the third week of work on this activity, you get a visit from the site cost accountant, who tells you that he has determined, based on the first 2 weeks of production, that you are trending far over budget. The cost estimate for this work package was 1 man-hour per unit, and the average cost per unit is already 65% higher than that, so this activity will end up over the cost budget and over the schedule too, unless you immediately add more workers (see Table 14.2). The cost accountant tells you that he is going to bring up the substandard performance of your two crews at the site superintendent's next review meeting tomorrow.

Per the estimate, the activity consists of installing a total of 7148 units, projected to last 20 weeks (5-day work weeks). Plot the marginal cost per unit (man-hours per unit installed in each day, by each crew) vs. the number of units installed, for the 2 weeks of actual data followed by the learning model forecast. Is there a difference between the performance of your two crews? What is the predicted average cost per unit at the completion of the activity? How confident are you in your projections?

Table 14.2 Problem data

Day	Crew 1 h paid	Crew 1 units installed	Crew 2 h paid	Crew 2 units installed
1	30	10	52	15
2	27	11	39	23
3	42	21	57	36
4	31	17	28	18
5	18	11	34	26
6	44	31	55	40
7	35	22	21	15
8	28	19	25	18
9	35	24	11	9
10	29	19	34	24
Total	319	185	356	224

What are you going to tell the site superintendent if the cost accountant raises the issue of your apparent overrun in tomorrow's project review meeting?

Problem 14.5.2 The U.S. Army is responsible for decontaminating a of site. The decontamination planning and execution was contracted to a major engineer-constructor. This project has been in operation for 18 months but is apparently over budget and behind schedule. Some facts about the project are given below. The CEO of the engineer-constructor has tasked your team with the job of re-estimating the cost and duration and developing a recovery plan, if possible (see below).

- Scope: Decontaminate 1690 containers of highly toxic waste
- Original contract: 31 month operational period
- Original plan: Ramp up decontamination over a 6 month period until reaching a steady-state of 60 containers per month. That is, increase processing rate reaching 60 per month, then constant at 60 per month until completion of all 1690 containers at 31 months
- Original price: $214.5 million.

The current project is 18 months in the operational phase. Project expenditures to date are $132 million. The project plan called for budgeted costs of $128 million at this point. The project has actually processed 522 containers, at an average rate of $522/18 = 29$ containers per month. The actual number of containers processed in each month to date is variable and shown in Table 14.3.

Site management estimates that the project is 31% complete, based on the number of containers completed to date: $522/1690 = 0.31$. According to the project accountant, costs on this project are largely fixed rather than variable; that is, the costs are not directly dependent on the number of containers processed; the total cost depends on how long it takes to completely process all the containers and shut down the facility. The fixed costs average $7.34 million per month ($132 million/18 months = $7.34 million per month. (Considering only the 18 month operational phase as having costs; hence, over-estimating cost per month.) The project

Table 14.3 Problem data

Month of project	Monthly containers processed	Cumulative containers
1	11	11
2	8	19
3	0	19
4	2	21
5	7	28
6	16	44
7	4	48
8	38	86
9	45	131
10	48	179
11	37	216
12	31	247
13	23	270
14	51	321
15	52	373
16	53	426
17	34	460
18	62	522

believes that project duration and operational costs would be minimized by processing the remaining containers in the shortest time possible.

The contractor's CEO at the home office is unhappy about the apparent overruns on this project in both duration and cost. The CEO sends your team to the site to provide a report to him on the forecast for this job, and your recommendations, to include, but not necessarily limited to:

- Is there any learning curve effects on this project?
- When do you believe this project will finish, if there are no changes to the project from the status quo? (Estimated Duration to Complete; Estimated Duration at Completion)
- What is your confidence level on the predicted duration?

References

Anzanello MJ, Fogliatto FS (2011) Learning curve models and applications: literature review and research directions. Int J Ind Ergon 41(5):573–583

Mosheiov G, Sidney JB (2003) Scheduling with general job-dependent learning curves. Eur J Oper Res 147(3):665–670

Index